Machine Learning for Sustainable Energy Solutions

Machine Learning for Sustainable Energy Solutions

Edited by

Zafar Said
United Arab University, Al Ain, UAE

Prabhakar Sharma
DSEU
New Delhi

WILEY

Published by JohnWiley & Sons, Inc., Hoboken, New Jersey.

Library of Congress Cataloging-in-Publication Data Applied for:

Hardback ISBN: 9781394267408

Cover Design: Wiley
Cover Image: © Jorg Greuel/Getty Images

Set in 9.5/12.5pt STIXTwoText by Straive, Pondicherry, India

Contents

List of Contributors

Tauseef Ahmed
Department of Mechanical Engineering
Universiti Teknologi PETRONAS
Bandar Seri Iskandar
Malaysia

Redhwan Almuzaiqer
Mechanical Engineering Department
College of Engineering
King Saud University
Riyadh
Saudi Arabia

Sundaraganesan Arivazhagan
Department of Mechanical Engineering
St. Joseph's College of Engineering
Chennai
India

Kathapillai Arun
Department of Mechanical Engineering
St. Joseph's College of Engineering
Chennai
India

Munusamy Arun
Department of Mechanical Engineering
Saveetha School of Engineering
Saveetha Institute of Medical and Technical Sciences (SIMATS)
Chennai
India

Vikas Attri
Mechanical Engineering Department
D.C.R. University of Science & Technology
Murthal (Sonepat)
India

H. Bahuruteen Ali Ahamadu
Department of Mechanical Engineering
Rajalakshmi Institute of Technology
Chennai
India

Debabrata Barik
Department of Mechanical Engineering
Karpagam Academy of Higher Education
Coimbatore
India

Niteen Bhirud
School of Engineering
Ajeenkya DY Patil University
Pune
India

Baiju S. Bibin
Department of Thermal and Energy Engineering, School of Mechanical Engineering
Vellore Institute of Technology (VIT)
Vellore
India

Bhaskor Jyoti Bora
Rajiv Gandhi Institute of Petroleum Technology
Sivasagar
India

Sreejesh S. R. Chandran
Department of Mechanical Engineering
College of Engineering Muttathara
CAPE, Government of Kerala
Thiruvananthapuram
India

Muhd Azimin bin Ab Ghani
Faculty of Engineering and Technology
Muscat University
Muscat
Oman

Hussein Safwat Hasan Hasan
Department of Information and Communications Engineering
Myongji University
Seoul
Republic of Korea

Lalhmingsanga Hauchhum
Department of Mechanical Engineering
National Institute of Technology Mizoram
Aizawl
India

Abida Hussain
Department of Fundamental and Applied Sciences
Universiti Teknologi PETRONAS
Bandar Seri Iskandar
Malaysia

Murid Hussain
Faculty of Engineering and Technology
Muscat University
Muscat
Oman

Akshay Jain
Department of Mechanical Engineering
National Institute of Technology
Hamirpur
India

Farrukh Jamil
Faculty of Engineering and Technology
Muscat University
Muscat
Oman

Parampreet Singh Jassal
Department of Mechanical Engineering
IIT Bombay
Mumbai
India

S. Kaliappan
Department of Mechatronics Engineering
KCG College of Technology
Chennai
India

Sanjeev Kumar
Department of Mechanical Engineering
Delhi Skill and Entrepreneurship University
Delhi
India

Neeraj Kumbhakarna
Department of Mechanical Engineering
IIT Bombay
Mumbai
India

Samuel Lalhriatpuia
Department of Mechanical Engineering
National Institute of Technology Mizoram
Aizawl
India

Nikhil S. Mane
School of Engineering
Ajeenkya DY Patil University
Pune
India

Sayantan Mukherjee
Mechanical Engineering Department
Gandhi Academy of Technology and Engineering
Brahmapur
India

Arun Munusamy
Department of Mechanical Engineering
Saveetha School of Engineering
Saveetha Institute of Medical and Technical Sciences (SIMATS)
Thandalam
India

Md. Gulam Mustafa
Department of Mechanical Engineering
Delhi Technological University
Delhi
India

Prabhu Paramasivam
Department of Research and Innovation
Saveetha School of Engineering
SIMATS
Chennai
India

Arni Gesselle M. Pornea
Center for Device Thermography and Reliability
University of Bristol
Bristol
UK

M. D. Rajkamal
Department of Mechanical Engineering
Velammal Institute of Technology
Chennai
India

Abdul Razak
Department of Mechanical Engineering
P. A. College of Engineering (Affiliated to Visvesvaraya Technological University, Belagavi)
Mangaluru
India

Abin Roy
Department of Thermal and Energy Engineering, School of Mechanical Engineering
Vellore Institute of Technology (VIT)
Vellore
India

Sridhar Sahoo
Department of Agricultural & Food Engineering
IIT Kharagpur
Kharagpur
India

Zafar Said
Mechanical and Aerospace Engineering Department
College of Engineering
United Arab Emirates University
Al Ain United Arab Emirat

Natarajan Sendhil Kumar
Department of Mechanical Engineering
National Institute of Technology
Karaikal
India

Saboor Shaik
Department of Thermal and Energy Engineering, School of Mechanical Engineering
Vellore Institute of Technology (VIT)
Vellore
India

Avdhesh K. Sharma
Mechanical Engineering Department
D.C.R., University of Science & Technology
Murthal (Sonepat)
India

Prabhakar Sharma
Department of Mechanical Engineering
Delhi Skill and Entrepreneurships University
Delhi
India

Farman Ullah
Department of Physics
University of Science and Technology
Bannu
Pakistan

Pethurajan Vigneshwaran
Department of Thermal and Energy Engineering, School of Mechanical Engineering
Vellore Institute of Technology (VIT)
Vellore
India

Mudasar Zafar
School of Mathematics, Actuarial and Quantitative Studies (SOMAQS)
Asia Pacific University of Technology & Innovation (APU)
Bukit Jalil
Kuala Lumpur
Malaysia

Preface

The transition toward a more sustainable, energy-secure future is no longer a conceptual aspiration but an urgent global imperative. Integration of sources of clean energy into mainstream energy systems has become both a scientific need and a policy goal as the world struggles with the twin issues of climate change and depleting fossil fuel stocks. Still, making this change is not easy. By nature, renewable energy systems are complex, nonlinear, and sensitive to a broad spectrum of operational and environmental conditions. Thus, maximizing their performance, raising their dependability, and lowering the related uncertainty need required for sophisticated computational methods. Machine learning, a subset of artificial intelligence, is among the most exciting of these methods.

The present book "Machine Learning for Sustainable Energy Solutions" emerges at the junction of sustainability and intelligent computation. This is a contemporary academic project combining data-driven innovation, smart modeling, and front-line green energy researcher contributions. The book explores how machine learning and associated technologies may provide deeper insights, better forecasts, and optimal operations across several sustainable energy sectors. This compilation primarily focuses on a wide spectrum of sustainable energy systems, ranging from biofuels and hydrogen to solar energy and nanofluid-based technologies. The book comprises fifteen meticulously selected chapters, each reflecting current advancements and highlighting promising future directions in the field. The first chapter, Green Energy Led Sustainable Development: Barriers and Opportunity, offers a wide and critical summary of how green energy projects are transforming sustainable development all around. It draws attention to technical, policy-level, and socioeconomic impediments as well as new chances for cross-sectoral synergy.

Chapters 2 through 6 focus on bio-based energy generation, paying especially close attention to biomass gasification and biohydrogen. These chapters show how well ML methods, including hybrid approaches, support vector machines, and neural networks, anticipate and maximize bioenergy production from challenging organic feedstocks. They not only show model accuracy but also underline in actual uses the need of strong validation, sensitivity analysis, and parameter tuning. Using response surface methodology (RSM) for multi-attribute optimization, the seventh chapter offers an extensive case study on hydrogen-powered dual-fuel engines. This shows how hybrid techniques may produce better-performing and more sustainable engine technologies, therefore bridging conventional statistical modeling with current intelligent systems.

Chapters 8 through 12 explore the growing and important use of nanotechnology in energy systems, particularly in engine emissions management, energy storage, and heat transfer improvements. Whether in biodiesel–diesel blends, biogas-doped fuels, or solar thermal systems, the application of nanoparticles provides a rich field for machine learning-driven invention. These chapters investigate the impacts of particle shape, concentration, and material compatibility using artificial intelligence methods for highly precise forecast of thermal conductivity, viscosity, and system performance. Chapters 10 and 11 should especially be mentioned as they address the junction of thermo-physical modeling and nanofluid optimization utilizing a combination of artificial intelligence and engineering principles. These chapters provide scalable, experimentally confirmed ML-based solutions for long-standing problems like sedimentation, pumping losses, and stability of nanofluids.

Chapter 13 is especially noteworthy as it offers a thorough review of several energy storage techniques. It prepares one to grasp the larger background in which chemical, thermal, electrical, mechanical, and machine learning technologies may be used to evaluate, categorize, and maximize different storage systems. The latter two chapters turn their emphasis toward anaerobic digestion and biogas generation, fields with a distributed and circular character. The confluence of the Internet of Things (IoT) and machine learning for real-time digestive parameter control and monitoring is described in Chapter 14. Chapter 15 broadens this subject by stressing the cooperative importance of mathematical modeling and artificial intelligence in maximizing reactor performance, gas output, and operational stability.

The editors of this book have stayed dedicated to variety in topic choice, methodological rigor, and relevance to practical applications above all. The book is useful for academic researchers, business leaders, and graduate students equally, as each chapter is based on modern research and enhanced by experimental or simulated data. Apart from offering individual insights, the chapters together provide a convincing story about the transforming power of machine learning in hastening the worldwide change to sustainable energy systems.

The editors, Dr. Zafar Said and Dr. Prabhakar Sharma, express their sincere gratitude to all contributing authors for their scholarly efforts and to the Wiley-Scrivener team for their professional support throughout the publication process.

Special thanks are extended to the research groups, funding bodies, and academic institutions that have supported the underlying work in each chapter. This book is envisioned as more than just a collection of technical writings; it is a blueprint for action. It invites readers to think critically, innovate responsibly, and act decisively in the shared mission of building a sustainable energy future powered by intelligent systems.

Zafar Said
UAE University
United Arab Emirates

Prabhakar Sharma
DSEU, Delhi, India

1

Green Energy-Led Sustainable Development: Barriers and Opportunities

Arni Gesselle M. Pornea[1] *and Hussein Safwat Hasan Hasan*[2]

[1] *Center for Device Thermography and Reliability, University of Bristol, Bristol, United Kingdom*
[2] *Myongji University, Seoul, Republic of Korea*

1.1 Introduction

The ever-growing energy demand and reliance on fossil fuels have incurred an increasingly severe environmental impact. With an expected future energy consumption escalation of a 3.5% annual increase from 2023 to 2030, the immediate development of sustainable energy is relatively essential for achieving the net-zero target [1]. Numerous green energy sources serve as viable alternatives to traditional energy sources, including solar, wind, hydropower, biomass, and geothermal energy harvesting, which can help meet and fulfill the rising energy requirements. This shift toward green energy is expected to open up new opportunities to accelerate economic growth. It is expected that clean energy consumption will increase by about ~67% between 2020 and 2030, covering about 80% of global energy consumption [2]. The transition to renewable energy has faced significant challenges that hinder progress, including issues with technological readiness, economic viability, and support across various sectors. As a result, the contribution of renewable energy remains frustratingly stagnant. It is imperative to expedite the transition to green energy sources and increase their energy yield. Hence, numerous barriers impede the dominance of sustainable energy in the global energy grid. For discussion simplification, the dominant barriers and obstructions to green energy implementation were mainly categorized into two main segments: (a) nontechnical hindrances and (b) technical challenges [3].

Barriers in the nontechnical category consist of financial constraints, policy stipulation, and public cooperation, which embodies the reliance of the sustainable development of green energy on the governmental implementation approach and programs. The lack of financial support is one of the hurdles to advancing

Machine Learning for Sustainable Energy Solutions, First Edition. Edited by Zafar Said and Prabhakar Sharma.

green energy; it is challenging to secure the initial investment necessary to develop and implement green energy solutions. Restrictions on policy are another obstacle to the green energy transition; most of the government's policies post inconsistent and ambiguous objectives, which restrain long-term private sector investment and interest. These obscure goals and regulations make implementation difficult, hindering the progress of renewable energy. Furthermore, the obligation to preserve the environment may occasionally be overshadowed by the government's consideration of economic growth. Analogously, social acceptance of alternate energy sources poses hurdles to the propagation of renewable energy as well. Certain schemes must be implemented with awareness promotion to obtain public support and literacy regarding green energy utilization [4].

Moreover, technical challenges are one of the primary obstacles to achieving sustainable development goals (SDGs). Viable green energy sources and options still require research and development to enhance their efficiency and economic viability. The advancement of sustainable energy is highly reliant on the material options (such as photovoltaic [PV] materials and green energy catalytic conversions) that must be selected from a myriad of material candidates and then synthesized efficiently to generate high yields and quality for utilization in energy-harvesting devices. The primary goal of the material investigation is to develop effective materials that can be readily utilized commercially. The discovery and development of these representative materials span approximately 15–20 years, with consideration given to improvement adjustments. This standard technological progression greatly hampers sustainable energy transformations [5].

Prompted by technological developments in machine learning (ML) and artificial intelligence (AI), these hold the potential to bring resolutions to both technical and nontechnical challenges that sustainable energy is facing [6]. The exploration of these techniques helps address the barriers to sustainable energy development. They can help predict specific material properties, thereby eliminating the lengthy material research and trial process. They can also postulate ideal material attributes with desired characteristics matching the application requirements. The acceleration of material discovery through these technological tools will amplify the immediate commercialization of sustainable energy sources, a phase that has never been seen before. It also explores new frontiers, such as hidden pattern identification and correlations, to generate hypotheses and solutions due to its immense data-handling capabilities. It can also be used to recognize patterns in renewable energy generation and utilization, which will expedite policy realization. Nonetheless, these green energy innovations not only promise great potential for achieving net-zero emissions and affirming the long-term viability of our planet but also imply enhanced economic growth and generate new job opportunities. This communication offers an overview of energy research development, its corresponding barriers and opportunities, and the role of ML and AI technology in directing a sustainable future [5, 7, 8].

1.2 The Current Landscape of Green Energy

1.2.1 Green Energy Types and Technologies

Green energy spans a wide array of sustainable energy sources, each with exceptional technologies and utilization has its advantages and hurdles:

- **Solar power:** This constitutes PV and solar thermal systems that directly transform sunlight into electrical and thermal energy. Solar energy-harvesting technology has advanced from the development of bifacial PVs that harvest energy from both sides of the panels to thin-film solar cells, boosting energy-harvesting efficiency while reducing installation costs; hence, it is still far from reaching its full potential. There is still more to explore regarding the implementation of solar energy [9].
- **Wind power:** Wind turbines harness wind energy by converting it into usable forms, such as electric energy. Technological innovations cover the turbine blades, offshore installations of wind turbines, and improvements in turbine efficiency, promoting higher energy harvesting even in areas with lower wind speeds. Innovations encompass larger turbine blades, floating offshore wind farms, and enhanced turbine efficiency, allowing for higher energy yield even in areas with lower wind speeds [1].
- **Hydropower:** Considered one of the oldest sustainable energy sources, it is now the major source of clean energy. It utilizes water flow to harvest energy. Improvements in hydropower generation focus on enhancing turbine efficiency and installing these facilities in remote areas, thereby reducing ecosystem disturbance [1, 10].
- **Bioenergy:** Utilizing organic materials, such as plant waste and manure, to generate heat or produce biofuels. The technological enhancement consists of gasification and anaerobic digestion, which raise energy output and reduce emissions [10].
- **Geothermal energy:** Geothermal systems utilize the Earth's internal heat to generate electricity and provide direct heating. Enhanced geothermal systems (EGS), which involve creating underground reservoirs in dry rock to extract heat, are expanding the potential of geothermal energy beyond natural volcanic regions [1].

1.2.2 Global Green Energy Usage Statistics

The global implementation and production of green energy have grown significantly over the past decade, reaching a record 30% of global renewable electricity generation in 2023. Green energy increased from 19% of overall global electricity in

2000, mainly driven by the expansion of solar and wind electricity, which contributed 0.2% in 2000, to a record high of 13.4% in 2023 [1].

- Solar energy is the dominating contributor to the green energy revolution. It has been the fastest-growing renewable energy source for 19 consecutive years. The energy produced through solar power has doubled compared to coal's electricity, accounting for 5.5% (1631 TWh) of the overall electricity generation in 2023. Obtaining a 23% increase in solar energy generation (from +307 TWh in 2023 to +307 TWh in 2022) compared to a 9.8% increase in 2022 (from +206 TWh to +227 TWh). This massive increment in solar energy harvesting can be attributed to the surge in solar power plant installations, such as in China, which contributed to 51% additional global solar energy generation [1].
- Wind energy was the new powerhouse of newly introduced clean energy. The combined efforts of sustainability-leading countries resulted in a power generation capacity exceeding 800 GW, with wind energy accounting for 7.8% (2374 TWh) of the world's electricity in 2023. Likewise, in solar energy, China also uplifts wind energy harvesting by 60% of the overall harvested wind energy [1].
- Given its popularity, hydropower production experienced a downward energy trend, accounting for 14% of global electricity in 2023. The falling trend can be attributed to the extensive droughts of the past few years. Under normal conditions, hydropower production can account for approximately 1.1% of fossil fuel energy generation. Hydropower recorded a fall in energy production (−88 TWh) for 2023, constituted by countries such as Mexico, China, India, and Vietnam [10].
- Global bioenergy electricity production increased by 21 TWh (+3.1%), primarily due to the expansion of bioenergy generation in China, which contributed 28 TWh (+15.6%), resulting in a 2.4% share (697 TWh) of the overall global energy capacity [1].
- Geothermal energy generally constitutes a very low energy harvesting due to the limited facilities that generate energy in this manner. Hence, it still obtained a 0.3% (90 TWh) global energy contribution [1].

1.3 Barriers to Green Energy Implementation

1.3.1 Economic and Financial Challenges

The implementation of green energy is often hindered by significant economic and financial challenges that impede its scalability and widespread adoption. Understanding these obstacles is crucial for developing effective solutions and strategies to address them.

1.3.1.1 High Initial Costs

One of the major hurdles impeded by the development and implementation of clean energy technologies is the high initial cost of developing and setting up renewable energy facilities. Solar energy-harvesting panels, wind turbines, and other renewable energy schemes commonly require large initial investments compared to typical energy sources, such as coal and natural gas. The difference in initial cost investment can hinder the overall adaptation and transition toward clean energy sources. For instance, the expenditure to build a residential PV can range from \$21,000 in the United States, considering government incentives [11]. While the cost can be compensated in the long run, the primary expense still poses a substantial barrier, particularly in regions that do not support financial incentives.

1.3.1.2 Investment Risks

Renewable energy projects are often considered risky investments compared to common energy projects. This insight stems from several factors, including market volatility, regulatory changes, and technological maturity. Clean energy harvesting can rely on environmental situations and intermittencies, such as solar radiation and wind. This can lead to inconsistent energy generation and obscured revenue forecasts, amplifying market volatility. Policy and regulatory change severely impact the implementation of green energy, influenced by political climates. Alterations to these tax incentives and subsidies can significantly sway the future of renewable energy. One of the primary hurdles to realizing these renewable energy technologies is their technological maturity. Most of these technologies are relatively new and require extensive testing and investigation to optimize the advancement of their capabilities. These posts uncertainty about the performance and reliability of the technology for economic gain.

1.3.2 Regulatory and Policy Frameworks

Government regulations and policies play a crucial role in implementing and adopting clean energy technologies, where effective policies significantly accelerate the transition. In most places worldwide, the advancement and enforcement of green energy are hindered by numerous factors, such as complex permitting processes, inconsistent policies, and insufficient incentive support. The complexity and duration of the permitting process are among the major obstacles to facilitating renewable energy infrastructure. For example, in the United States, it will take a long review and approval process to facilitate renewable energy projects, considering the approval of multiple governmental sectors, such as local, state, and federal agencies. Similar to policy approval, inconsistencies in policies hinder the advancement of clean energy implementation. Sudden changes in policies hinder the continued development of clean energy, leading to market instability and a reduction in investor confidence. Moreover, financial support from the government plays a big role in making renewable energy projects viable, which can be implemented through tax credits, rebates, and subsidies. In regions that do not invoke or lack sufficient incentives, slow growth in renewable energy adoption is realized. For example, in India and Indonesia, progress has been made in renewable energy through the implementation of targeted policies by Jawaharlal Nehru National Solar Mission and the Solar Energy Corporation of India (SECI). Nevertheless, these countries are still facing issues in legislative, funding, and facility dilemmas that tariff incentives, subsidies, tax credits, and regulatory adjustments can resolve [2]. This thereby proves the importance of regulations and policy frameworks in promoting the propagation of clean energy, particularly in developing nations.

1.3.3 Social Acceptance and Cultural Factors

The adoption of clean energy technologies is influenced by technological and economic barriers, as well as social and cultural acceptance. Societal acceptance plays a significant role in the deployment of green energy technologies. Public perception considerably influences the adoption and acceptance of green energy. Misunderstandings and insufficient knowledge about the advantages and potential of renewable energy may cause doubt and opposition among communities. For example, a widespread misconception is that solar and wind energy cannot consistently power a modern economy, which is contradicted by advancements in grid management and energy storage technologies that mitigate the fluctuations of these energy sources [12].

1.3.4 Technological Barriers

A variety of technological challenges must be addressed to facilitate a swift transition to renewable energy. Effective renewable technological assessment and development, as well as a successful strategy for utilizing natural resources, involves promoting the development of wind and solar energy. The rising demand for electricity in various aspects of life, such as heating, electric vehicle charging, and residential and industrial requirements, requires meticulous regulation to ensure an adequate amount of power is produced while attempting to achieve a net-zero state. Additionally, incorporating a significant amount of collected clean energy into the power grid requires additional effort to ensure the proper amount of energy is harnessed, thereby avoiding an energy crisis. This demands careful process planning and flexible grid connections due to the variability of renewable energy sources. It is essential to ensure that both renewable energy installations and fossil fuel plants can adjust in response to fluctuations in energy production.

Several significant technological challenges often impede the transition to green energy sources:

- **Efficiency issues:** Numerous green energy technologies still encounter hurdles related to efficiency. For instance, the typical efficiency range of solar panels is between 15% and 22%, and although higher efficiency options exist, they typically come at a greater expense. Enhancing the efficiency of these technologies is vital for making them more competitive with fossil fuels.
- **Scalability and infrastructure:** A number of renewable energy solutions require substantial modifications to the current infrastructure, which can be expensive and technologically complex.
- **Storage and intermittency:** One of the primary obstacles faced by renewable energy sources like solar and wind is their inconsistency, as energy generation fluctuates based on weather conditions.
- **Energy grid integration hurdle:** The intermittency of most renewable energy sources creates a notable challenge for grid integration since the energy supply needs to align with demand reliably. Energy storage solutions like batteries are crucial for addressing this problem.

1.4 ML and AI in Green Energy

Investing in research and development for the cultivation of ML and AI utilization in renewable energy, while intentionally focusing on factors that have the most significant impact, such as sophisticated forecasting models, intelligent grids, and advanced renewable energy solutions. Efficient technology transfer in knowledge and expertise through international alliances is implemented through joint research programs and technological exchange engagements. These initiatives progressively contribute to developing industry regulations and standard frameworks to initiate the seamless integration of ML and AI in the renewable energy sector [6].

1.4.1 Technological Assessment and Optimization

The integration of ML and AI into renewable energy fields, such as wind power, solar power, and comprehensive net-zero efforts, is driving significant technological advancements. These advancements are designed to tackle current challenges and improve the efficiency and sustainability of facilities that harvest renewable energy [5, 6].

- In terms of general renewable energy harvesting, AI and ML technologies aid in the technical design and evaluation of various renewable sources. Utilizing the ML and AL techniques increases the efficiency and predictability of the output of these renewable sources, which is vital for their industrialization and integration into the energy grid system. AI-empowered renewable energy technology developers analyze and optimize energy generation, improve the probability of providing solutions, and dynamically adapt to energy market demand [3].

- In solar energy, ML and AI play essential roles in improving the energy forecasting efficiency and optimal energy harvesting of PV systems [2]. AI-driven predictive performance can identify possible failures or inefficiencies in solar panel design and utilization [13, 14]. This functionality greatly minimizes the initial system design assessment duration, ensuring optimal arrangement and architecture for energy harvesting is obtained [15]. Multiple ML techniques can also be implemented to predict the solar energy-harvesting capacity in certain areas, thereby providing insight into the feasibility of PV installation [14]. Through the utilization of AI algorithms such as Google Sunroof, solar companies can enhance the positioning of panels, reduce shading, and guarantee the best tilt angles for maximum energy harvesting [16]. In terms of the technological aspect, ML was also utilized to detect and localize PV panel defects, reducing the required time to identify flawed cell parts [2].
- In wind energy systems, ML and AI have projected advantageous contributions to energy prediction as well as system design optimization. ML techniques have been used to maximize wind farm forecasts by taking advantage of the predictive and optimization ability, thereby substantially improving wind energy harvesting and relatively increasing its contribution to the global energy grid [2]. AI can also be applied to optimize the wind turbine's overall design, including the blade architecture and articulation in response to wind speed, thereby improving overall farm performance [17].
- In the bioenergy industry, ML and AI are transforming optimization, waste minimization, and sustainability efforts to tackle challenges associated with traditional bioenergy methods. With their ability for smart decision-making, predictive analytics, and adaptive management to enhance bioenergy operations. ML algorithms play a critical role in predicting biomass characteristics, optimally selecting feedstock, and refining energy conversion processes overall. AI systems, on the other hand, can forecast biomass yield from different sources by analyzing environmental conditions and crop status. This technologically advanced tool can initiate the rapid production of clean energy and its relative adaptation [7].

1.4.2 Predictive Net-Zero Initiative

The idea of net zero stems from the aim to attain a balance between the number of greenhouse gases emitted and removed from the atmosphere, preventing additional damage to our planet. It proposes strategies for minimizing carbon emissions and offsetting the remaining emissions through carbon harvesting. Achieving net zero is essential to controlling global warming and managing climate change. Hence, multiple challenges still hinder the pursuit of net-zero emissions, such as accurate measurement and tracking of emissions across all sectors. AI and ML are tools that can help advance net-zero efforts, and by leveraging predictive algorithms, they can enhance real-time energy consumption across different systems and sectors to align energy demand with the most reasonable and ideal net-zero condition. For instance, DeepMind, an ML algorithm developed by Google, has utilized ML to forecast wind power output 36 hours ahead of actual generation, thereby improving the value and efficiency of wind energy integration into the grid [3, 18].

1.4.3 Enhancing Energy Storage Systems

AI plays a crucial role in enhancing energy storage, which is essential for managing the variability of renewable energy. By forecasting times of high energy demand and streamlining energy storage operations, AI algorithms can successfully balance energy supply and decrease dependence on fossil fuel backup sources. An example of this is Tesla's Autobidder AI software designed for battery storage systems, where ML is utilized to autonomously trade energy in real-time markets, thereby optimizing profits and efficiency [16].

1.4.4 Energy Demand and Supply Forecasting

Incorporating renewable energy into current grid systems poses several difficulties, especially in handling the electricity generated from diverse and fluctuating sources. ML and AI play a critical role in enhancing renewable energy adoption, not only by improving the performance and reliability of energy sources but also through the aid of predictive analysis. Smart grids, which connect electric infrastructure, storage facilities, and consumers, ensure energy efficiency and have implemented ML and AI to estimate price variation, capacity prediction, and failure detection [5]. Accurate predictions are crucial for successfully integrating renewable energy sources into the grid, especially because of their natural fluctuations. ML models that examine weather patterns can be utilized to assess the potential output of wind and solar energy, thereby improving energy generation [19]. With the aid of weather forecasting, grid management has been enhanced, facilitating a better supply of renewable energy. ML and AI enhance grid management by processing vast amounts of information from smart meters and grid sensors. Analyzing previous and current global energy consumption using ML and AI helps identify patterns and trends in areas that can be improved to achieve optimal energy-harvesting efficiency. Investigation of the variables that influence energy consumption, including population increase, industrialization, and economic growth [5]. This allows for real-time management of energy flows and the immediate identification of anomalies. This technology promotes a more dependable energy distribution and aids in averting energy outages [20].

Moreover, ML- and AI-assisted decision-making capabilities have been developed, reinforcing numerous models, including supervised and unsupervised learning, reinforcement learning, Bayesian methods, probabilistic ML models, logistic regression, and deep neural networks to facilitate core energy technology and energy distribution automation. These modes facilitated effectiveness in energy forecasting, distribution, and management, correlated with the assimilation of renewable energy. These models demonstrate effective load and data forecasting, stabilized energy distribution, and issue management, all of which are correlated with clean energy grid integration. The proposed smart ML automation is estimated to yield savings of $237 billion to $813 billion [2]. These reliable models will initiate accelerated policy establishment, effective incentive mechanisms, and reduction in technical barriers before deployment [8].

1.4.5 Setting Ambitious Goal

Determining a systematic strategy and ambitious targets for expanding the portion of renewable energy in the global energy market, analogous with international treaties and climate aims. Integrating ML and AI develops renewable energy objectives considering the probable maximum performance, cost reduction, and reliability improvement. Foster international collaboration to align national and regional targets with global objectives [21].

1.4.6 Activate Support and Financial Investment

Establish financial incentives and policies, such as tax credits, grants, and low-interest loans, to encourage investments in utilizing ML and AI for green energy projects, with the promotion of green financial packages, private–public collaboration, and international funding assistance to collect capital for renewable energy projects. Another strategy that can be implemented is investment reduction in fossil fuels to be relocated to ML- and AI-mitigated renewable energy projects [19]. Numerous agencies worldwide have taken the initiative to advance the clean energy agenda through policies and programs. The United Nations has proposed SDGs to initiate a sustainable future by 2030, focusing on the accessibility and affordability of clean energy [2]. The International Energy Agency (IEA) estimated that more than $1 trillion in annual investment is needed to satisfy the world's energy demand sustainably [21].

1.5 Challenges in the Integration of ML and AI in Renewable Energy

While ML and AI have opted to provide solutions for mitigating renewable energy adaptation and objectives, several hurdles need to be addressed to obtain their optimal impact.

- **High preliminary cost:** The assimilation of ML and AI technology requires considerable upfront financing for the hardware, software, and training. This will incur overhead costs for installing and establishing the renewable energy facility. Which might be inaccessible for developing countries with inadequate resources [19].
- **Date security and privacy infringement:** The utilization of ML and AI in the renewable energy domain requires accumulation, processing, and massive amounts of data. This can be a critical concern for data privacy and security, involving sensitive information such as energy consumption trends and grid flaws [19].
- **Technological constraints:** Current AI functions might not yet be capable of handling the uncertainty and complexities of renewable energy techniques, such as unprecedented weather conditions or sudden grid conditions. Therefore, it is essential to continuously develop ML and AI algorithms that can better tackle these issues and provide more robust strategies for renewable energy adaptation [6].
- **Job displacement and expertise breaches:** It is evident that the integration of ML and AI into any field will result in job displacement, as most functions will be automated. Moreover, there will be skill gaps in personnel, as the industry requires employees knowledgeable in both renewable energy and AI technology. This requires training and education to help the employees adapt to the job requirements.
- **Data quality reliance:** ML and AI algorithms depend on quality data to perform optimally.
- **Ethical standards monitoring:** The utilization of ML and AI algorithms in renewable energy processes may pose ethical and regulatory issues, such as ensuring that the decision-making will not be influenced and exploited, that can be imposed on the algorithm biases. Postulating pronounced ethical guidelines and monitoring frameworks is essential to resolve these concerns, thereby ensuring the sensible use of these emerging technologies [19].

1.6 Directive in ML and AI Improvement Toward Its Application

ML and AI play a critical role in enhancing renewable energy adoption by improving the performance and reliability of various energy sources, achieving net zero. But it is nowhere near perfect; there is still room for improvement to achieve its full potential. There are still limitations that still need to be addressed and improved [6].

1.6.1 Workforce Capacity Increase

Implement technical support, training systems, and educational sources to establish workforce capacity for integrating ML and AI into the energy grid, particularly in developing countries. The development of programs that address the skills gap and displaced personnel ensures that workers are equipped to mitigate ML and AI in the green energy economy. Collaboration between academia, industry, and government is essential to simultaneously support the innovation of the renewable energy sector while creating opportunities and jobs with it.

1.6.2 Large-Scale Project Implementation

Planning and implementation of projects assisted with ML and AI technologies showcase their potential and advantages when assimilated with green energy projects mainly applied to the major performance parameters such as smart grids, microgrids, and advanced energy storage solutions. These initial projects will testify to the advantages of integrating ML and AI into the energy system grid; they will also provide insight into problems and limitations of ML and AI, which will be important for future initiatives. Supervising and evaluating the project's performance is necessary to fortify and improve the contribution of ML and AI to the clean energy grid, maximizing its ecological sustainability and net-zero contribution [1, 2].

1.6.3 Public Awareness

Initiation of public awareness and educational programs to promote ML and AI technologies' contribution to renewable energy and address misconceptions. Engaging the community in the projects guarantees reasonable contributions and benefits of ML and AI integration onto the energy grid. Transparency regarding the role and coverage area of the application of ML and AI in clean energy frameworks fosters trust and public confidence [1].

1.6.4 Continuous Progress Monitoring and Strategies Adjustment

Monitoring energy expenditure trends and progress on integrating ML and AI into the energy grid to assess its relative contribution to sustainability and the net-zero objectives. Continuously assessing and updating objectives, approaches, and schemes to establish alignment with international treaties, technological progress, and rising global energy demands. Sharing lessons learned is the best strategy to approach progress; increasing renewable energy and reducing the traditional energy sources on the global energy grid with ML and AI is one way to approach overall net zero emissions. Considering the apparent energy demand increase in the long run, ML and AI make the renewable energy supply adaptable to the change in the energy requirement through statistical calculations. These make the system flexible on the anticipated future energy consumption increase with a relative estimation of the required infrastructure and investment requirement on different clean energy sources [5].

The utilization of ML and AI allows statistical projections assuming a relative increase in renewable energy and a reduction in conventional energy production, considering population expansion, economic growth, and technological development. The relative cost was also estimated in correlation with the expansion capacity of different clean energy forms. This overall plan elucidates the corresponding shift from traditional energy sources toward renewable energy sources in the upcoming years [12].

Moreover, identifying sections where ML and AI technologies have a significant impact is crucial to their effective utilization in the energy sector. This not only boosts the implementation of renewable energy but also makes it more viable compared to conventional energy sources.

1.7 Conclusion

The assimilation of ML and AI into the development of green energy technologies presents a significant opportunity to address the challenges hindering the growth of renewable energy. ML and AI possess significant potential to improve the efficiency and effectiveness of renewable energy systems, from enhancing the accuracy of energy production forecasting and optimizing grid connections to facilitating more intelligent management of energy storage solutions. By leveraging these advanced technologies, the energy sector can improve renewable energy sources' competence, dependability, and profitability. These opportunities can also be benefited through advocacies for policies through a more reliable research and development renewable energy scheme, thereby boosting their investments, policies, and subsidies, which play a key role in building a sustainable, efficient, and resilient energy ecosystem. Additionally, AI-driven analysis can yield deeper insights into energy consumption behaviors, resulting in more effective demand response tactics and further decreasing dependence on fossil fuels. Combining blockchain technology with ML and AI can potentially improve the decentralization of energy markets, enabling more transparent and efficient transactional energy exchanges. Together, these technological innovations could greatly speed up the advancement of net-zero objectives, rendering collaborations between the government and private sector and ensuring the implementation of renewable energy policies, supplementing a more dependable, cost-efficient, and widely available energy source globally.

References

1 Wiatros-Motyka, M., Jones, D. and Fulghum, N. (2024). Global Electricity Review 2024.

2 Bin Abu Sofian, A.D.A., Lim, H.R., Munawaroh, H.S.H. et al. (2024). Machine learning and the renewable energy revolution: exploring solar and wind energy solutions for a sustainable future including innovations in energy storage. *Sustainable Development* 32: 3953–3978:https://doi.org/10.1002/sd.2885.

3 Olawade, D.B., Wada, O.Z., David-Olawade, A.C. et al. (2024). Artificial intelligence potential for net zero sustainability: current evidence and prospects. *Next Sustainability* 4: 100041.

4 Al Ali, H.M., Ajmal, M.M., Gunasekaran, A., and Helo, P.T. (2019). Managing change in constant evolving sustainable energy market: an exploratory study of motivators, enablers, and barriers. *Sustainable Production and Consumption* 20: 84–97.

5 Rangel-Martinez, D., Nigam, K.D.P., and Ricardez-Sandoval, L.A. (2021). Machine learning on sustainable energy: a review and outlook on renewable energy systems, catalysis, smart grid and energy storage. *Chemical Engineering Research and Design* 174: 414–441.

6 Zhu, J.J., Yang, M., Jiang, J. et al. (2024). Enabling GPTs for expert-level environmental engineering question answering. *Environmental Science & Technology Letters* https://doi.org/10.1021/acs.estlett.4c00665.

7 Anbarasu, K., Thanigaivel, S., Sathishkumar, K. et al. (2025). Harnessing artificial intelligence for sustainable bioenergy: revolutionizing optimization, waste reduction, and environmental sustainability. *Bioresource Technology* 418: 131893.

8 Zhenpeng Yao, Y.L., Lum, Y., Johnston, A. et al. (2022). Machine learning for a sustainable energy future. *Nature Reviews Materials* 8: 202–215.

9 Special Report on Solar PV Global Supply Chains. www.iea.org/t&c/.

10 Wiatros-Motyka, M., Jones, D., Broadbent, H., et al. 2023. Global Electricity Review 2023. Ember. https://ember-energy.org/app/uploads/2024/11/Global-Electricity-Review-2023.pdf.

11 PVCASE TEAM. (2024). The biggest problems with solar power today, and how to solve them.

12 Ma, Z., Mei, G., and Piccialli, F. (2023). An attention-based cycle-consistent generative adversarial network for IoT data generation and its application in smart energy systems. *IEEE Transactions on Industrial Informatics* 19: 6170–6181.

13 Safwat Hasan, H. (2018). Beamforming technique assisted by machine learning algorithm for next location prediction. *International Journal of Wireless Communications and Mobile Computing* 6: 37.

14 Hasan, H.S. and Hwang, H. (2021). Performance assessment of motorized solar photovoltaic louvers system using PVSYST software. *Electrical Science & Engineering* 3: 30–39.

15 Bezrukov, O. *Optimizing Solar Panel Efficiency and Placement Using AI-Driven Project Management Tools.* (2024). https://huggingface.co/blog/omniaiml/ai-helping-solar-panel-optimization.

16 Boza, P. and Evgeniou, T. (2021). Artificial intelligence to support the integration of variable renewable energy sources to the power system. *Applied Energy* 290: 116754.

17 Kabir, E., Kumar, P., Kumar, S. et al. (2018). Solar energy: potential and future prospects. *Renewable and Sustainable Energy Reviews* 82: 894–900. https://doi.org/10.1016/j.rser.2017.09.094.

18 Asif, R. (2020). Deep neural networks for future low carbon energy technologies: potential, challenges and economic development. *Proceedings - 2020 IEEE 6th International Conference on Big Data Computing Service and Applications, BigDataService 2020* 136–141. Institute of Electrical and Electronics Engineers Inc. https://doi.org/10.1109/BigDataService49289.2020.00028.

19 Wu, Y., Xu, M., and Liu, S. (2024). Generative artificial intelligence: a new engine for advancing environmental science and engineering. *Environmental Science and Technology*. https://doi.org/10.1021/acs.est.4c07216.

20 Mazhar, T., Irfan, H.M., Haq, I. et al. (2023). Analysis of challenges and solutions of IoT in smart grids using AI and machine learning techniques: a review. *Electronics (Switzerland)* 12: 242.

21 International Energy Agency (IEA). Perspectives for the Energy Transition: The Role of Energy Efficiency. www.iea.org/t&c/.

2

Machine Learning-Driven Valorization of Organic Waste for Sustainable Bio-Hydrogen Production

Munusamy Arun[1], Debabrata Barik[2], and Sreejesh S. R. Chandran[3]

[1] *Department of Mechanical Engineering, Saveetha School of Engineering, Saveetha Institute of Medical and Technical Sciences (SIMATS), Thandalam, India*
[2] *Department of Mechanical Engineering, Karpagam Academy of Higher Education, Coimbatore, India*
Centre for Energy and Environment, Karpagam Academy of Higher Education, Coimbatore, India
[3] *Department of Mechanical Engineering, College of Engineering Muttathara, CAPE, Government of Kerala, Thiruvananthapuram, India*

2.1 Introduction

Valorization of organic waste for sustainable energy solutions driven by machine learning (ML) encounters multiple obstacles [1]. First, taking into account the changes in waste composition is essential for accurate energy yield prediction from varied organic sources [2]. This calls for models that are strong and adaptive [3]. Improving model generalizability across various waste streams is challenging due to the absence of consistent datasets [4]. Furthermore, state-of-the-art sensors and monitoring systems are required for continuous data input in order to optimize energy output in real time [5]. Potential biases in decision-making algorithms and other ethical considerations are important to keep in mind while thinking about how to apply ML to trash valorization [6]. Finally, there are practical issues with these models' scalability for broad deployment and their incorporation into pre-existing waste management systems [7]. In order to fully utilize ML for the conversion of organic waste into renewable energy sources, it is essential to resolve these difficulties [8]. Valorization of organic waste for sustainable energy solutions led by ML has seen the emergence of several methodologies [9]. For the purpose of forecasting energy production from various forms of organic waste, supervised learning models like classification and regression are utilized [10]. Patterns in varied waste compositions can be better identified with the use of unsupervised learning methods like dimensionality reduction and grouping [11]. The optimization of waste-to-energy conversion operations is aided by reinforcement learning. Still,

Machine Learning for Sustainable Energy Solutions, First Edition. Edited by Zafar Said and Prabhakar Sharma.

problems remain [12]. It is difficult to generalize models to different types of garbage due to the absence of complete and standardized datasets [13]. Predicting and optimizing in real time are challenging due to environmental conditions and the dynamic composition of waste [14]. It is also important to pay close attention to ethical considerations, such as the possibility of bias in ML algorithms, and how new technologies will interact with current waste management systems [15]. Promoting sustainable energy solutions through ML-driven organic waste valorization requires a delicate balancing act between the advantages and disadvantages of these techniques [16].

2.1.1 Objectives

- Construct and verify ML-BHV to maximize the flexibility and efficiency of bio-hydrogen generation from organic waste, taking into account changing waste compositions and real-time optimization of process parameters.
- Investigate and illustrate the many ways ML-BHV may be used to improve sustainability initiatives, waste management, and renewable energy, highlighting its ability to transform many industries by optimizing intelligent processes.
- Demonstrate the improved performance, adaptability, and cost-effectiveness of ML-BHV in real-world renewable energy projects by conducting thorough simulation assessments to determine its scalability and feasibility.

The remaining section of the research follows the identical format as the literature review in Section 2.2, which looked at past efforts to drive the valorization of organic waste for renewable energy solutions. Machine Learning-Enhanced Bio-Hydrogen Valorization (ML-BHV) is the subject of Section 2.3. Section 2.4 provides an in-depth description of the results and analysis, while Section 2.5 provides a concise overview and final thoughts.

2.2 Literature Review

According to a recent study, machine learning (ML) and mechanistic methods have been integrated into waste conversion technologies and biorefinery processes. This integration promises a synergistic approach to improved modeling and understanding. Li et al. present a framework for modeling wet waste conversion technologies (WCTs) [17] that combine ML with mechanistic methods such as computational fluid dynamics (CFD) and process simulation (PS). Complex processes, molecular kinetics, and reaction kinetics may all be better understood with the help of this integrated framework, which will also make techno-economic evaluations and sensitivity studies much easier. The results, which combine the best of mechanistic and data-driven modeling techniques, foretell improvements in waste management and sustainable, clean energy production. The review by Tsui et al. assesses the use of ML [18] in biorefinery processes such as thermochemical conversions, anaerobic digestion, fermentation, and composting. In comparison to more conventional mechanistic methods, this analysis takes into account algorithmic decisions, modeling difficulties, and possible improvements. Sustainable digital transformation in line with development goals is the aim of the suggested method, which seeks to direct joint efforts in this direction. A recent review by Ascher et al. examines the use of machine learning techniques (MLTs) [19] to model biomass and waste gasification/pyrolysis, focusing on their benefits, drawbacks, and recent spike in popularity. With ML, and especially with ANNs, the prediction accuracy is great (R2 > 0.90). Based on the requirements of the application, several ML approaches provide flexibility. Better accuracy, non-numerical parameter inclusion, and solution diversity generation are the potential areas. In order to gain a better understanding of the processes under investigation, the study recommends placing an emphasis on model interpretability. Particularly noteworthy is the fact that ML-BHV, a developing technology for ML-BHV, promises better results than current methods.

2.3 Proposed Method

This paper gives a progressive approach referred to as Machine Learning-Enhanced Bio-Hydrogen Valorization (ML-BHV) in response to the crucial need for sustainable energy options. With a focal point on natural waste repurposing for bio-hydrogen production, ML-renewable electricity, waste control, and sustainability are only some of the areas that could benefit from ML-BHV's innovative approach to organic waste usage, which in turn leads to progressive system parameter optimization, dynamic management of operations, and predictive protection, among other things. This paper seeks to validate the viability and versatility of ML-BHV in realistic conditions via thorough simulation analysis.

Figure 2.1 shows the entire procedure for the use of machine learning to reap renewable energy resources by way of valorizing natural waste. The process starts off with data collection, which includes amassing vital information at the homes and environmental elements, which affect natural waste. This dataset is now ready for machine learning research by undergoing comprehensive data preprocessing steps such as cleaning, transformation, and feature extraction. Intelligent ML models shape the inspiration of the system; one such version is the trash valorization model, which is responsible for figuring out the best way to valorize waste consistent with its unique properties. The optimization algorithm is predicated closely on this version's input to refine the valorization method in an iterative cycle of improvement. Valorization is the process of reworking organic waste into beneficial products like biogas and biofuels. A renewable strength environment is powered in this electricity-generating phase, which powers several applications.

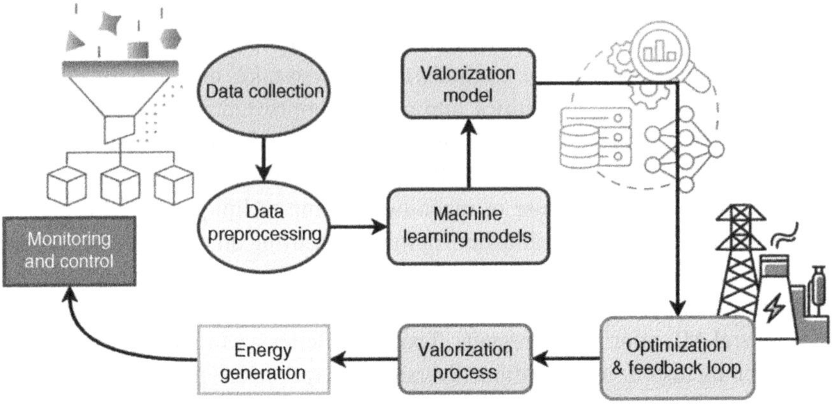

Figure 2.1 Organic waste valorization driven by machine learning.

At the same time, a complicated system for monitoring and controlling using sensors and the Internet of Things devices ensures that the valorization process is optimized and overseen in real time. Environmental impact assessments evaluate emissions and sustainability measures, incorporating environmental elements into the procedure. The evaluation's findings direct the optimization technique as a whole. Stakeholders can easily understand the device's power output, environmental effect, and performance through the reporting and visualization. The interrelated levels of this innovative approach are illustrated in Figure 2.1, which demonstrates the synergy among natural waste valorization and system gaining knowledge for a greater sustainable and environmentally friendly destiny.

$$Q = \frac{\beta \cdot F \cdot (1-\gamma)}{\delta \cdot \alpha} \cdot \left(1 - \frac{\text{RMSE}}{\eta \cdot \vartheta}\right) + \frac{\varepsilon}{\theta} \cdot \left(\frac{\lambda}{\mu}\right)^{\nu} - \frac{w \cdot (\rho + \sigma)}{\varphi \cdot Y} \cdot \left(\frac{\tau}{\omega}\right)^{w} \tag{2.1}$$

The performance of ML-BHV is evaluated by way of thinking about every variable in equation (2.1). The effectiveness of bio-hydrogen technology is represented by using (β), the power recovered from natural waste is denoted with the aid of (F), and the impact of fluctuations in waste composition is considered through (γ). The real-time model factor is denoted via (δ), and the optimization of bio-hydrogen manufacturing methods is connected to (α). In device studying, (η) represents dynamic management, and (ϑ) represents predictive upkeep, while (RMSE) represents the root mean squared error, which shows the accuracy of the predictions. (ε) denotes the full complexity of the machine, while (Θ) indicates how well ML-BHV scales. (λ) represents the discount of environmental effect, (μ) stands for the practicability from a monetary attitude, and (ν) is an operator that indicates how distinct components work collectively. Additional complexity is introduced by means of the variables (w), (ρ), and (σ), while (φ) and (Y) function as elements for balancing and normalization, respectively. A new player on the field is symbolized via (τ), and its impact is contemplated by using an exponent (ω). All of these elements work together to assess ML-BHV efficiency in the sustainable the bio-hydrogen era from each angle.

$$B = \frac{\beta \cdot (X + Y) \cdot \alpha}{\gamma \cdot (U + Z) \cdot \eta^2 + \delta \cdot (Q + A) \cdot \theta^2 + \mu \cdot (X + Y) \cdot (U + Z + A)} \tag{2.2}$$

In equation (2.2) for adaptability, (B) represents the total adaptability of the ML-BHV machine, which includes its potential to respond dynamically to converting situations and optimize for efficiency, affordability, and environmental impact, in addition to its potential to respond to various and evolving situations. The mastering capability of the system is verified through the variable (β), which shows the impact of gaining knowledge of algorithms on optimizing process parameters. Both internal and external elements can affect bio-hydrogen era, and the weights attributed to improving system efficiency and external environmental variables, respectively, are denoted by means of (X) and (Y), respectively. The effect of converting waste composition is considered through (γ), (U), and (Z), which emphasize flexibility throughout timescales by using thinking about the temporal dynamics of managing sports and waste composition. (δ) represents the position that device mastering performs in forecasting the need for maintenance. The fee-effectiveness element and additional prices due to surprising events are addressed by way of (Q) and (A), which pertain to economic issues. The addition of multipliers and factors, inclusive of (α, η, θ) and (μ), provides intricacy to equation (2.2) and represents the complex nature of elements impacting adaptation within the ML-BHV system. These factors indicate cumulative results, dynamic interactions, nonlinear correlations, and synergies.

$$\mathrm{DF} = \frac{\sum_{u=1}^{U} F_{\mathrm{output}}(u) \cdot Q_{\mathrm{market}}(u) - D_{\mathrm{total}}(u)}{\int_0^U \left[\gamma \cdot F_{\mathrm{output}}(u) \cdot Q_{\mathrm{market}}(u) - \delta \cdot D_{\mathrm{total}}(u)\right] e^{-s \cdot u} du} \tag{2.3}$$

In the cost-effectiveness equation (2.3) for ML-BHV, the (DF) stands for the monetary efficiency metric that adjusts with time. The entire time span inside the evaluation is denoted by (U). The bio-hydrogen power produced is represented through ($F_{\mathrm{output}}(u)$) at a certain time (u), and the market fee of bio-hydrogen is represented via ($Q_{\mathrm{market}}(u)$) at the same time. The basic value ($D_{\mathrm{total}}(u)$) of ML-BHV at time (u) consists of all charges (setup, working, and preservation prices). The monetary fee of bio-hydrogen generation and the price details are represented by means of the parameters (δ) and (γ), respectively, which function as weighting elements. (s) is the discount rate that is used to account for the time value of money while applying it to destiny coin flows. The economic feasibility of ML-BHV is

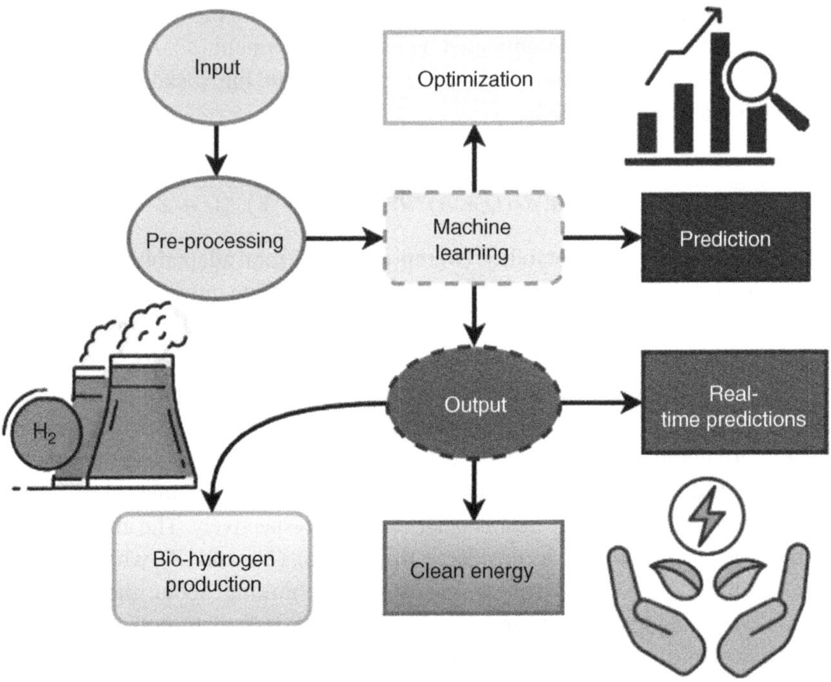

Figure 2.2 Bio-hydrogen production system.

thoroughly evaluated by including all relevant variables in equation (2.3), which incorporates bio-hydrogen generation, market dynamics, and costs over time.

An incorporated framework for changing waste substances into sustainable bio-hydrogen energy using progressive technology is shown in Figure 2.2 as the bio-hydrogen production system. The input segment is where the machine takes in data and natural waste. These by-products, which come from a number of sources, are used as the foundation for bio-hydrogen production. At the same time, the system's activities are informed by input data, which includes information about the waste's composition and environmental conditions. To get the garbage within a feasible circumstance for the subsequent steps, this process uses progressive techniques consisting of sorting and shredding. Critical to the machine's operation is the incorporation of system learning, which allows adaptive and dynamic optimization in response to information gathered in real time. ML algorithms improve common efficiency through gaining knowledge from past data, predicting perfect situations, and continuously refining technique parameters. The system boils down to producing bio-hydrogen gas, a clean energy by-product, during the bio-hydrogen production stage. Here, microorganisms use anaerobic digestion to decompose the subtle waste. The end result is this renewable energy that has many capabilities, as well as real-time forecasts that provide light on the potential for bio-hydrogen manufacturing in the future. By tackling environmental issues and contributing to the production of renewable energy, the bio-hydrogen production system showcases an environmentally friendly method for waste control.

$$\eta_{\text{biohydrogen}} = \frac{\left(Q_{\text{bio}-\text{hydrogen}} - \beta \cdot F_{\text{NM}}\right)\left(1 - e^{-\gamma \cdot X_{\text{organic waste}}}\right) \cdot e^{-\delta \cdot U_{\text{reaction}}}}{\left(1 + \frac{\alpha \cdot F_{\text{NM}}}{\epsilon \cdot X_{\text{oraganic waste}}}\right) \cdot \left(\frac{1}{1 + e^{-\vartheta \cdot U_{\text{reaction}}}}\right)} \tag{2.4}$$

The dynamics of the bio-hydrogen manufacturing method is captured by way of every variable within equation (2.4) for bio-hydrogen production efficiency ($\eta_{\text{bio}-\text{hydrogen}}$). The real price of bio-hydrogen production is denoted by means of ($Q_{\text{bio}-\text{hydrogen}}$), and the effect of algorithmic mastering on the method is measured with the aid of (F_{NM}). The amount of natural waste used, denoted as ($X_{\text{organic waste}}$), affects efficiency through an exponential term. Efficiency is affected by a decaying exponential time period in the reaction length of bio-hydrogen synthesis, denoted as (U_{reaction}). The impact of machine mastering, natural waste, and reaction time can be first-class tuned, and the interaction between those elements is captured via their coefficients (β, γ, δ, α, ϵ, and ϑ). The inclusion of exponential and logistic elements provides complexity to equation (2.4), which represents the complex web of interactions that govern the performance of bio-hydrogen manufacturing in the framework of techniques better via system learning.

An integrated approach to optimizing bio-hydrogen (BHV) production through the application of ML techniques is shown in Figure 2.3 because of the ML-BHV simulation and validation framework. The process starts with simulation input, which includes all of the many parameters that affect BHV production. The ML-BHV simulation model, a sophisticated system that makes use of ML algorithms designed for the complexities of the BHV production process, makes use of these facts as its foundation. To optimize manufacturing parameters and expect consequences, these algorithms scour the input statistics for complex linkages. In the validation section, the ML-BHV simulation model is examined notably by comparing its results with real facts. This evaluates the accuracy and dependability of the simulation by studying the model's effects and providing tips for improvements. The feedback loop, the vital part of the system, allows for steady enhancement. The ML-BHV simulation model undergoes an iterative procedure of adjustment in response to validation effects, allowing the model to trade and develop as time goes on. The remaining step is to generate simulation output, which incorporates essential metrics for BHV manufacturing, statistics on the feasibility of the technique, and beneficial records for making selections. By constantly optimizing BHV manufacturing through the use of system studying,

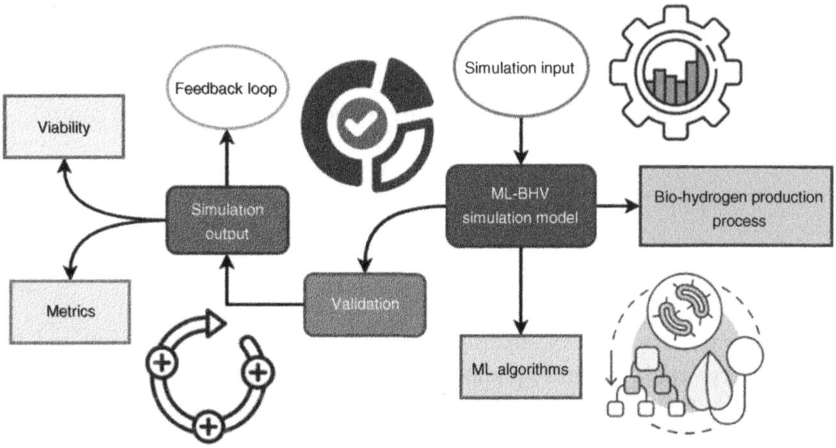

Figure 2.3 ML-BHV simulation and validation.

the ML-BHV simulation and validation system reveal a sparkling and dynamic approach.

$$\mathrm{SG}(u) = f\left(\sum_{j=1}^{o} x_j \cdot \sigma\left(\sum_{k=1}^{n} V_{jk} \cdot Q_k(u) + \sum_{l=1}^{n} w_{jl} \cdot \mathrm{SG}(u-1)\right) + c\right) \tag{2.5}$$

The machine's potential to conform to changing situations is represented by actual-time flexibility at time (u) in equation (2.5) for real-time flexibility in the ML-BHV machine. The variety of input variables that impact real-time flexibility is represented by way of the variable (o), with each parameter certain through ($Q_k(u)$) and (j) starting from 1 to (o). The weight of the (jth) input measurement, denoted as (x_j), determines the way it influences the overall flexibility. A concealed layer of (n) neurons, connected to input parameters through weights (V_{jk}), captures elaborate interactions in equation (2.5) that resembles a neural network. At the preceding point in time ($\mathrm{SG}(u-1)$), (w_{jl}) represents weights that link the neurons in the hidden layer to the actual-time flexibility, thereby incorporating a recurrent neural network aspect. The activation characteristic introduces nonlinearity, and the bias issue (c) modifies the full output. The output function, denoted as (f), is accountable for calculating the final real-time flexibility prediction using the activated and weighted inputs. Equation (2.5) is a reflection of a quite smart model that could research and adapt to the time-structured nature of bio-hydrogen generation.

One crucial approach to reusing organic waste for renewable electricity is machine learning, more advantageous bio-hydrogen valorization or ML-BHV. With the assistance of the most advanced device-studying techniques, ML-BHV improves bio-hydrogen technology strategies in real time and successfully handles the complexity of changing waste composition. This leap forward gives a new way of considering bio-hydrogen valorization and has far-reaching effects for sustainable power, the disposal of waste, and renewable energy. By exhibiting advanced performance, adaptability, and value effectiveness in lots of actual global settings, the dependence on simulation evaluation validates the practicality and versatility of ML-BHV.

2.4 Results and Discussion

The ground-breaking ML-BHV technique, which aims to transform organic waste conversion, is the focus of this investigation. Here, we investigate the most important results from all of these studies to show how ML-BHV might revolutionize the development of green power.

2.4.1 Bio-Hydrogen Production Efficiency Analysis

Evaluating the practicality of the Machine Learning-Driven Valorization of Organic Waste for Sustainable Energy Solutions requires an efficiency analysis of bio-hydrogen production. With an emphasis on real-time adaptability and predictive maintenance capabilities, ML-BHV presents novel methods to maximize the conversion of organic waste for bio-hydrogen. Energy yield, process flexibility, and cost-effectiveness are some of the important metrics that are evaluated in this analysis. ML is the key to ML-BHV's goal of improving bio-hydrogen production efficiency while dealing with different types of wastes. Findings from the study shed light on how ML-BHV optimizes process parameters, which in turn boosts energy production and sustainability, as shown in Figure 2.4, as 93.5%. This analysis sheds light on the revolutionary possibilities of ML-BHV in improving the efficiency of bio-hydrogen production, which could lead to a greener energy future and is therefore crucial for those involved in renewable energy, waste management, and sustainability.

2.4.2 Performance Analysis

An essential part of determining how well this novel method works is the performance evaluation of Machine Learning-Driven Valorization of Organic Waste for Sustainable Energy Solutions. The use of ML methods to optimize the conversion of organic waste into sustainable energy is the focus of this study. Considerations including overall process efficiency, prediction accuracy, and adaptability to changing waste compositions are taken into account. The research dives into the ways

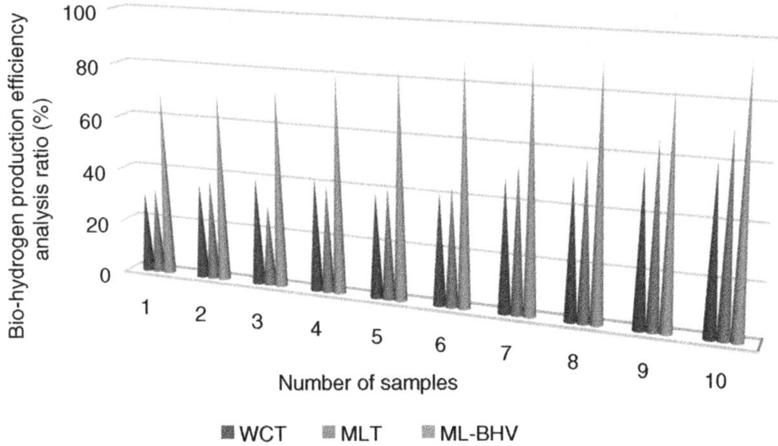

Figure 2.4 Bio-hydrogen production efficiency analysis.

in which ML allows for predictive maintenance and dynamic control of processes in real time, thus increasing flexibility as shown in Figure 2.5, as 94.2%. Increasing energy yield and improving cost-effectiveness are the goals of these strategies, which optimize process parameters. The results of this performance analysis provide important information for those involved in renewable energy, waste management, and sustainability. They show how ML can be used to utilize organic waste more efficiently and how sustainable energy solutions can be advanced.

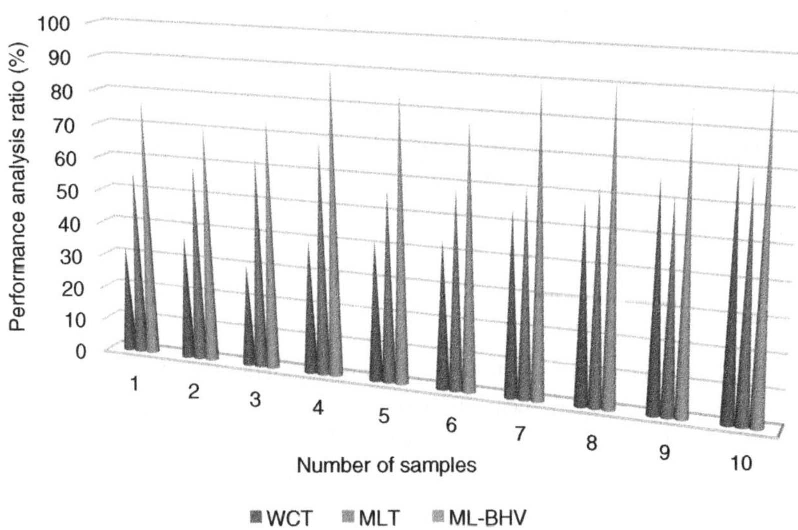

Figure 2.5 Performance analysis.

2.4.3 Adaptability Analysis

Machine Learning-Driven Valorization of Organic Waste for Sustainable Energy Solutions relies heavily on the adaptability analysis to evaluate how well the system responds to different environments. This study assesses the efficacy of ML approaches in making organic waste conversion processes more flexible, with the goal of producing sustainable energy as shown in Figure 2.6, as 98.2%. The model's adaptability to changing waste compositions, operational control changes in real time, and response to different input circumstances are some of the parameters that are examined. The study dives into the ways in which the system may efficiently adapt to changes in organic waste properties through ML-driven valorization, which leads to real-time flexibility. These approaches strive to improve adaptability by dynamically optimizing process parameters. Their goal is to ensure efficient and consistent energy generation from a variety of organic sources. This adaptability analysis sheds light on the revolutionary possibilities of ML to tackle the problems linked to organic waste variability in sustainable energy systems, and its results are highly relevant to stakeholders.

To assess the financial viability of Machine Learning-Driven Organic Waste Valorization for Sustainable Energy Solutions, a cost-effectiveness analysis must be carried out as shown in Figure 2.7, as 99.2%. In light of the potential monetary and environmental costs, this research evaluates the efficacy of ML methods for improving the conversion of organic waste into energy. Operating costs, maintenance charges, and total investment returns are some of the key parameters that are examined. Using process parameter optimization, energy yield enhancement,

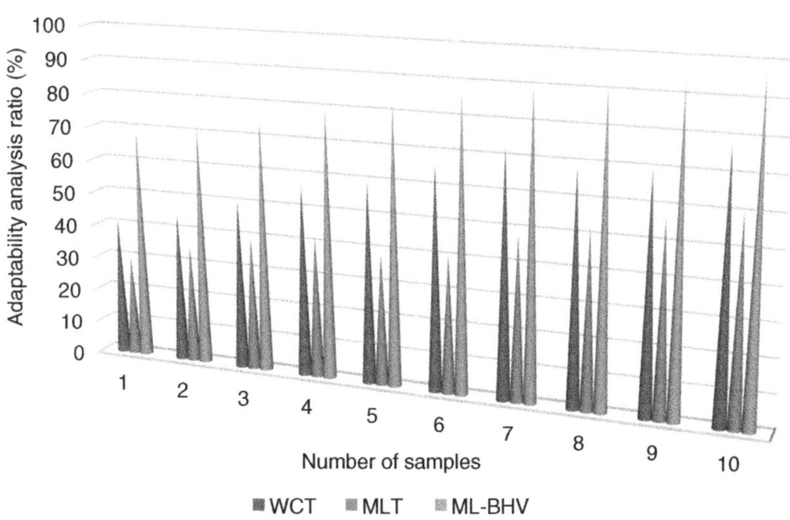

Figure 2.6 Adaptability analysis.

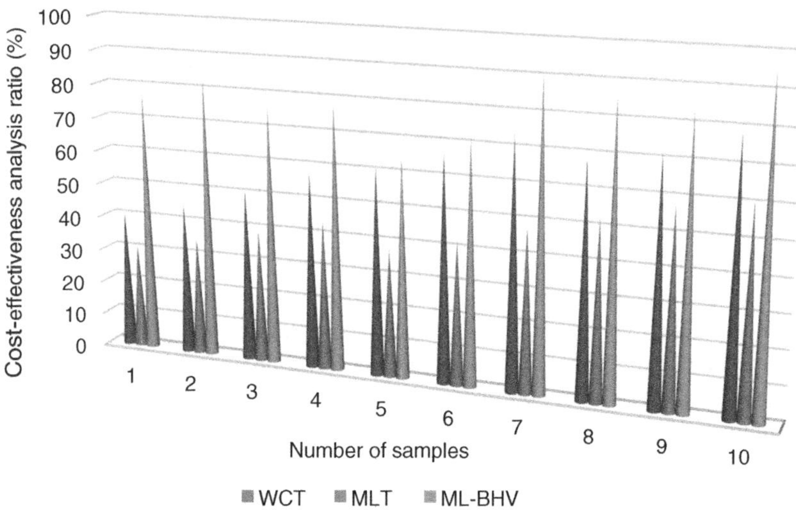

Figure 2.7 Cost-effectiveness analysis.

and resource usage optimization, the investigation delves into how ML improves cost-effectiveness. Overall operational expenses can be reduced with these technologies, which dynamically adapt to changing waste compositions and predict maintenance needs. Sustainable energy solutions that prioritize financial viability can be advanced with the help of the results of this cost-effectiveness analysis, which show how incorporating ML into organic waste valorization processes could be economically beneficial. All things considered, the novel features of ML-BHV are brought to light by these evaluations. It offers a more environmentally friendly and financially feasible future for sustainable energy generation by combining ML with organic waste valorization, which improves efficiency, performance, and adaptability while also positioning itself as a cost-effective option.

2.5 Conclusion

The research concludes by presenting ML-BHV, or ML-BHV, as a game-changing method for producing bio-hydrogen from organic waste in a sustainable manner. In order to optimize bio-hydrogen production processes, manage changes in waste composition, and efficiently harness energy from organic waste, ML-BHV is adopted. The suggested approach improves bio-hydrogen generation, increases efficiency, and achieves real-time flexibility by utilizing modern ML algorithms. A number of industries stand to benefit from ML-BHV's application, including those dealing with waste management, renewable energy, and environmental initiatives. Machine learning-based BHV opens up new possibilities for sustainable energy solutions by optimizing process parameters, dynamically controlling operations, and predicting maintenance needs. Because it provides a more flexible and economical option, this novel method may cause a significant change in the bio-hydrogen valorization industry. To guarantee adaptability and applicability in real-world contexts, validation is based on simulation studies. Findings provide light on ML-BHV's potential for widespread usage in renewable energy projects by showing how it enhances performance and cost-effectiveness in actual scenarios and datasets. In the quest for greener energy sources, ML-BHV stands out as a revolutionary approach that combines ML with bio-hydrogen production.

Author Contributions

"Conceptualization, methodology, data collection and analysis, resources, writing-original draft preparation, writing-review, and editing—M. Arun, Debabrata Barik, and Sreejesh S R. Chandran; Supervision—Debabrata Barik."

Acknowledgment

The authors sincerely thank Karpagam Academy of Higher Education and Saveetha Institute of Medical and Technical Sciences, India, for providing research facilities to carry out the research work.

Data Availability Statement

The data will be available upon request.

Funding Statement

This research is not funded by any funding agency.

Conflict of Interest

The authors declare that they have no conflicts of interest.

References

1 Shahbeik, H., Rafiee, S., Shafizadeh, A. et al. (2022). Characterizing sludge pyrolysis by machine learning: towards sustainable bioenergy production from wastes. *Renewable Energy* 199: 1078–1092.

2 Shafizadeh, A., Shahbeik, H., Rafiee, S. et al. (2023). Machine learning-based characterization of hydrochar from biomass: implications for sustainable energy and material production. *Fuel* 347: 128467.

3 Said, Z., Sharma, P., Nhuong, Q.T.B. et al. (2023). Intelligent approaches for sustainable management and valorisation of food waste. *Bioresource Technology* 377: 128952.

4 Velidandi, A., Gandam, P.K., Chinta, M.L. et al. (2023). State-of-the-art and future directions of machine learning for biomass characterization and for sustainable biorefinery. *Journal of Energy Chemistry* 81: 42–63.

5 Kabir, M.M., Roy, S.K., Alam, F. et al. (2023). Machine learning-based prediction and optimization of green hydrogen production technologies from water industries for a circular economy. *Desalination* 567: 116992.

6 López-Guajardo, E.A., Delgado-Licona, F., Álvarez, A.J. et al. (2022). Process intensification 4.0: a new approach for attaining new, sustainable and circular processes enabled by machine learning. *Chemical Engineering and Processing: Process Intensification* 180: 108671.

7 Shahbeig, H. and Nosrati, M. (2020). Pyrolysis of biological wastes for bioenergy production: thermo-kinetic studies with machine-learning method and Py-GC/MS analysis. *Fuel* 269: 117238.

8 Ge, S., Shi, Y., Xia, C. et al. (2021). Progress in pyrolysis conversion of waste into value-added liquid pyro-oil, with focus on heating source and machine learning analysis. *Energy Conversion and Management* 245: 114638.

9 Zhu, X., Liu, B., Sun, L. et al. (2023). Machine learning-assisted exploration for carbon neutrality potential of municipal sludge recycling via hydrothermal carbonization. *Bioresource Technology* 369: 128454.

10 Sharma, P., Jain, A., Bora, B.J. et al. (2023). Application of modern approaches to the synthesis of biohydrogen from organic waste. *International Journal of Hydrogen Energy* 48 (55): 21189–21213.

11 Velvizhi, G., Jacqueline, P.J., Shetti, N.P. et al. (2023). Emerging trends and advances in valorization of lignocellulosic biomass to biofuels. *Journal of Environmental Management* 345: 118527.

12 Wang, Z., Peng, X., Xia, A. et al. (2022). The role of machine learning to boost the bioenergy and biofuels conversion. *Bioresource Technology* 343: 126099.

13 Shahbeik, H., Shafizadeh, A., Nadian, M.H. et al. (2023). Using evolutionary machine learning to characterize and optimize co-pyrolysis of biomass feedstocks and polymeric wastes. *Journal of Cleaner Production* 387: 135881.

14 Löfgren, J., Tarasov, D., Koitto, T. et al. (2022). Machine learning optimization of lignin properties in green biorefineries. *ACS Sustainable Chemistry & Engineering* 10 (29): 9469–9479.

15 Prioux, N., Ouaret, R., Hetreux, G., and Belaud, J.P. (2023). Environmental assessment coupled with machine learning for circular economy. *Clean Technologies and Environmental Policy* 25 (2): 689–702.

16 Li, J., Li, L., Tong, Y.W., and Wang, X. (2023). Understanding and optimizing the gasification of biomass waste with machine learning. *Green Chemical Engineering* 4 (1): 123–133.

17 Li, J., Suvarna, M., Li, L. et al. (2022). A review of computational modeling techniques for wet waste valorization: research trends and future perspectives. *Journal of Cleaner Production* 367: 133025.

18 Tsui, T.H., van Loosdrecht, M.C., Dai, Y., and Tong, Y.W. (2023). Machine learning and circular bioeconomy: building new resource efficiency from diverse waste streams. *Bioresource Technology* 369: 128445.

19 Ascher, S., Watson, I., and You, S. (2022). Machine learning methods for modelling the gasification and pyrolysis of biomass and waste. *Renewable and Sustainable Energy Reviews* 155: 111902.

3

Application of Neural Networks for Model Prediction of Combustion and Emissions in Diesel Engines

Parampreet Singh Jassal[1], Sridhar Sahoo[2], and Neeraj Kumbhakarna[1]

[1] *Department of Mechanical Engineering, IIT Bombay, Mumbai, Maharashtra, India*
[2] *Department of Agricultural & Food Engineering, IIT Kharagpur, Kharagpur, West Bengal, India*

List of Abbreviations

Variable	Definition
AI	Artificial Intelligence
ANFIS	Adaptive Neuro-fuzzy Inference System
ANN	Artificial Neural Network
BTE	Brake Thermal Efficiency
CA50	Crank Angle for 50% Mass of Fuel Burnt
CFD	Computational Fluid Dynamics
CGP	Conjugate Gradient with Polak-Rieber Updates
CNG	Compressed Natural Gas
CNN	Convolutional Neural Networks
CR	Compression Ratio
DBT	Dry Bulb Temperature
EGT	Exhaust Gas Temperature
EGR	Exhaust Gas Recirculation
EVT	Exhaust Valve Temperature
FBNN	Feedback Neural Networks
FFNN	Feedforward Neural Network
GA	Genetic Algorithms
GRNN	General Regression Neural Network
HCCI	Homogeneous Charge Compression Ignition
HHO	Hydroxyl Gas
LM	Levenberg–Marquardt
LS-SVM	Least Square Support Vector Machine

Machine Learning for Sustainable Energy Solutions, First Edition. Edited by Zafar Said and Prabhakar Sharma.

MAPE	Mean Absolute Percentage Error
MIMO	Multi-input Multi-output
MISO	Multi-input Single-output
PCCI	Premixed Charge Compression Ignition
PM	Particulate Matter
PSO	Particle Swarm Optimization
RBFNN	Radial Basis Function Neural Networks
RMSE	Root Mean Square Error
RNN	Recurrent Neural Networks
RSM	Response Surface Methodology
SCG	Scaled Conjugate Gradient
SFC	Specific Fuel Consumption
SOC	Start of Combustion
SPL	Sound Pressure Level
SVM	Support Vector Machines
TA	Training Algorithm
TDC	Top Dead Center
UHC	Unburnt Hydrocarbons
WCO	Waste Cooking Oil

3.1 Introduction

Artificial intelligence (AI) is the execution of human intelligence with the help of machines and digital tools, mainly computers, to solve complex relationships [1, 2]. The advent of AI has penetrated different sectors, from research and development to goods and services. The automotive industry, one of the largest sectors, also implements AI in different applications to improve driving comfort, road safety, and engine efficiency to achieve environmentally friendly solutions. As the backbone of the automobile, the engine is now utilizing AI to model the in-cylinder performance and complex combustion phenomena to increase the engine's overall efficiency [3].

This chapter showcases the application of AI in diesel engine performance predictions using different input and output features. The initial section deals with the AI fundamentals, classifications, and applications for engine performance predictions. The parameters that affect the diesel engine performance, such as injection timing, pressure, and blend ratios, are reviewed and highlighted considering neural network models [3]. Thus, a brief review of the different neural network characteristics and their applicability to different parameters is presented. The existing research community has focused on implementing different strategies such as premixed charge compression ignition (PCCI), homogeneously charge compression ignition (HCCI), biodiesel blend, and nano-additive application to reduce emissions and improve the performance of the diesel engine [4]. However, carrying out the experiments requires time and resources before reaching any conclusions. Advanced computing facilities in recent times have helped reduce the burden of emerging 2D and 3D computational fluid dynamics (CFD) simulations [5].

Though these tools can handle large computational problems, neural networks have become widespread due to their generalization capability and their ability to solve nonlinear problems across all domains without mathematical modeling [6, 7]. However, optimizing the neural network model is necessary to obtain an optimal result with high accuracy [8–10]. Utilizing ANN models offers a time- and cost-effective alternative to experimental testing, particularly in scenarios where traditional methods like computational fluid dynamics (CFD) simulations or kinetic modeling may be impractical.

3.2 Artificial Neural Networks

Artificial intelligence (AI) is the general term for a wide range of technologies designed to automate tasks that traditionally require human efforts and intelligence. Artificial neural networks (ANNs) represent a subset of AI that analyses complex data by emulating the construction and functionality of the human brain. ANNs can address nonlinear problems, and because of their generalization ability, they are widely popular in different scientific and technical fields. ANNs tend to enhance the prediction performance through data training and validation by reviewing the input data without explicit mathematical modeling.

Figure 3.1 illustrates the different components and variables associated with the neural networks in a flowchart. ANNs have replaced conventional statistical and regression modeling due to their wide adaptability and exceptional versatility in different domains. ANNs can handle large amounts of data, avoiding issues with scalability, latency, and processing speeds. Due to its self-learning, nonlinearity, and adaptability, it is widely accepted for function approximation in different paradigms. ANNs consist of three interconnected layers: input, hidden, and output, to enter the data, process them, and produce output vectors [11].

To develop robust models, ANNs require careful consideration during the input and output parameter selection and normalization of large datasets. The choice of activation functions, the number of hidden layers, and training algorithms impact the developed model's efficiency. ANNs iteratively adjust the network parameters, such as weight and bias, to achieve the desired prediction accuracy. Various back-propagation training algorithms, such as train-bfg, –lm, and –rp, are used to optimize the ANN model.

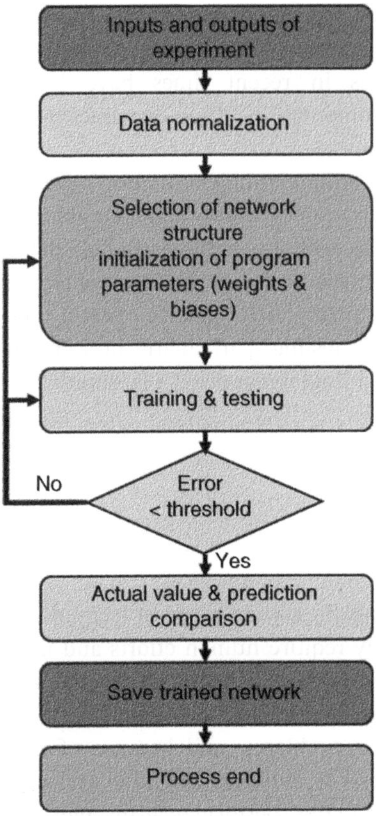

Figure 3.1 Outline of the application progress for ANN.

Subsequent testing and validation phases evaluate the effectiveness of ANN prediction models using metrics such as correlation coefficient (R) and coefficient of determination (R^2), root mean square error (RMSE), and mean absolute percentage error (MAPE), providing a holistic assessment of model efficiency and robustness, wherein the ranges of these calculations have been accepted as $R > 0.98$, $MSE < 0.001$, and $MAPE < 5\%$ [11]

$$R^2 = \frac{\left[n\sum_{i=1}^{n}\hat{y}_i y_i - \sum_{i=1}^{n}\hat{y}_i\sum_{i=1}^{n}y_i\right]^2}{\left[n\sum_{i=1}^{n}\hat{y}_i^2 - \left(\sum_{i=1}^{n}\hat{y}_i\right)^2\right]\left[\left[n\sum_{i=1}^{n}y_i^2 - \left(\sum_{i=1}^{n}y_i\right)^2\right]\right]}$$

$$\text{RSME} = \sqrt{\frac{1}{n}\sum_{i=1}^{n}(\hat{y}_i - y_i)^2}$$

$$\text{MAPE} = \frac{100\%}{n}\sum_{i=1}^{n}\left|\frac{(y_i - \hat{y}_i)}{y_i}\right|$$

where n is the number of samples; $\bar{y}_i$ is the prediction value; y_i is the actual experimental value.

There are numerous challenges associated with using ANNs to model diesel engines. First, collecting input data that accurately represents a range of engine variables and operating conditions. The intricate nonlinear characteristics of diesel engines, which are governed by variables such as fuel quality, load, and speed, make it challenging to create models without extensive experimental data.

Furthermore, reducing uncertainty, enhancing extrapolation capacity, and resolving convergence must be considered. Overcoming challenges related to noise quantization, variable difficulties, and continuous gradient enigmas is critical for advancing ANN modeling approaches and their practical use in engine modeling.

3.2.1 Types of Artificial Neural Networks

Various ANN types are depicted in Figure 3.2, particularly the feed-forward neural network (FFNN). FFNNs comprise well-ordered layers similar to the neurons processing inside the human brain [13]. These layers contain input units sending data through schematically planned network connections to the hidden layers and then to the output units. The network's capacity is determined by the strength of these connections, known as weights. FFNNs function in a one-way manner, sending data from input-to-hidden-to-output nodes, and that's why the name is feed-forward. A single-layer FFNN contains input and output layers; however, multilayer FFNNs consist of at least two hidden layers of neurons, thus allowing more complicated interconnections and computations [14, 15]. The capability of the ANN network to estimate the nonlinearity between the input and the output can be enhanced using more hidden layers in the network. Thus, it is specifically advantageous when working on large datasets. Radial basis function neural networks (RBFNNs) signify another category of ANN models known particularly for reducing the training time attributed to the simplified structure of the network. RBFNNs are efficient in diminishing extrapolation errors, thus improving dependability. In contrast, recurrent neural networks (RNNs) include feedback pathways, thus permitting the modeling of nonlinear systems with high computational efficiency [12]. Feedback neural networks (FBNNs) employ internal state memory to compute sequential inputs of data, thus enabling them to perform tasks like pattern recognition. Formulating ANN models for predicting the combustion behavior in diesel engines requires the transfer of input signals via the hidden layers, and the activation functions are used to estimate the nonlinear behavior of the data [16]. The training of the ANN model requires learning in which the given network acquires knowledge from a set of training commands and expected outputs. This step is generally supervised; however, there is a problem of overfitting when the algorithm results in an illogical response to new data. This is the primary

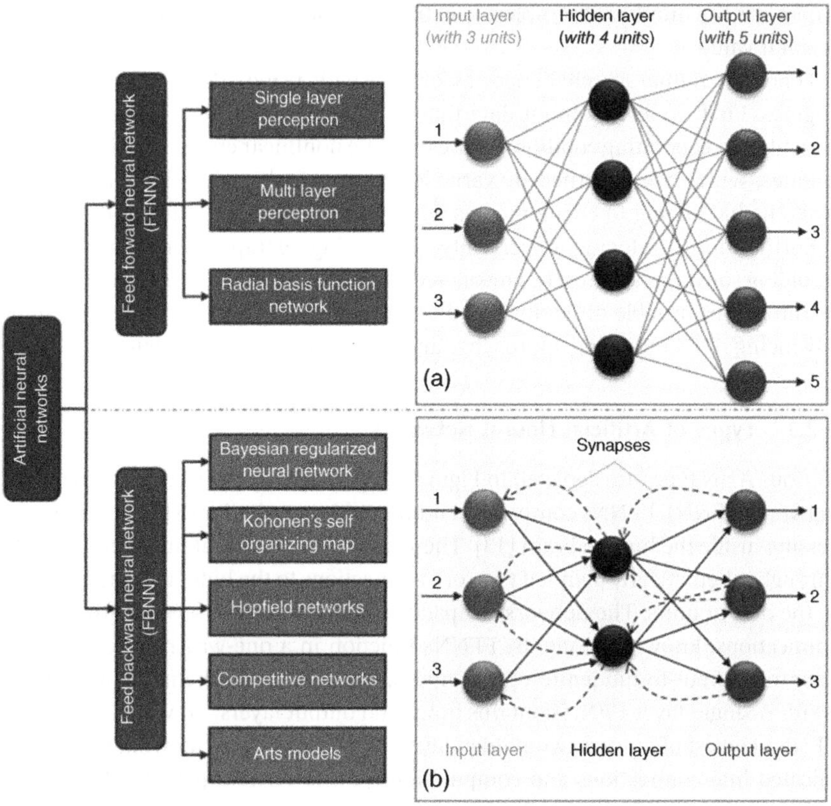

Figure 3.2 Classification of ANN. Hoang et al. [12]/with permission of Elsevier.

concern, and careful validation is needed to confirm that the model can be generalized [15]. The training tools like MATLAB have different training algorithms such as Levenberg–Marquardt (trainlm), gradient descent possessing adaptive learning (traingda), resilient back-propagation (trainrp), gradient descent having momentum and adaptive back-propagation (traingdx), scaled conjugate gradient back-propagation (trainscg), conjugate gradient back-propagation via Fletcher-Reeves updates (traincgf), and BFGS quasi-Newton back-propagation (trainbfg). Along with this, the transfer functions used are usually a hyperbolic tangent sigmoid (tansig), a logarithmic sigmoid (logsig), and a linear (purelin) function. ANN models pose hopeful abilities and skills to predict complex engine behaviors. ANNs comprehend different unique features and thus need to be carefully chosen as per the application requirement. The fundamental type of neural network, the FFNN, passes the data in one direction suitable for pattern recognition and regression analysis. In the case of RNNs, sequential data processing is done by feeding the output of the previous step as input to the current step using loops. RNNs are suitable for language modeling, time series analysis, and speech recognition. Convolutional neural networks (CNNs) can perform image classification and object detection using grid-like data utilizing convolutional layers for feature extraction.

The neural network comprises a set of interconnected layers with specific weights and biases. The weight and bias are adjustable parameters that are used to improve the network predictions. The weight represents the strength of the network between two neurons, whereas bias represents the offset from the output to match the data better. During training, these parameters are adjusted iteratively to increase the prediction accuracy and minimize errors. The activation function does the nonlinear transformation of the input data, enabling the network to learn complex data relationships. Various activation functions include Linear, Sigmoid, Tanh, ReLU, and Softmax. The sigmoid function is generally used for binary classifications, squashing input values between 0 and 1. The Tanh function extends the range to $[-1, 1]$, facilitating symmetric representation. The rectified linear unit (ReLU) is popular due to its simplicity and effectiveness, eliminating the vanishing gradient problem. Each activation function serves specific purposes, influencing network performance and training dynamics based on the task's requirements.

3.3 AI and ANN in Internal Combustion Engines

The modern-day scenario points toward a prominent surge in the development of artificial intelligence (AI) procedures, highlighting the beginning of a third upsurge after the previous ones in the decades of the 1970s and 2000s. Within the orthodox approaches directed at improving engine performance, the intricate challenges and cut-throat competition among various manufacturers pushed the researchers working in this field to research the utilization of AI algorithms in the engineering of internal combustion engines [8–10, 12, 14, 15]. These complicated challenges include different aspects related to the mapping of engine performance characteristics under varying operating conditions, the nonlinear dynamics of combustion reactions, and the control of particulate matter [12].

AI algorithms present a favorable prospect for assessing engine performance by avoiding these intricacies and exclusively depending on empirical data. Moreover, traditional engine evaluation techniques often neglect the effect of random cyclic variations, thus compromising the technique's robustness. AI models, on the other hand, permit the formulation of strong and effective control on ICE combustion analysis, encompassing the assessment of homogeneous charge compression ignition or reactivity-driven compression ignition [15]. With substantial innovations made in the computational management of big data, engine design has advanced even further.

With the support of comprehensive data obtained from engines, such as velocity or air, engine load, power, density and viscosity of fuel, fuel spray, and atomization characteristics, AI models enable the conception of sophisticated and intelligent control systems. AI procedures are progressively favored because of their competence, swiftness, precision, and cost compliance. AI models can even extract the data from interconnected engines to formulate intelligent control systems for engine management. The interlinked ICE network of similar configurations acts as a vital resource, delivering important insights and steering toward incessant improvements in ICE technology and engine performance [17]. ANNs being applied to assess the performance of internal combustion engines is a relatively recent development, and the main aim of this application is to gauge engine performance and predict emissions. ANN has been applied to various sectors; however, the implementation of ANN in the automotive sector is becoming increasingly popular due to its capability to model nonlinear combustion dynamics and intricate engine behaviors. In contrast to conventional computer simulations, the computational resources required by ANN are relatively low, and at the same time, the results obtained are quicker and more accurate.

3.4 ANN in Diesel Engines

In the context of diesel engines, technical breakthroughs made in AI have driven us into a new age of brainy and conscious vehicle systems. This comes in with a blend of knowledge obtained from machine learning and deep learning models and algorithms. There are a wide range of methods comprising supervised, unsupervised, and reinforcement learning methods. Sensors incorporated at various engine component locations provide data for AI methods such as detection, regression, classification, clustering, and segmentation. This is very important in decoding parameters that affect the engine performance, emissions, fault detection, maintenance, assistance required by the driver, etc. [18]. For a self-driving automotive, quick autonomous selection of complex decisions is crucial and thus demands reinforced learning, deep learning, and semi-supervised learning. These systems are eventually set out to route environmental conditions, aid in path forecasting, detect objects, and perform decision-making control tactics. Scheming a neural network model for estimating combustion performance and emissions in diesel engines demands thoughtful deliberation of the architecture, data processing, and training.

3.4.1 ANN for Different Fuel Properties

The fuel properties that affect engine combustion and performance can be extensively correlated using various modeling techniques. Razavi et al. [19] used the least square support vector machine (LS-SVM) method to predict the biodiesel characteristics as a function of the fatty acid composition. The model's performance was evaluated using the coefficient of determination to evaluate the pour point, cloud point, iodine value, and kinematic viscosity. Further, the existing model was compared with the adaptive neuro-fuzzy inference system (ANFIS) model, concluding that the LS-SVM method has better prediction ability. Sharon et al. [20] employed ANN and used the developed correlation to develop a Simulink model to predict engine performance and emissions. The biodiesel blend composition and engine brake power were input parameters to predict the fuel consumption, thermal efficiency, NO_x, HC, CO, and smoke density. The developed ANN model predicted well for biodiesel blends with a correlation coefficient higher than 0.99. However, the study does not detail the training criteria and select the performance metrics.

A similar study was conducted by Rao et al. [21] to predict the engine performance and emission parameters as a function of the engine load and fuel blend type. The error percentage of the predicted values was within the range of 0.01–0.03. The optimum blend percentage with a higher performance output was achieved using the ANN. However, the details about the data normalization training-validation splits are not mentioned. Dey et al. [22] employed ANN to forecast the performance and emissions of a single-cylinder diesel engine operating with different biodiesel blends. They found that the Levenberg–Marquardt back-propagation algorithm with logistic sigmoid activation function is predicted to have MAPE in the 2.32–4.54% range.

Gul et al. [23] applied the ANN model to optimize the engine parameters (engine load, speed, and fuel properties) after optimizing them using the Grey-Taguchi method. They found that the ANN model accurately predicted the optimal engine conditions, significantly improving the response factors. Najafi et al. [24] combined GA for one neuron with ANN to optimize biodiesel and CNG usage in dual-fuel CNG engines. They found a reduction in NO_x, CO, and PM emissions, maintaining a power output similar to that of the diesel engine. These studies show the versatile application of ANN models to predict and optimize engine performance with high efficiency and lower emissions.

3.4.2 ANN for Diesel Engine Performance

There are different studies available where diesel engine behavior for different running conditions has been studied using diesel fuel. Some of these studies have particularly used a multi-input-one-output modeling approach. Celik and

Arcaklioglu [25] used different ANN models to estimate parameters such as specific fuel consumption (SFC), air-to-fuel ratio, and exhaust gas temperature (EGT) while working on an eight-cylinder diesel engine.

The inputs to the model included engine rpm, power, and water temperature. The ANN architecture consisted of one hidden layer containing four neurons. Back-propagation through the SCG and LM algorithms achieved high accuracy, with R^2 values approaching 0.99 and RMSE lesser than 0.03. Irdemoosa et al. [26] focused on predicting the fuel consumption in a mining truck using ANNs. Input parameters included loading time, loaded travel time, idled empty time, empty travel time, payload, and idle time to load. The ANN 6.9.9.1 architecture included two hidden layers carrying nine neurons individually. Bietresato et al. [27] assessed the use of ANNs to indirectly evaluate farm tractor performance. Predictions for torque and brake-specific fuel consumption (BSFC) were made from the known temperature of the exhaust gas (EGT) and motor oil. The authors concluded that the Gaussian function outperformed the sigmoidal function for their application. Goudarzi et al. [28] formulated novel ANN methods to compute the exhaust valve temperature (EVT) and the convection heat transfer coefficient. The inputs required were temperatures at different regions, and inverse heat conduction procedures were then used to estimate the convective coefficient. The ANN model employed different neuron quantities in their hidden layer, and the LM model portrayed the best convergence. The EVT estimation model needed two hidden layers with 15 and 7 neurons, whereas the convective coefficient model, which comprised two hidden layers, required 10 and 5 neurons, respectively.

Arcaklioğlu et al. [29] examined the impact of injection pressure, throttle position, and speed on engine parameters and emissions. ANNs were used to predict the performance and emissions separately. A single hidden layer with 5–15 neurons was utilized to predict performance parameters.

In order to estimate the emissions, two hidden layers have been used, and different numbers of neurons, that is, 8-7, 9-5, 9-7, and 10-7, were used in these layers. These configurations have been used to estimate different parameters. Three training functions (LM, CGP, and SCG) have been used in conjunction with the back-propagation model and the logsig transfer function. The capability of the ANN models has been determined using statistical metrics, namely, R^2 (coefficient of determination), RSME (root mean square error), and MAPE (mean absolute percentage error). Roy et al. [37] developed an ANN model to scrutinize the common rail direct injection engine for different amounts of exhaust gas recirculation. Their model estimated various parameters such as brake thermal efficiency (BTE), specific fuel consumption (SFC), nitrogen oxides (NO_x), CO_2 emissions, and particulate matter.

The architecture utilized for the model consisted of four input nodes, five output nodes, and two hidden layers, each containing ten neurons. The transfer function employed was logsig, with the LM algorithm, and the least MSE being the criterion for the loss function. Nikzadfar and Shamekhi [30] utilized the CFD-generated data as input for their ANN model. The model predicted engine parameters such as brake-specific fuel consumption (BSFC), torque, NOx, and soot. The engine input parameters correspond to ten variables, namely, the rate of exhaust gas recirculation, the thermophysical properties of air, rpm, the pilot and main fuel amount, the exhaust and rail pressure, the injection crank angle, and the pilot retard. Perturbation sensitivity analysis was carried out to assess the individual impact of engine inputs on outputs. Table 3.1 summarizes the studies conducted for diesel engines using ANN.

3.4.3 ANN for Diesel Engines Using Biodiesel Blends

Kshirsagar et al. [38] used ANN to forecast the performance and emission attributes of stationary four-stroke engines operating on diesel blended with *Calophyllum inophyllum*-based biodiesel. They varied the engine loads and biodiesel blending percentages while investigating the impact of the time of fuel injection and pressure. ANN techniques were formulated to predict diesel engines' performance parameters like BTE, EGT, and BSFC, as well as emission parameters like CO, HC, CO_2, NO_x, and soot formation tendency. The architecture of the ANN model employed comprised single-hidden and two-hidden layer models containing different numbers of neurons. These neurons were optimized based on the minimization of mean square error (MSE). The hidden layers employed the tan-sig transfer function and trainlm as the training algorithm. The input parameters were also normalized to keep them constrained in the $[-1, 1]$ range. For different engine parameters being scrutinized, the results obtained indicated notable accuracy of prediction, having R^2 value of almost equal to 1.

Likewise, Sharma et al. [34] used the back-propagation method to estimate the emissions and performance of a single cylinder four-stroke diesel engine blended with polanga-based biodiesel. The model being used portrayed significant accuracy of prediction for performance parameters like BTE, BSFC, EGT, peak pressure, and emissions parameters like CO, HC, and NO_x. Canakci et al. [33] studied five different ANN models to investigate the performance and emissions of diesel blends made from waste-frying palm oil (WFPO). Correlations were developed among different input and output parameters.

The selection of relevant input variables and ANN architectures was emphasized to augment the accuracy of prediction. The logsis transfer function was used for all indicated networks, and one hidden layer was employed. For uniformity, the input and output data have been normalized in the range of 0 to 1. Oğuz et al. [39] used different fuel blends in a diesel engine and modeled the performance using ANN methods for BSFC, torque, and power. Their model had input

Table 3.1 Summary of recent studies.

Fuel(s)	Training algorithm	Transfer function	Network	Input	Output	Reference
Diesel	Back-propagation	–	6.9.9.1	Payload, loading and idling times	Fuel flow rate	Irdemoosa and Dindarlo [26]
Diesel	Back-propagation	Gaussian	2.26.10.1	Exhaust gas temperature and temperature of the motor oil	Brake-specific fuel consumption	Bietresato et al. [27]
Calophyllum inophyllum methyl ester	Back-propagation, Levenberg–Marquardt	Tangent sigmoid	4.16.3, 4.14.14.6	Engine load, fuel blend proportion, time of injection and pressure	Brake thermal efficiency, brake-specific energy consumption, exhaust gas temperature; CO, CO_2, HC, O_2, NO, Soot	Kshirsagar and Anand [30]
Jatropha oil, Hydrogen	Back-propagation, Levenberg–Marquardt.	Logarithmic sigmoid and Tangent sigmoid	3.16.8	Load, biodiesel blend, H_2	Brake thermal efficiency, brake-specific fuel consumption, NO_X, HC, exhaust gas temperature, O_2, CO, CO_2	Javed et al. [31]
WCO	Levenberg–Marquardt, Back-propagation	Logarithmic sigmoid and Tangent sigmoid	2.25.4	Blend proportion, engine rpm	Engine torque, specific fuel consumption, CO, HC	Ghobadian et al. [32]
Bioethanol-biodiesel	Back-propagation	Tangent sigmoid	2.28.4	Engine speed and the type of fuel	Torque, engine power, fuel flow rate, and brake-specific fuel consumption	Oguz et al. [33]

Waste Palm Oil	Back-propagation, Levenberg–Marquardt, scaled conjugate gradient	Logarithmic sigmoid	8. 8.11	Lower calorific value, pressure, density, viscosity, cetane number, humidity, DBT, engine rpm	Torque, load, cylinder pressure, thermal efficiency, CO, NO_x, CO_2, EGT, HC, smoke, flow rate of fuel, and injection pressure	Canakci et al. [34]
CNG	Back-propagation, gradient Descent.	Logarithmic sigmoid and linear	3.5.4.1, 3.3.5.1, and 3.5.3.1	Biogas flow, methane percentage, power	Exhaust gas temperature, pilot fuel and airflow rate	Akkouche et al. [35]
Diesel CNG	Back-propagation, Levenberg–Marquardt	Logarithmic sigmoid and linear	4.4.1, 4.5.1	Engine rpm, cetane number, CNG flow rate and density	Vibration and sound pressure level	Celebi et al. [36]

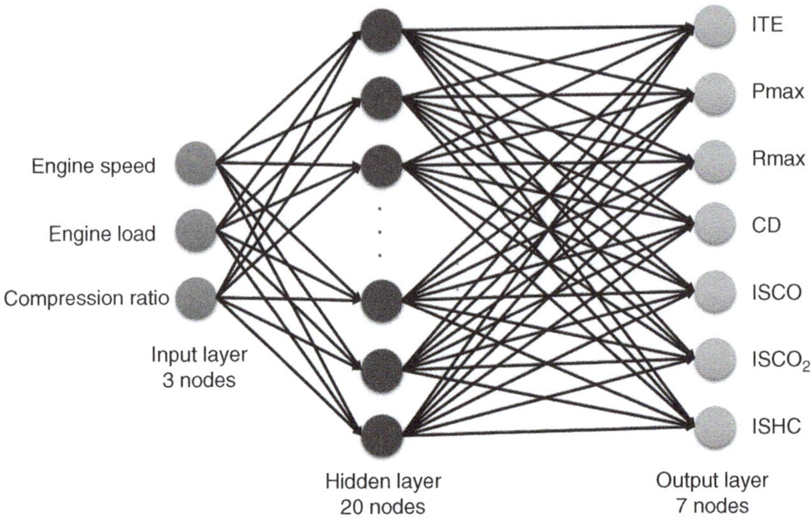

Figure 3.3 Structure of an ANN model. Sahoo et al. [31]/with permission of Elsevier.

data comprising biodiesel-diesel blends and engine speeds, and it was noticed that the value of the correlation coefficient was quite high. In a separate research work, Iscan et al. [40] estimated the capability of a diesel engine running on biodiesel fuels. The FFBP procedure with the mean square error function was used, where the neural network comprised three layers having ten neurons. They were able to predict engine performance and emission-related parameters, highlighting the ruggedness of their ANN model for high accuracy of prediction.

Furthermore, Venugopal et al. [41] applied ANN to estimate the performance characteristics and emissions of a single-cylinder four-stroke diesel engine running on biodiesel blends at different compression ratios (CRs) and engine loads. The NN based on different hidden layers contained two input parameters, namely, CR and load and multiple output data, as illustrated in Figure 3.3. The ANN model successfully predicted parameters such as the brake thermal efficiency, mechanical efficiency, brake-specific fuel consumption, and emissions of HC, CO_2, CO, and NO_x. The experimental research results showed that a compression ratio (CR) of 18:1 improved the engine performance.

In the context of emissions, it was found that a CR of 17:1 is more suitable. Karthickeyan et al. [32] used ANN to deduce the effects of different compression ratios on the engine performance characteristics and emissions. ANN models formulated using back-propagation algorithms could estimate the characteristics of a diesel engine running on biodiesel blends at varying values of CR. Their study utilized a set of seven training algorithms comprising four different functions. For all the models studied, the trainlm training function with log and tan sigmoidal functions was the most optimum configuration of the ANN architecture. The Trainlm algorithm is a back-propagation training function frequently employed in ANN training. It is effective in managing the weights of the network, such that the error can be minimized during the training of the ANN model.

While working on a diesel engine based on a biodiesel blend prepared from waste cooking oil (WCO), Ghobadian et al. [42] used ANN to predict the engine behavior. They accurately estimated the BSFC, engine torque, and HC and CO emissions at different engine rpm values and blending ratios. They particularly emphasized the effect of tansig and logsig activation functions containing a single hidden layer. Moreover, they also studied the effect of variation in the training algorithm (TA), namely, traingdx, trainlm, trainscg, and trainrp, and varied the neurons in the range of 22–28 in the hidden layer. The input and output parameters were regulated within −1 and 1 through normalization. This study investigated the effect of using tansig (hyperbolic tangent) and logsig (logistic sigmoid) activation functions. It was observed that tansig maps the input-to-output in the range of −1 and 1, whereas logsig does the same in the range of 0 and 1.

By switching between these two activation functions, the researchers aimed to gauge their effect on the performance of the ANN model. Javed et al. [36] explored the use of ANN for predicting a diesel engine's emission and performance characteristics fueled with hydrogen and biodiesel blends.

The importance of the selection of suitable activation functions and training procedures was emphasized in this research work. Many activation functions and training algorithms were tested for each hidden layer. In a separate study, Aydin et al. [35] employed ANN and response surface methodology (RSM) to estimate the behavior of a diesel engine based on biodiesel fuel blends. It was noticed that the degree of correlation between the ANN prediction and the actual data is quite strong.

The studies summarized above highlight the importance of ANN models in estimating engine performance with a high degree of confidence. ANN estimates pose a cost-complaint and competent substitute to traditional experimental methods, thus making researchers aware of the complicated relationships between inputs and outputs. Further exploration and refinement in this direction will certainly advance the ANN model's capability and potential to develop clean and efficient engines.

3.4.4 ANN for Gaseous Fuels

Many researchers have utilized ANN models to study the applicability of gaseous fuels in dual-fuel mode engines. Some focus on multi-input single-output (MISO) models, whereas others stress using multi-input multi-output (MIMO) models for studying engine behaviors.

3.4.4.1 MISO Model Studies

Celebi et al. [43] studied the engine working on biodiesel and natural gas blends for its propensity to vibrations. The ANN architecture comprising 4-4-1 and 4-5-1 signatures was used to estimate the vibrations and sound pressure level (SPL). Transfer functions, namely, purelin and logsig, were utilized, and the results obtained were quite satisfactory.

Akkouche et al. [44] constructed models aimed at forecasting airflow, pilot fuel flow, and exhaust temperature within a microcombined heat unit functioning in the CNG-diesel dual-fuel mode. They employed various network architectures and utilized gradient descent with momentum as the learning function. The outcomes demonstrated high R^2 values close to 1 for all the models.

Javed et al. [45] predicted engine noise levels in dual-fuel mode using biodiesel blended with zinc oxide nanoparticles and H_2 induction. They achieved a regression coefficient of 0.99 with a 4-8-1 network architecture.

3.4.4.2 MIMO Model Studies

Javed et al. [36], while working on various blends of biodiesel with H_2 gaseous fuel, examined the performance of the engine. A three-layered neural network architecture with three input and eight output layers interlinked via 16 nodes in the hidden layer. The Levenberg–Marquardt (LM) algorithm was employed along with tangent sigmoid and logarithmic sigmoid transfer functions to learn the ANN model. It was observed that the regression coefficient value is very high; thus, the model was recognized as effective in predicting engine behavior.

Yusuf et al. [46] formulated an ANN model and used it for a CNG engine running in the dual-fuel mode. They studied different output variables as a function of engine rpm and the flow rate of the CNG fuel. High correlation coefficients spanning over 0.92 to 0.99 have been attained.

Uludamar et al. [47] studied engine vibrations fueled with hydroxyl (HHO) gas-biodiesel, achieving satisfactory results with an ANN model considering the cetane number, heating value, HHO percentage, and speed as inputs.

3.4.4.3 Comparative Studies

Syed et al. [48] compared seven learning algorithms for an ANN model developed for an engine running on a dual fuel, namely, hydrogen-diesel. They found that the quasi-Newton back-propagation and tansig-tansig transfer function combination yielded the best results.

Taghavifar et al. [49] estimated the convective coefficient of a diesel engine running on H_2 fuel through a CFD model to generate data for learning. The

ANN model utilized for the study attained RMSE magnitudes close to 9.13 for a 3-17-3 configuration. Thus, it can be noticed from the above-mentioned studies that ANN models are highly efficient and effective in estimating different variables for engine characteristics, giving a sound comprehension of the parameters that need to be optimized. These, in a nutshell, aid in reducing engine emissions and improving engine performance.

Kenanoglu et al. [50] studied the performance of a two-cylinder, four-stroke diesel engine using hydroxy (HHO) gas fuel enrichment in a biodiesel blend. They employed the Levenberg–Marquardt (LM) learning algorithm to fine-tune the weights within the cascade-forward network to construct the ANN architecture. Cascade-forward networks resemble feed-forward networks, with the only difference being that they include the connections from the input and all preceding layers to the succeeding layers. Input variables comprised motor speed, fuel type, and fuel flow rate; outputs were brake power, torque, and NO_x emissions. There were two hidden layers in the architecture comprising 10 and 15 neurons, respectively, which were determined randomly using trial and error. Figure 3.4 describes the mean square error as a function of the number of epochs. Epochs represent the

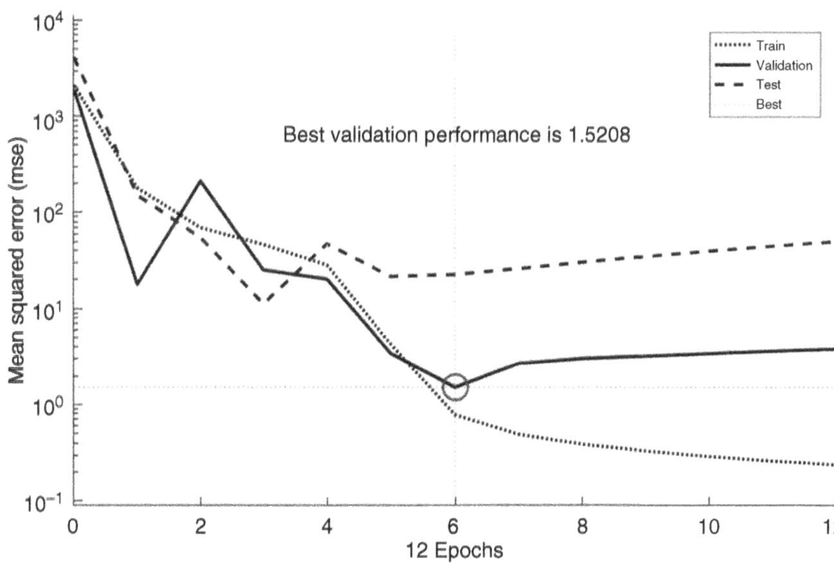

Figure 3.4 Mean square error vs number of epochs for estimating the credibility of ANN. Kenanoglu et al. [50]/with permission of Elsevier.

number of forward and backward runs carried out during the training of the ANN model. The mean square error of 1.5208 justifies only a small value of error and, thus, high accuracy in the model estimation.

3.4.5 ANN for HCCI Engines

Homogeneous charge compression ignition (HCCI) engines epitomize encouraging progress in combustion expertise because of their potential for high efficacy and low pollutant discharges, particularly in low-temperature combustion modes. Controlling HCCI engines poses a significant challenge, particularly considering the impracticality of experimental testing. Developing an efficient and accurate model for HCCI combustion control is imperative. Compared to computational fluid dynamics (CFD) simulations or kinetic modeling, using an ANN model offers a faster and cheaper alternative. This is particularly relevant considering that traditional thermo-kinetic HCCI models require a substantial amount of time, sometimes taking several hours to days for the simulations of only one engine cycle, which typically lasts from 40 to 120 ms. There are three major categories of HCCI models, namely, empirical, thermos-kinetic, and ANN models, as highlighted in Figure 3.5. Although empirical models demand a large amount of experimental data and thermokinetic models are limited by computational requirements that are more than actual engine capabilities, ANN models on the contrary work as a bridge among accuracy, resources,

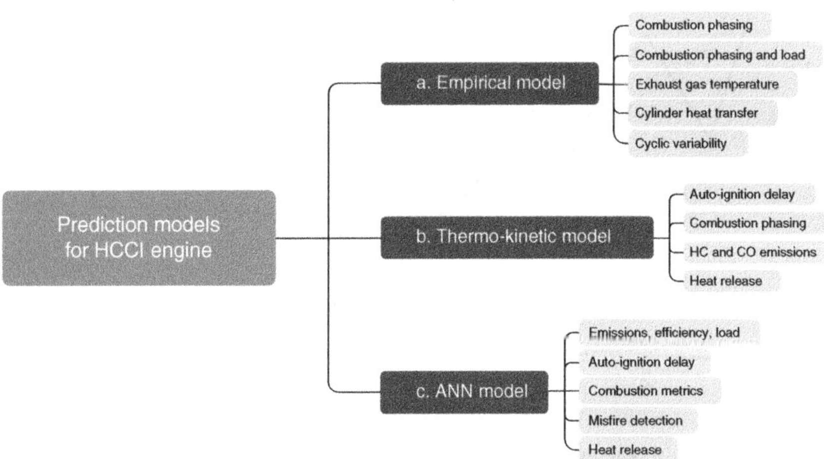

Figure 3.5 ANN model used in the HCCI engines. Adapted from [12].

and data requirements. Compared to internal combustion engines, the use of ANN in HCCI engines is quite limited to date.

Rezaei et al. [51] developed an ANN model capable of real-time operation to predict the performance and emissions characteristics of HCCI engines using oxygenated fuels such as butanol and ethanol. Maurya and Kumar [52] devised a real-time model to discern the intensity of ringing in an HCCI engine powered by hydrogen. After formulating an ANN model, they found a good degree of correlation (with a value as large as 0.99) between the ringing operation and CA50. Furthermore, they could correlate that the ringing intensity tends to intensify with alterations like advanced combustion phasing, higher intake air temperature, and higher value of equivalence ratio. Likewise, Bahri et al. [9], while working on the HCCI engines, established an ANN model to predict the ringing action in real time. They could develop precise estimations of ringing intensity and recognize the operating points. RI was controlled by varying CA50, and the ANN model was developed with the values of in-cylinder pressure at 5, 10, and 15 °CA aTDC (P5, P10, and P15) and its maximum pressure (Pmax).

To settle the issue of the long training needed for the ANN models and the possibility of local minima problems in conventional back-propagation neural networks, Bendu et al. [53] developed a general regression neural network (GRNN) model to estimate the behavior of the HCCI engine. The GRNN approach is established around a probabilistic functional network with a layout as described in Figure 3.6. This model portrayed a maximum error of just 2% and needed reduced training data compared to other models. Additional improvements comprise the amalgamation of GRNN with particle swarm optimization (PSO) to enhance the smooth and efficient operation of the HCCI engine. Moreover, some studies also used ANSYS simulations in combination with ANN models to study the behavior of HCCI engines and obtained satisfactory results in estimating combustion performance and emissions.

Nazoktabar et al. [54] formulated a gray-box HCCI model by combining physical and ANN models to predict complex non-linear characteristics such as HCCI emissions and its reliance on the mixture strength. This gray-box model curtailed the dependence on the huge experimental data required otherwise while optimizing the engine's performance. The physical model outputs were the inputs to the gray-box model, and these subsequent gray-box outputs were the main inputs for the emissions model. Moreover, ANN has also been used to envision HCCI start of combustion (SOC) and locate misfires through dynamic measurements. Thus, these studies highlight the prospects of ANN models in augmenting the understanding and control of HCCI engines.

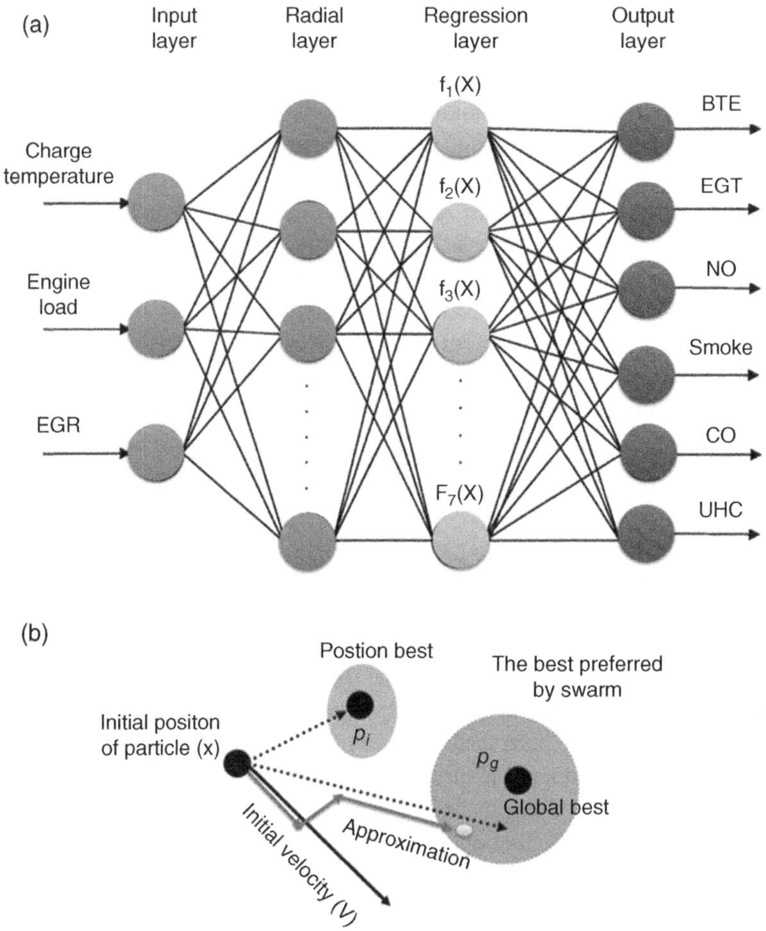

Figure 3.6 (a) Typical GRNN architecture and (b) PSO structure. Bendu et al. [53]/with permission of Elsevier.

3.5 Conclusions

AI presents a new, innovative method for predicting and optimizing engine properties and performances. Conventional methods have difficulty dealing with nonlinearity and stochastic variability, but AI models handle these issues efficiently. With the availability of large amounts of data and advanced computational power, AI helps in engine optimization, real-time engine management, and performance predictions, deciding the future of autonomous vehicle development. Also, the individual effect of each fuel and engine parameter has shown promising results in predicting engine output parameters using ANN.

ANN offers a cost-effective approach to predict engine performance, combustion, and emission characteristics to optimize the complete engine. This can be further enhanced by combining other optimization methods, such as Genetic Algorithms (GA) and support vector machines (SVMs), to increase its versatility and operating regimes. The literature referred to in the chapter also reveals the accurate prediction capability of ANN in predicting engine performance and emissions for various fuel blends, engine load, and operation criteria. This enables the researchers to expand their insights into the complex engine behavior. In this way, ANN promises the development of advanced and more efficient engine technologies, contributing to environmental sustainability and energy efficiency.

ANN models require low computational power and time to predict and optimize engine design parameters compared to their counterparts, such as CFD and chemical kinetic models. Recent technology uses the design of experiments to choose the least number of experiments or simulations and further extrapolate the data throughout the domain using ANN models. This enables the efficient utilization of ANN models in designing the intricate shapes of the combustion chamber, intake manifolds, and combustion modeling coefficients. Combining experiments or simulations with ANN modeling increases predictive capabilities and reduces the time and computational effort needed to design modern engines.

References

1 Chowdhary, K.R. (2020). *Fundamentals of Artificial Intelligence*, 603–649. New Delhi: Springer India.

2 Müller, V.C. (ed.) (2016). *Fundamental Issues of Artificial Intelligence*, vol. 376, 520. Berlin: Springer.

3 Khanzode, K.C.A. and Sarode, R.D. (2020). Advantages and disadvantages of artificial intelligence and machine learning: a literature review. *International Journal of Library & Information Science* 9 (1): 3.

4 Hall, W. and Pesenti, J. (2017). *Growing the Artificial Intelligence Industry in the UK*. UK, London: Department for Digital, Culture, Media and Sport (United Kingdom).

5 Makridakis, S. (2017). The forthcoming artificial intelligence (AI) revolution: its impact on society and firms. *Futures* 90: 46–60.

6 Agwu OE, Akpabio JU, Alabi SB, Dosunmu A (2018) Artificial intelligence techniques and their applications in drilling fluid engineering: a review. *Journal of Petroleum Science and Engineering* 167:300–315. https://doi.org/10.1016/j.petrol.2018.04.019

7 Haleem A, Javaid M, Khan IH (2019) Current status and applications of Artificial Intelligence (AI) in medical field: an overview. *Current Medicine Research and Practice* Volume 9, Issue 6, Pages 231–237. https://doi.org/10.1016/j.cmrp.2019.11.005

8 Drews, P., Albin, T., HeBeler, F.J. et al. Fuel-Efficient Model-Based Optimal MIMO Control for PCCI Engines. *Preprints of the 18th IFAC World Congress Milano (Italy),* August 28 – September 2, 2011

9 Bahri B, Shahbakhti M, Aziz AA (2017) Real-time modeling of ringing in HCCI engines using artificial neural networks. *Energy* 125:509–518. https://doi.org/10.1016/j.energy.2017.02.137

10 Anarghya, A., Rao, N., Nayak, N. et al. (2018). Optimized ANN-GA and experimental analysis of the performance and combustion characteristics of HCCI engine. *Applied Thermal Engineering* 132: 841–868. https://doi.org/10.1016/j.applthermaleng.2017.12.129.

11 Uslu, S. and Celik, M.B. (2018). Prediction of engine emissions and performance with artificial neural networks in a single cylinder diesel engine using diethyl ether. *Engineering Science and Technology, an International Journal* 21 (6): 1194–1201. https://doi.org/10.1016/j.jestch.2018.08.017.

12 Hoang AT, Nizetic S, Ong HC, Tarelko W, Pham VW, Le TH, Chau MQ, Nguyen XP, A review on application of artificial neural network (ANN) for performance and emission characteristics of diesel engine fueled with biodiesel-based fuels. *Sustainable Energy Technologies and Assessments* 47 (2021) 101416. https://doi.org/10.1016/j.seta.2021.101416

13 Hertz, J., Krogh, A., and Palmer, R.G. (2018). *Introduction to the theory of neural computation.* Boca Raton: CRC Press https://doi.org/10.1201/9780429499661.

14 Huang Guang-Bin, Chen Yan-Qiu, Babri HA. Classification ability of single hidden layer feed-forward neural networks. *IEEE Transactions on Neural Networks* 2000; 11 (3): 799–801. https://doi.org/10.1109/72.846750.

15 Bin, H.G., Zhu, Q.Y., Siew, C.K. (2004). Extreme learning machine: A new learning scheme of feed-forward neural networks. *IEEE International Joint Conference on Neural Networks.* https://doi.org/10.1109/IJCNN.2004.1380068.

16 S.V. Channapattana, Abhay A. Pawar, Prashant G. Kamble, Optimisation of operating parameters of DI-CI engine fueled with second generation Bio-fuel and development of ANN based prediction model, *Applied Energy* 187 (2017) 84–95. https://doi.org/10.1016/j.apenergy.2016.11.030

17 Turkson, R.F., Yan, F., Ali, M.K.A., and Hu, J. (2016). Artificial neural network applications in the calibration of spark-ignition engines: an overview. *Engineering Science and Technology, an International Journal* 19 (3): 1346–1359.

18 Kakati, D., Roy, S., and Banerjee, R. (2019). Development of an artificial neural network based virtual sensing platform for the simultaneous prediction of emission-performance stability parameters of a diesel engine operating in dual

fuel mode with port injected methanol. *Energy Conversion and Management* 184: 488–509.

19 Razavi, R., Bemani, A., Baghban, A. et al. (2019). An insight into the estimation of fatty acid methyl ester based biodiesel properties using a LSSVM model. *Fuel* 243: 133–141. https://doi.org/10.1016/j.fuel.2019.01.077.

20 Sharon, H., Jayaprakash, R., M, K.s. et al. (2012). Biodiesel production and prediction of engine performance using SIMULINK model of trained neural network. *Fuel* 99: 197–203.

21 Prasada Rao, K., Victor Babu, T., Anuradha, G., and Appa Rao, B.V. (2017). IDI diesel engine performance and exhaust emission analysis using biodiesel with an artificial neural network (ANN). *Egyptian Journal of Petroleum* 26: 593–600.

22 Dey S, Reang NM, Majumder A, Deb M, Das PK. A hybrid ANN-Fuzzy approach for optimization of engine operating parameters of a CI engine fueled with diesel palm biodiesel-ethanol blend. *Energy* 2020; 202: 117813. https://doi.org/10.1016/j.energy.2020.117813.

23 Gul M, Shah AN, Aziz U, Husnain N, Mujtaba MA, Kousar T, et al. Grey-Taguchi and ANN based optimization of a better performing low-emission diesel engine fueled with biodiesel. *Energy Sources Part A: Recovery, Utilization, and Environmental Effects* 2019: 1–14. https://doi.org/10.1080/15567036.2019.1638995.

24 Najafi B, Akbarian E, Lashkarpour SM, Aghbashlo M, Ghaziaskar HS, Tabatabaei M. Modeling of a dual fueled diesel engine operated by a novel fuel containing glycerol triacetate additive and biodiesel using artificial neural network tuned by genetic algorithm to reduce engine emissions. *Energy* 2019; 168: 1128–1137. https://doi.org/10.1016/j.energy.2018.11.142.

25 Çelik, V. and Arcaklioğlu, E. (2005). Performance maps of a diesel engine. *Applied Energy* 81: 247–259. https://doi.org/10.1016/j.apenergy.2004.08.003.

26 Siami-Irdemoosa E, Dindarloo SR. Prediction of fuel consumption of mining dump trucks: a neural networks approach. *Applied Energy* 2015; 151: 77–84. https://doi.org/10.1016/j.apenergy.2015.04.064

27 Bietresato, M., Calcante, A., and Mazzetto, F. A neural network approach for indirectly estimating farm tractors engine performances. *Fuel* 143: 144–154. https://doi.org/10.1016/j.fuel.2014.11.019.

28 Goudarzi K, Moosaei A, Gharaati M. Applying artificial neural networks (ANN) to the estimation of thermal contact conductance in the exhaust valve of internal combustion engine. *Applied Thermal Engineering* 2015; 87: 688–697. https://doi.org/10.1016/j.applthermaleng.2015.05.060

29 Arcaklioğlu, E. and Çelıkten, I. (2005). A diesel engine's performance and exhaust emissions. *Applied Energy* 80: 11–22. https://doi.org/10.1016/j.apenergy.2004.03.004.

30 Nikzadfar, K. and Shamekhi, A.H. (2014). Investigating the relative contribution of operational parameters on performance and emissions of a common-rail diesel engine using neural network. *Fuel* 125: 116–128. https://doi.org/10.1016/j.fuel.2014.02.021.

31 Sahoo S, Kumar VNSP, Srivastava DK. Quantitative analysis of engine parameters of a variable compression ratio CNG engine using machine learning. *Fuel* 2022; 311: 122587. https://doi.org/10.1016/j.fuel.2021.122587

32 Karthickeyan V, Balamurugan P, Rohith G, Senthil R. Developing of ANN model for prediction of performance and emission characteristics of VCR engine with orange oil biodiesel blends. *Journal of the Brazilian Society of Mechanical Sciences and Engineering* 2017; 39(7): 2877–2888. https://doi.org/10.1007/s40430-017-0768-y.

33 Canakci M, Necati A, Arcaklioglu E, Erdil A. Expert systems with applications prediction of performance and exhaust emissions of a diesel engine fueled with biodiesel produced from waste frying palm oil. *Expert Systems with Applications* 2009; 36: 9268–9280. https://doi.org/10.1016/j.eswa.2008.12.005

34 Sharma A, Sahoo PK, Tripathi RK, Meher LC. Artificial neural network-based prediction of performance and emission characteristics of CI engine using polanga as a biodiesel. *International Journal of Ambient Energy* 2016; 37(6): 559–70. https://doi.org/10.1080/01430750.2015.1023466.

35 Aydın, M., Uslu, S., and Bahattin, Ç.M. (2020). Performance and emission prediction of a compression ignition engine fueled with biodiesel-diesel blends: a combined application of ANN and RSM based optimization. *Fuel* https://doi.org/10.1016/j.fuel.2020.117472.

36 Javed S, Satyanarayana Murthy YVV, Baig RU, Prasada Rao D. Development of ANN model for prediction of performance and emission characteristics of hydrogen dual fueled diesel engine with Jatropha Methyl Ester biodiesel blends. *Journal of Natural Gas Science and Engineering* 2015; 26: 549–557. https://doi.org/10.1016/j.jngse.2015.06.041

37 Roy, S., Banerjee, R., and Bose, P.K. (2014). Performance and exhaust emissions prediction of a CRDI assisted single cylinder diesel engine coupled with EGR using artificial neural network. *Applied Energy* 119: 330–340.

38 Kshirsagar, C.M. and Anand, R. (2017). Artificial neural network applied forecast on a parametric study of Calophyllum inophyllum methyl ester-diesel engine out responses. *Applied Energy* 189: 555–567. https://doi.org/10.1016/j.apenergy.2016.12.045.

39 Oguz H, Sarıtas I, Baydan HE. Prediction of diesel engine performance using biofuels with artificial neural network. *Expert Systems with Applications* 2010; 37(9): 6579–6586. https://doi.org/10.1016/j.eswa.2010.02.128

40 Iscan, B. (2020). ANN modeling for justification of thermodynamic analysis of experimental applications on combustion parameters of a diesel engine using diesel and safflower biodiesel fuels. *Fuel* 279: 118391. https://doi.org/10.1016/j.fuel.2020.118391.

41 Venugopal P, Kasimani R, Chinnasamy S. Developing a smart fuel using artificial neural network for compression ignition engine fueled with Calophyllum-inophyllum diesel blend at various compression ratio. *Environmental Progress & Sustainable Energy* 2020. https://doi.org/10.1002/ep.13356.

42 Ghobadian B, Rahimi H, Nikbakht AM et al. Diesel engine performance and exhaust emission analysis using waste cooking biodiesel fuel with an artificial neural network. *Renewable Energy* 2008; 34: 976–982. https://doi.org/10.1016/j.renene.2008.08.008

43 Celebi K, Uludamar E, Tosun E et al. Experimental and artificial neural network approach of noise and vibration characteristic of an unmodified diesel engine fuelled with conventional diesel, and biodiesel blends with natural gas addition. *Fuel* 2017; 197: 159–173. https://doi.org/10.1016/j.fuel.2017.01.113

44 Akkouche N, Loubar K, Nepveu F et al. Micro-combined heat and power using dual fuel engine and biogas from discontinuous anaerobic digestion. *Energy Conversion and Management* 2020; 205: 112407. https://doi.org/10.1016/j.enconman.2019.112407

45 Javed S, Baig RU, Murthy YVVS. Study on noise in a hydrogen dual-fuelled zinc-oxide nanoparticle blended biodiesel engine and the development of an artificial neural network model. *Energy* 2018; 160: 774–782. https://doi.org/10.1016/j.energy.2018.07.041

46 Yusaf, T.F., Buttsworth, D.R., Saleh, K.H., and Yousif, B.F. (2009). CNG diesel engine performance and exhaust emission analysis with the aid of artificial neural network. *Applied Energy* 87: 1661–1669. https://doi.org/10.1016/j.apenergy.2009.10.009.

47 Uludamar E, Tosun E, Tüccar G et al. Evaluation of vibration characteristics of a hydroxyl (HHO) gas generator installed diesel engine fuelled with different diesel–biodiesel blends. *International Journal of Hydrogen Energy* 2017; 42: 23352–23360. https://doi.org/10.1016/j.ijhydene.2017.01.192

48 Syed J, Baig RU, Algarni S et al. Artificial Neural Network modeling of a hydrogen dual fueled diesel engine characteristics: an experiment approach. *International Journal of Hydrogen Energy* 2017; 42: 14750–14774. https://doi.org/10.1016/j.ijhydene.2017.04.096

49 Taghavifar H, Taghavifar H, Mardani A et al. On the modeling of convective heat transfer coefficient of hydrogen fueled diesel engine as affected by combustion parameters using a coupled numerical-artificial neural network approach. *International Journal of Hydrogen Energy* 2015; 40: 4370–4381. https://doi.org/10.1016/j.ijhydene.2015.01.140

50 Kenanoglu, R., Baltacıoglu, M.K., Demir, M.H., and Ozdemir, M.E. (2020). Performance & emission analysis of HHO enriched dual-fuelled diesel engine with artificial neural network prediction approaches. *International Journal of Hydrogen Energy* 45: 26357–26369. https://doi.org/10.1016/j.ijhydene.2020.02.108.

51 Rezaei, J., Shahbakhti, M., Bahri, B., and Aziz, A.A. (2014). Performance prediction of HCCI engines with oxygenated fuels using artificial neural networks. *Applied Energy* 138: 460–473. https://doi.org/10.1016/j.apenergy.2014.10.088.

52 Maurya RK, Saxena MR. Characterization of ringing intensity in a hydrogen-fueled HCCI engine. *International Journal of Hydrogen Energy* 2018; 43: 9423–9437. https://doi.org/10.1016/j.ijhydene.2018.03.194

53 Bendu, H., Deepak, B.B.V.L., and Murugan, S. (2017). Multi-objective optimization of ethanol fuelled HCCI engine performance using hybrid GRNN-PSO. *Applied Energy* 187: 601–611.

54 Nazoktabar, M., Jazayeri, S.A., Arshtabar, K., and Ganji, D.D. (2018). Developing a multi-zone model for a HCCI engine to obtain optimal conditions using genetic algorithm. *Energy Conversion and Management* 157: 49–58.

4

Enhanced Energy Storage with Hybrid Nanoparticles and Machine Learning for Energy Sustainability

Arun Munusamy[1], *Debabrata Barik*[2,3], *and Sreejesh S.R. Chandran*[4]

[1] *Department of Mechanical Engineering, Saveetha School of Engineering, Saveetha Institute of Medical and Technical Sciences (SIMATS), Thandalam, India*
[2] *Department of Mechanical Engineering, Karpagam Academy of Higher Education, Coimbatore, India*
[3] *Centre for Energy and Environment, Karpagam Academy of Higher Education, Coimbatore, India*
[4] *Department of Mechanical Engineering, College of Engineering Muttathara, CAPE, Government of Kerala, Thiruvananthapuram, India*

4.1 Introduction

A revolutionary approach to the problems limiting the efficiency of biohydrogen production from solar energy is proposed in this pioneering study as a result of combining hybrid nanoparticles and machine learning [1]. Problems with energy storage, variations in solar input, and the need for adaptive control systems are some of the obstacles encountered by the traditional biohydrogen production process [2]. Using modern materials science and AI in combination, this research suggests a novel strategy called hybrid-ESML, which stands for energy storage and machine learning optimization using hybrid nanoparticles [3].

Making the most of the complementary capabilities of hybrid nanoparticles and machine learning algorithms is crucial for optimizing biohydrogen production through solar energy use [4]. In doing so, the study hopes to find ways around the constraints and inefficiencies of current sustainable energy alternatives [5]. To address all of the problems with renewable energy in an environmentally responsible way, the hybrid-ESML approach improves intelligent systems while simultaneously increasing energy storage capacity [6].

Some fields may benefit from the suggested approach, such as renewable energy storage, biohydrogen synthesis, and sustainable energy solutions [7]. The goal of this study is to show that hybrid-ESML may improve energy storage, system flexibility, and overall performance [8]. It will do this by analyzing simulations and

Machine Learning for Sustainable Energy Solutions, First Edition. Edited by Zafar Said and Prabhakar Sharma.

validating real-world scenarios [9]. This innovative strategy can transform renewable energy projects, offering more efficient and environmentally friendly options for the future [10].

The main contribution of this chapter is as follows:

Improve the efficiency of biohydrogen production with hybrid-ESML. It combines machine learning for intelligent system optimization with hybrid nanoparticles for enhanced energy storage. Harness the combined power of hybrid nanoparticles and machine learning to overcome obstacles like poor energy storage, unpredictable solar input, and the need for adaptive control. Through simulation analyses, validate the hybrid-ESML technique and demonstrate its real-world application in various renewable energy scenarios by enhancing power storage, system adaptability, and overall performance.

In light of the worldwide need to wean ourselves off fossil fuels and toward renewable energy sources, this perspective highlights the critical importance of developments in materials, technologies, and systems that facilitate effective energy collection, storage, conversion, and administration. The perspective emphasizes machine learning (ML) to accelerate these advances [11]. To evaluate the efficacy of ML-accelerated workflows in energy research, it presents KPIs for measuring the impact of smart grids, photovoltaics, batteries, and electrocatalysis on recent advances. The perspective continues by identifying that the energy sector might benefit from ML applications, demonstrating ML's crucial role in sustainable energy solutions.

In the endeavor to improve energy storage options for modern electronic and electrical systems, this study investigates the possibilities of high-density polymer nanocomposites supplemented with high-dielectric-constant ceramic nanoparticles [12]. Traditional methods are hindered by ceramic nanoparticle loading, which reduces the breakdown strength. The paper presents a unique approach to solve this problem by distributing spherical ceramic nanoparticles in a gradient inside the polymer matrix. Gradient-structured nanocomposites have ultrahigh breakdown strength and energy density due to their increased mechanical and electrical characteristics.

This work develops hydrogel-based piezoresistive sensors with mechanical resilience, consistent conductivity, and environmentally acceptable manufacturing processes. Alkaline lignin and silver nanoparticles (Ag NPs) are used to synthesize lignin-silver hybrid nanoparticles (Lig-Ag NPs). Hybrid nanoparticles minimize Ag NPs discharge and create dynamically stable lignin semi-quinone radicals [13]. A hydrogel with strong sacrificial hydrogen bonding is formed using a polyvinyl alcohol (PVA) matrix for electrical delivery. Ammonia-induced pore formation makes the PVA/Lig-Ag NPs hydrogel compressible, pressure sensitive, and signal response stable.

4.2 Materials and Methods

Due to the need for long-term, cost-effective power solutions, the research proposes hybrid-ESML or hybrid nanoparticle-integrated energy storage and machine learning optimization. This technology uses hybrid and device-learning nanoparticles to change solar power-based biohydrogen production. A complete solution to power storage issues and intermittent solar entry is the hybrid-ESML system. This novel technique improves smart systems and strengthens storage for more sustainable and dependable biohydrogen generation. This generation is likely to be used in environmentally friendly energy and renewable storage, as shown by rigorous simulation assessment and validation of real international situations.

Figure 4.1 shows the integrated system, which uses hybrid nanoparticles and machine learning to transform strength storage capacities. All sections of this machine work together to make energy storage systems efficient and durable. This system starts with solar panels, which provide power. These solar panels start power accumulation. The energy conversion unit converts collected solar power to electricity. The next phase is introducing innovative hybrid nanoparticles. Due to their excellent power storage, these nanoparticles boost the machine's storage potential. It ensures the system's ability to acquire and maintain strength by serving as a complex electric strength storage medium. The hybrid nanoparticle energy storage system saves electricity. With the unique properties of hybrid

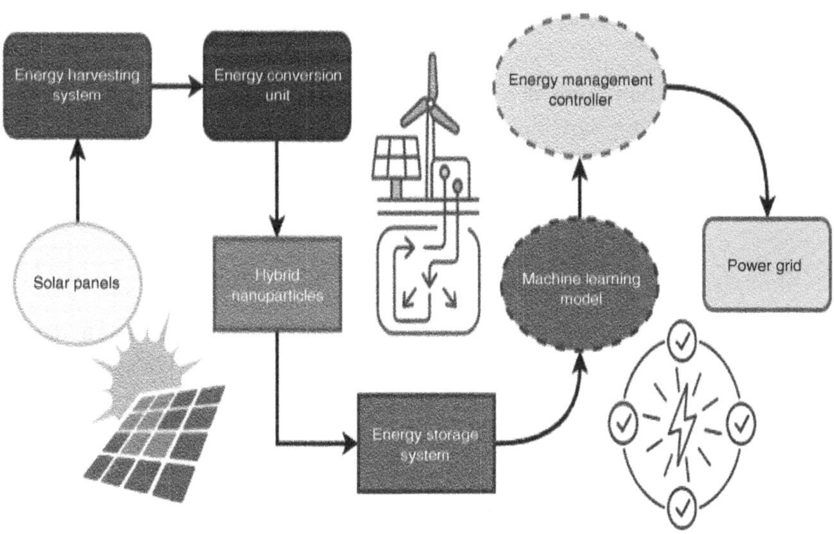

Figure 4.1 Enhanced energy storage using an integrated system.

nanoparticles, this element stores and regulates electricity. This machine learning model enhances and calibrates the electricity storage device's performance.

Using predictive analytics, this technique improves energy storage unit charging and discharge cycles by continually mastering and responding to system features. Due to system learning, the power storage system performs well under different situations. This makes a solution sustainable and responsive. Energy management processors are essential for device-wide power and path control. It talks with the version that utilizes machine learning to make real-time decisions to distribute power to meet immediate needs or store it for later use. Connecting the system to the power grid completes the integration of conserved energy into the larger energy networks.

$$F = D \times W \times \int_{u_1}^{u_2} J(u) \cdot du - \frac{1}{2} n(\Delta W)^2 + \sum_{j=1}^{O} \rho_j \cdot F_{NM,j} \tag{4.1}$$

The standard strength storage capability of the machine is denoted by (F) in equation (4.1). The capacitance, represented by using the (D), is a degree of an energy storage system's potential to save charge. The voltage throughout the device is denoted with the aid of (W), and the integral of the modern over the years, which represents the cumulative price provided or withdrawn from the system, is proven by means of ($\int_{u_1}^{u_2} J(u) \cdot du$). The weight of hybrid nanoparticles, denoted by the amount (n), is a hallmark of how material features have an effect on power storage at some stage in the technique. The voltage exchange that happens throughout storage is represented by way of (ΔW). The effect of artificial intelligence on optimizing the dynamics of energy storage is reflected throughout the entire duration ($\sum_{j=1}^{O} \rho_j \cdot F_{NM,j}$), which provides information on the blended strength contribution from diverse systems gaining knowledge of algorithms ($F_{NM,j}$) with efficiency variables (ρ_j). Sustainable biohydrogen generation is proposed by the hybrid-ESML machine, which takes into consideration all interrelated elements. The equation (4.1) offers a comprehensive view of these factors.

$$B = \frac{\beta \cdot e^{\gamma \cdot \sin(IO)} + \delta \cdot \ln(1 + NM) + \alpha \cdot \tan(IO) \cdot NM}{\epsilon \cdot \sqrt[3]{FT} + \vartheta \cdot \cos i(IO \cdot \tan i(NM))^2 - \eta \cdot \mathrm{COS}(FT)} \tag{4.2}$$

In the system adaptability equation (B), a couple of variables are used to depict the machine's adaptability in trendy. (IO) refers back to the impact of hybrid nanoparticles, which is contained inside a hyperbolic cosine operation, a hyperbolic tangent characteristic, a sine function, and a tangent function. The effect of system mastering, which is engaged in tangent features, hyperbolic tangent capabilities, and natural logarithm capabilities, is denoted with the aid of (NM). The cosine feature encapsulates the power potential for storage, denoted as (FT). The coefficients that affect the weights of each time period in equation (4.2) and their contributions to the overall system adaptability are (β, γ, δ, α, ϵ, ϑ, η). To replicate the complex dynamics that affect the power of smart systems inside the framework of sustainable power answers, equation (4.2) explains the intricate hyperlinks and interactions among hybrid nanoparticles, algorithms for mastering, and electricity storage.

$$Q_{\text{overall}} = \beta \cdot \left(\frac{\rho . D_{\text{storage}} \cdot U_{\text{charge}}^2}{\tau_{\text{discharge}} - \tau_{\text{charge}}} + \gamma \cdot \left(1 - e^{-\frac{\alpha . \theta^2}{\sigma^2}} \right) + \delta \cdot \int_0^M \left(\frac{\partial F_{\text{efficiency}}}{\partial y} \right) \text{dy} \right) \tag{4.3}$$

Multiple variables are used to represent the hybrid-ESML technique within equation (4.3) for normal performance (Q_{overall}). The weight of hybrid nanoparticles is represented through (rho), and the unique heating capability of the strength storage mechanism is denoted with the aid of (D_{storage}). The charging time is represented via the time period (U_{charge}^2), while the release and rate time constants, ($\tau_{\text{discharge}}$) and (τ_{charge}), seize the dynamic approaches of charging and discharging, respectively. The tuning elements that have an effect on the adaptability factor are (γ) and (α). (θ) is a variable that is related to machine flexibility, and (σ) controls the adaptability feature. A crucial factor, with (δ) appearing as a weighting thing, represents the geographic variation in strength utilization ($F_{\text{efficiency}}$) during the machine period (M). Equation (4.3) represents the total performance of the proposed generation in energy from renewable assets packages by integrating complex features of hybrid nanoparticle properties, charging factors, flexibility, and spatial power utilization version.

Figure 4.2 displays the hybrid-ESML system architecture, which outlines the essential components and methods of cautioned storage for solar-powered biohydrogen production. This current method, hybrid-ESML, uses hybrid nanoparticles and machine learning optimization to improve energy storage in sun-powered biohydrogen generation. Solar power, a crucial yet variable component of renewable energy plants, is the first input in Figure 4.2. Solar energy provides energy for storing procedures. Hybrid nanoparticles concentrate on this strength from substance technology. These nanoparticles help solve the strength storage issue in biohydrogen production. The next stage in the system design is the energy storage system, which uses hybrid nanoparticles to advance. This component stores and manages solar power, making it easier to cope with challenges caused by the solar's variable production.

Machine learning model activation promotes machine optimization and adaptability. Using real-time data and simulations, the model optimizes energy use, adapts to changing environmental conditions, and boosts performance. This optimization is necessary in the biohydrogen age due to version complexity and significance. The hybrid-ESML integrated intelligent energy generation gadget

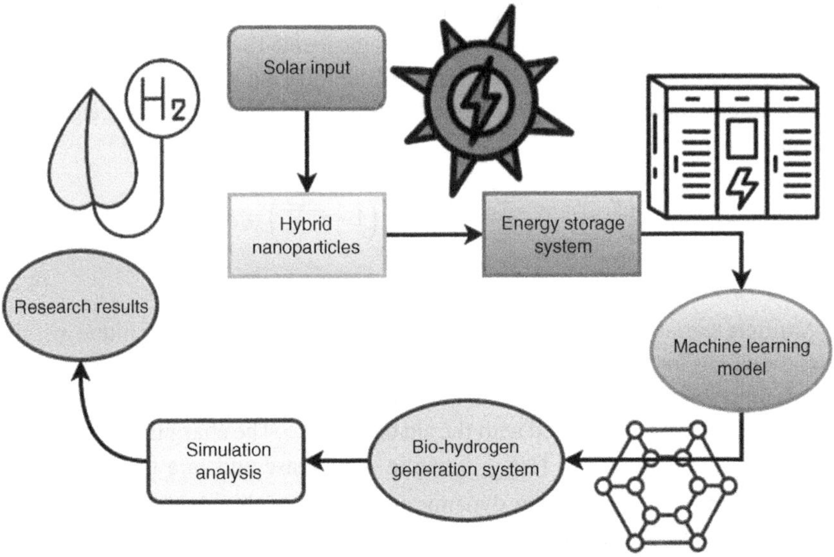

Figure 4.2 Architecture for a hybrid-ESML system.

combines an energy storage system and machine learning model findings for strong biohydrogen generation. Combining such aspects creates a comprehensive and practical sustainable energy plan. Simulation analysis is crucial for testing the proposed hybrid-ESML technique's performance and scalability. Actual-existence activities and information are utilized to assess whether the storage may be used for renewable energy initiatives.

$$T = \frac{\beta.\left(F_{\text{storage}} + F_{\text{NM}}\right)}{\gamma \cdot \left(D_{\text{material}} + D_{\text{algorithm}} + \delta \cdot \left(W_{\text{system}} \cdot S_{\text{adaptability}} \cdot G_{\text{reliability}}\right)\right)} \tag{4.4}$$

Multiple variables have separate, however related, features in the scalability equation (T) as it pertains to the hybrid-ESML approach of storing energy and device mastering. How properly the integration captures the performance of the combined nanoparticles and synthetic intelligence in optimizing strength storage is represented by using the issue of scaling (β). The common value, which incorporates spending on material (D_{material}) and algorithms ($D_{\text{algorithm}}$), is scaled via (γ). To account for system adaptability (W_{system}), adaptability to changing solar input ($S_{\text{adaptability}}$), and device reliability ($G_{\text{reliability}}$), the extra scaling issue (δ) is introduced. The (F_{storage}) and (F_{NM}) calculate the upgrades in strength storage capacity due to hybrid nanoparticles and machine mastering, respectively. When

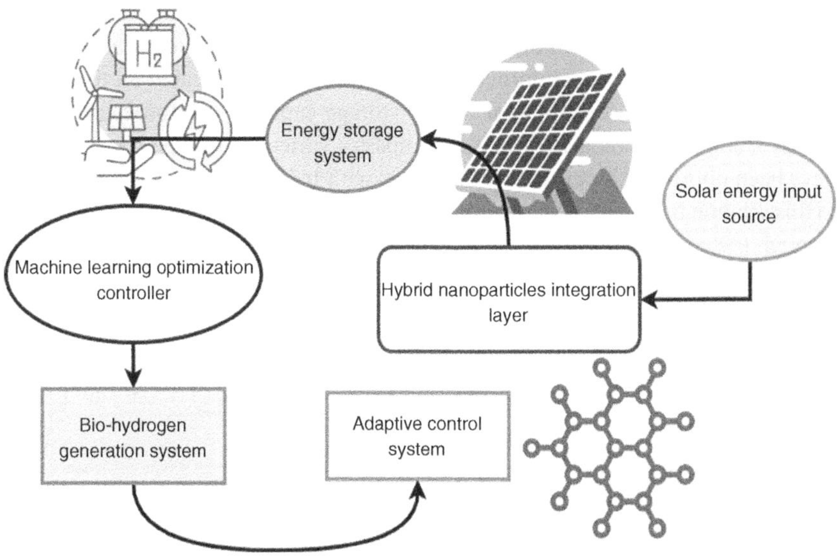

Figure 4.3 Advanced bio-hydrogen generation using a hybrid-ESML integration scheme.

comparing the hybrid-ESML approach's scalability in renewable energy initiatives, equation (4.4) emphasizes the equilibrium among electricity-related advanced profits, overall fees, and the vital capabilities of system flexibility and reliability.

Figure 4.3 shows the complex hybrid-ESML storage, which is aimed to enhance solar power biohydrogen generation. It combines hybrid nanoparticles with power storage and device optimization studies. The hybrid-ESML integration scheme for advanced bio-hydrogen generation, which sums up this all-encompassing design that makes use of contemporary technology to address issues like strength storage inefficiencies and solar enter variability. The solar-based biohydrogen generation system is an important element of the discern three. It makes use of solar strength as its main enter. To produce hydrogen sustainably, this system coordinates a couple of tactics, including the electrolysis of water and biohydrogen production, to use solar energy. Incorporating the hybrid-ESML module, however, brings approximately the actual innovation. Hybrid nanoparticles are brought within the Materials Science field as an innovation in substances engineering that enhances energy storage potential.

As a vital part of producing biohydrogen, these nanoparticles are reshaping the strength storage device's performance and overall performance. At the identical time, the device's intelligence is augmented using machine learning (ML) algorithms. Applying ML procedures permits the storage to maximize its performance, responding in actual time to inefficiencies and adjusting to numerous solar inputs. Achieving a green, improved biohydrogen era is predicated on this cognitive regulation. The device's total energy storage capacity is improved by using energy storage technologies, which encompass a variety of storage strategies that have been optimized by using them together with hybrid nanoparticles. Essential to this scheme is the adaptive control system, which ensures dynamic adaptability through the consistent analysis of information from machine learning algorithms and the kingdom of the storage.

$$G_{\text{feasebility}} = \frac{\beta \cdot \left(\gamma \cdot \left(\frac{\varepsilon \cdot F^2_{\text{storage}}}{\eta \cdot \theta}\right) + \delta \cdot \left(\frac{\vartheta \cdot B^3_{\text{control}}}{K}\right)\right) \cdot \alpha}{\sqrt{\epsilon \cdot \rho \cdot \sigma}} \times \left(1 + \frac{\lambda \cdot \mu}{v \cdot w^2}\right) \tag{4.5}$$

When comparing the practicability of the counseled hybrid-ESML technique to biohydrogen manufacturing, each of the equation's variables is extensively essential. The overall feasibility, as laid low with a component of scale (β) that takes machine efficiency into consideration, is represented as ($G_{\text{feasebility}}$). They have an impact on the hybrid nanoparticles (ε) on stepped forward electricity storage (F^2_{storage}), considering the homes of the material (η) and stability capabilities (θ), is illustrated through the time period $\gamma \cdot \left(\frac{\varepsilon \cdot F^2_{\text{storage}}}{\eta \cdot \theta}\right)$. They have an impact on system getting to know on adaptive control structures, with computational complexity taken into account, is represented via the term $\delta\left(\frac{\vartheta \cdot B^3_{\text{control}}}{K}\right)$. Variables ($\epsilon, \rho, \sigma$) characterizing material attributes, solar enter variability, and machine adaptability are covered in the denominator ($\sqrt{\epsilon \cdot \rho \cdot \sigma}$). Controlling for sun variability (μ), adaptability (v), and system edition below numerous situations (w), the remaining term includes nonlinear outcomes λ related to the interactions among nanoparticles, which are hybrid and machine learning. Equation (4.5) encapsulates several factors that make the hybrid-ESML approach viable for long-term biohydrogen manufacturing.

With the groundbreaking hybrid-ESML technology, improving solar-powered biohydrogen production has never been easier. The work overcomes energy storage issues in sun-powered biohydrogen synthesis using hybrid and algorithmic nanoparticles. The unique technology not only solves difficulties caused by strength storage and variable solar input but also enables flexibility via smart systems. This breakthrough ensures a safer and more efficient biohydrogen synthesis, addressing sustainable power challenges holistically. Hybrid-ESML proves its feasibility and scalability via simulations and real-world validations. Renewable power storage, biohydrogen synthesis, and sustained strength are conceivable uses. The results make it a viable solution for many renewable power projects.

4.3 Result and Discussion

Researchers are experimenting with hybrid nanoparticles and machine learning to find sustainable energy storage options. This research compares machine learning (ML), polyvinyl alcohol (PVA), and hybrid-ESML energy storage technologies for performance, flexibility, and scalability. Tables 4.1 and 4.2 and Figures 4.4 and 4.5 summarize the experiments. These results suggest that hybrid systems might revolutionize energy storage technology for a more sustainable and adaptive future.

Table 4.1 explains the interesting results from the hybrid nanoparticle-machine learning energy storage project. In the experiment, 1–10 samples were taken. The machine learning model showed energy storage values from 45.5% to 57%. PVA (polyvinyl alcohol) ranged from 66% to 75%, whereas hybrid-ESML improved from 77% to 97.7%. These findings suggest that machine learning and hybrid nanoparticles may improve energy storage. Sustainable and efficient energy storage systems are possible with the hybrid technique, which outperformed ML and PVA methodologies.

Figure 4.4 shows the results of a ten-sample study on system flexibility as it relates to energy storage performance. Curiously, the data show that the system components are quite adaptable. Values ranging from 48% to 58% were shown by the ML model, indicating its variable flexibility. Varieties of polyvinyl alcohol

Table 4.1 Analysis of energy storage.

Number of samples	ML	PVA	Hybrid-ESML
1	45.5	72	77
2	49	68.2	91
3	55	66	83
4	47	71	94.4
5	53.2	69	98
6	56	74	93.3
7	48	67.7	79
8	52.2	75	85
9	58	73.8	95.8
10	57	67	97.7

Table 4.2 Overall performance.

Number of samples	ML	PVA	Hybrid-ESML
1	49.7	64	76.5
2	48	63.5	82
3	57	67	78.9
4	60.5	72	84
5	47	63	77
6	52	61.7	92
7	55	65.5	88
8	45	73	94
9	51	71	89
10	53	74	97.8

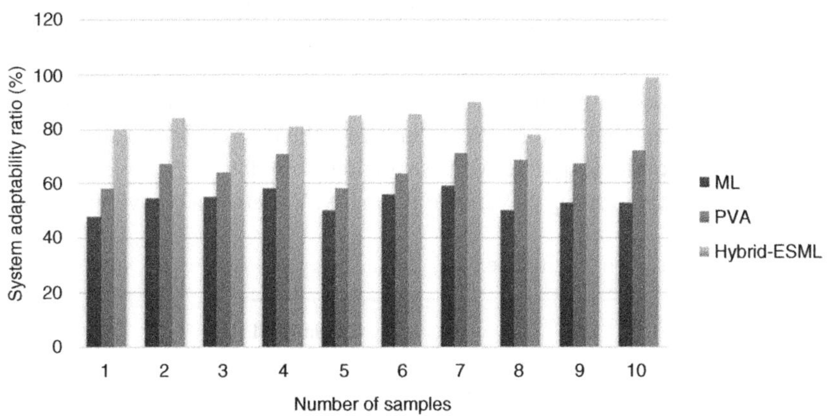

Figure 4.4 Analysis of system adaptability.

(PVA) consistently displayed adaptability scores ranging from 58% to 72%. With a remarkable range of 78–98.9%, the hybrid-ESML method stood out as a top performer. This figure highlights the hybrid-ESML system's strong flexibility, indicating that it might be a flexible and efficient option for meeting changing energy storage needs.

Table 4.2 gives a useful evaluation of the overall performance of energy storage devices, with an emphasis on ML, PVA, and the hybrid-ESML method, across ten samples. The ML model's versatility was on display with values ranging from 45% to 60.5%. Values between 61.7% and 73% demonstrated consistent performance

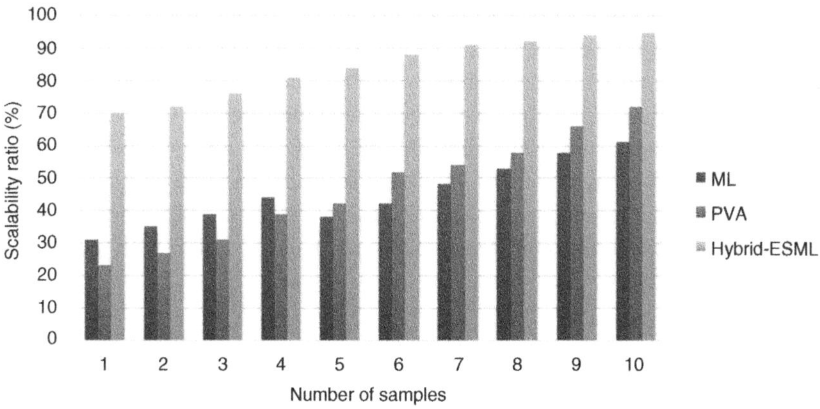

Figure 4.5 Analysis of scalability.

from PVA. The hybrid-ESML method outperformed ML and PVA in terms of efficiency, with values ranging from 76.5% to a staggering 97.8%. These findings highlight the hybrid-ESML approach's capacity to provide sustainable solutions by offering higher overall performance in energy storage.

Figure 4.5 shows the results of the scalability study for three different energy storage approaches: hybrid-ESML, machine learning (ML), and polyvinyl alcohol (PVA). These systems' scalability may be better understood with the help of the data, which spans 10 data points. ML's capacity to manage rising demands was proven by its incremental scalability, which progressed from 31% to 61%. From 23% to 72%, PVA increased steadily, suggesting constant scalability. With values increasing from 70% to an astounding 94.8%, hybrid-ESML demonstrated remarkable scalability. The hybrid-ESML system has the capacity to efficiently and dependably handle a wide range of energy storage needs, as seen in this figure, due to its beneficial scalability.

This suggests that it can suit many energy storage needs. Taken as a whole, these findings highlight the potential of renewable and efficient energy storage technologies that combine hybrid nanoparticles with machine learning. One promising strategy for meeting the ever-changing problems in energy storage is the hybrid-ESML method, which provides a scalable, adaptable, and effective alternative.

4.4 Conclusion

Overall, the study's results show that hybrid nanoparticles combined with machine learning might revolutionize energy storage. Scalability, versatility, and efficiency across varied energy storage needs are shown by the hybrid-ESML approach, which emerges as a beacon of hope. Its persistent superior performance compared to standalone machine learning and polyvinyl alcohol approaches points to a solid and trustworthy answer to the ever-changing problems in the energy storage industry. Insights from the research indicate that the hybrid-ESML method might make a substantial contribution to efficient and green energy storage technologies. The hybrid-ESML approach seems like a good long-term bet in light of the growing need for sustainable energy policies. An energy storage system that can adapt to the future's unpredictable needs may be possible with the help of such novel techniques.

Author Contributions

"Conceptualization, methodology, data collection and analysis, resources, writing-original draft preparation, writing-review, and editing—M. Arun, Debabrata Barik, and Sreejesh S.R Chandran; Supervision—Debabrata Barik".

Acknowledgment

The authors sincerely thank Karpagam Academy of Higher Education and Saveetha Institute of Medical and Technical Sciences, India, for providing research facilities to carry out the research work.

Data Availability Statement

The data will be available upon request.

Funding

This research is not funded by any funding agency.

Conflict of Interest

The authors declare that they have no conflicts of interest.

References

1 Said, Z., Sharma, P., Aslfattahi, N., and Ghodbane, M. (2022). Experimental analysis of novel ionic liquid-MXene hybrid nanofluid's energy storage properties: model-prediction using modern ensemble machine learning methods. *Journal of Energy Storage* 52: 104858.

2 Sharma, P., Said, Z., Kumar, A. et al. (2022). Recent advances in machine learning research for nanofluid-based heat transfer in renewable energy system. *Energy and Fuels* 36 (13): 6626–6658.

3 Jia, Y., Hou, X., Wang, Z., and Hu, X. (2021). Machine learning boosts the design and discovery of nanomaterials. *ACS Sustainable Chemistry and Engineering* 9 (18): 6130–6147.

4 Kanti, P.K., Shrivastav, A.P., Sharma, P., and Maiya, M.P. (2024). Thermal performance enhancement of metal hydride reactor for hydrogen storage with graphene oxide nanofluid: model prediction with machine learning. *International Journal of Hydrogen Energy* 52: 470–484.

5 Ma, T., Guo, Z., Lin, M., and Wang, Q. (2021). Recent trends on nanofluid heat transfer machine learning research applied to renewable energy. *Renewable and Sustainable Energy Reviews* 138: 110494.

6 Shen, Z.H., Liu, H.X., Shen, Y. et al. (2022). Machine learning in energy storage materials. *Interdisciplinary Materials* 1 (2): 175–195.

7 Ramachandran, T., Sana, S.S., Kumar, K.D. et al. (2023). Asymmetric supercapacitors: unlocking the energy storage revolution. *Journal of Energy Storage* 73: 109096.

8 Fan, M., Hu, P., Dan, Z. et al. (2020). Significantly increased energy density and discharge efficiency at high temperature in polyetherimide nanocomposites by a small amount of Al_2O_3 nanoparticles. *Journal of Materials Chemistry A* 8 (46): 24536–24542.

9 Hui, X., Ge, X., Zhao, R. et al. (2020). Interface chemistry on MXene-based materials for enhanced energy storage and conversion performance. *Advanced Functional Materials* 30 (50): 2005190.

10 Ye, J., Zhai, X., Chen, L. et al. (2021). Oxygen vacancies enriched nickel cobalt based nanoflower cathodes: mechanism and application of the enhanced energy storage. *Journal of Energy Chemistry* 62: 252–261.

11 Yao, Z., Lum, Y., Johnston, A. et al. (2023). Machine learning for a sustainable energy future. *Nature Reviews Materials* 8 (3): 202–215.

12 Jiang, Y., Zhang, X., Shen, Z. et al. (2020). Ultrahigh breakdown strength and improved energy density of polymer nanocomposites with gradient distribution of ceramic nanoparticles. *Advanced Functional Materials* 30 (4): 1906112.

13 Han, X., Lv, Z., Ran, F. et al. (2021). Green and stable piezoresistive pressure sensor based on lignin-silver hybrid nanoparticles/polyvinyl alcohol hydrogel. *International Journal of Biological Macromolecules* 176: 78–86.

5

Model Prediction of Biomass Gasification Using Support Vector Machines

Arun Munusamy[1] *and Debabrata Barik*[2,3]

[1] *Department of Mechanical Engineering, Saveetha School of Engineering, Saveetha Institute of Medical and Technical Sciences (SIMATS), Thandalam, India*
[2] *Department of Mechanical Engineering, Karpagam Academy of Higher Education, Coimbatore, India*
[3] *Centre for Energy and Environment, Karpagam Academy of Higher Education, Coimbatore, India*

5.1 Introduction and Literature Survey

Support vector machines (SVMs) for biomass gasification prediction provide many constraints, which make the gasification prediction extra difficult [1]. Due to its nonlinear and dynamic nature, biomass gasification involves complex thermo-chemical tactics laid low with the composition of the feedstock, operating situations, and reactor layout [2]. SVMs efficiently handle limitless interactions; however, the complexity of biomass and dynamic dynamics can be very difficult due to a couple of variables and high stages [3]. High-quality education information is scarce and heterogeneous, which is a complex problem [4]. Biomass gasification approaches vary depending on the feedstock type, moisture level, and gasification conditions [5]. The problem of gathering a large dataset that reflects this variability makes training and generalization models with confined information questionable [6]. SVMs cannot research and adapt to actual-world conditions as biomass gasification databases are not regular [7]. Moreover, SVMs are very sensitive to hyperparameter modifications, which can be an actual problem [8]. The default parameter, the kernel coefficients, and the kernel feature are regularly changed to locate the best working space. Choosing these hyperparameters requires trial and error, which will increase the time and performance of the modeling method [9]. Due to the complex calibration technique and the risks of overfitting and inefficiency, the model might not predict the effects for particular and untested biomass gasification configurations [10]. Biomass gasification is time established, making the model prediction complicated. SVMs are static and do

Machine Learning for Sustainable Energy Solutions, First Edition. Edited by Zafar Said and Prabhakar Sharma.

not show time-dependent modifications in gasification and temporal responses [11]. SVMs may not be bendy enough to deal with a huge transfer of pieces of biomass feedstock. This can cause inaccurate estimates, especially for feedstock variations or technique parameter adjustments [12]. SVM modeling is every other problem in biomass gasification. Understanding the mechanics and variables of the gasification method is crucial for version validation, optimization, and process management [13]. SVMs cannot predict the physical and chemical reactions of biomass due to the loss of transparency [14]. This limits the actual use of the version and its reliability. SVMs are useful in a few systems that study situations where biomass gasification prediction is difficult. These issues require a multidisciplinary approach that includes area expertise, new feature engineering, hyperparameter modification, and record availability [15]. SVM-based total biomass gasification methods may be more accurate and dependable if those demanding situations are addressed, and people can examine further and improve this essential renewable power technique. There have been several attempts to improve biomass gasification model predictions using SVMs, but problems remain. Feature engineering is frequently used to optimize the entry parameters of the SVM. Features are decided on and extracted to separate the most crucial varieties of biomass gasification for SVM education [16]. When capturing unlawful connections, the choice of kernel features is vital. Researchers are experimenting with kernel capabilities, including distribution polynomials and basic capabilities with a radial foundation, to illustrate the complicated gasification dynamics. Clustering techniques use multiple SVM models to improve the prediction accuracy. Mixing techniques, intensification, and bagging reduce variation in reaction and gasification. Researchers are studying hybrid models integrating SVMs with machine learning (ML) techniques. These hybrid approaches can combine SVMs with neural networks or fuzzy logic structures to enhance prediction. Variable feedstock, running situations, and reactor designs robustly affect biomass transfer. The lack of various and representative datasets makes training SVMs difficult, leading to growing biases and wrong predictions. Because of its effect on the SVM overall performance, hyperparameter tuning remains a critical difficulty. A complicated restoration manner, which calls for much computing power and know-how, hinders adoption. The SVM estimates are affected by the time dependence of the biomass gasification. SVMs may additionally leave out gas-associated changes due to their static nature. SVM models are often not smooth to recognize, which restricts their utility for gasification studies. The lack of transparency of SVM can hinder real-world applications and decision-making. Although various attempts exist to optimize biomass gasification using SVM-based estimation, there are still problems. To develop SVMs for biomass gasification prediction, we must overcome data constraints, simplify parameter change, and make the models temporal.

- The current research aims to develop an ISVM-BGP model that improves the prediction of biomass gasification results. Complex biomass feedstock, operating parameters, and different types of combustion cycles must be managed.
- This research aims to optimize biomass gasification processes for better resource usage, less environmental impact, and energy efficiency. This project improves the forecast model to help switch to less-polluted, more sustainable energy.
- ISVM-BGP could assist in bioenergy production, waste-to-energy conversion, and sustainable power generation, among other industries. This paradigm is powerful for process control and decision-making in many industrial scenarios. This provides effective solutions for many situations.

This paper follows a similar approach to Section 5.2 literature review of classifying SVM models for biomass gasification prediction. Section 5.3 explains the mathematical foundations of the integration SVMs for biomass gasification prediction (ISVM-BGP). Section 5.4 covers the research findings and discussion.

Machine learning (ML) techniques have emerged as useful tools for predicting important outputs and improving processes in recent studies on biomass gasification. These techniques have been developed as a result of recent research. The intriguing new technique to generate power from waste and renewable resources is biomass gasification, and Elmaz et al. [17] are particularly concerned with forecasting its key outputs, including carbon monoxide, hydrogen, methane, carbon dioxide, and syngas. Due to their assumptions, traditional models like kinetic and equilibrium models are difficult to apply in real-life situations. To avoid this, the author uses a multilayer perceptron, a decision tree, a polynomial, and a support vector regression. Principal component analysis (PCA) improves the computing efficiency of downdraft fixed-bed gasifier data. The decision tree and multilayer perceptron regression outperformed the competition with an R-squared value of 0.9 for most outputs after k-fold cross-validation. Ayodele et al. [18] investigate the co-gasification of oil palm and coconut wastes utilizing Portland cement, dolomite, and limestone as catalysts to obtain zero emissions. Process uncertainties are reduced using 12 ML methods. Support vector machines with linear, quadratic, and cubic kernels; nonlinear response quadratic models (NRQMs) with sequential quadratic programming (SQP) and Levenberg–Marquardt (LM) algorithms; and Gaussian process regression (GPRs) with rotational, squared, Matern 5/2, and exponential kernels. NLRQM-LM and NLRQM-SQP scored best in the syngas composition prediction ($R^2 > 0.97$). Ozbas et al. [19] tested the clean hydrogen generation prediction of biomass gasification in a steel fixed-bed updraft gasifier. Time, temperature, CO, CO_2, CH_4, O_2, and heating value were inputs into the linear regression (LR), K-nearest neighbors (KNN), support vector machine regression (SVMR), and decision tree regression (DTR) algorithms to predict hydrogen concentration. The maximum hydrogen concentration was 35%

vol. with an HHV of 3400 kcal/m^3. Ascher S et al. [20] found 70 papers in the 2000s and 549 in the 2010s using ML to predict biomass and waste gasification/pyrolysis processes. The talk covers the most popular ML methods and their pros and cons. Regression, tree-based, ANN, and SVM methods are examples. Although R^2 values vary, most research had prediction accuracy above 0.90. The focus on ML models' adaptability, especially for non-numerical factors, suggests that the gasification and pyrolysis process modeling accuracy may increase. Model interpretability is a priority from our perspective. Yucel et al. [21] employed ANNs and a nonlinear autoregressive exogenous (NARX) neural network model with two distinct sets of features to represent fixed-bed downdraft gasification. Two datasets were collected: ER, AF, temperature distribution (T0–T5), final or near-final biomass values, ER, AF, and reduction temperature (T0). Calculations predicted the calorific value of syngas and composition containing H_2, CO_2, CO, and CH_4. The NARX time series model outperformed the other designs, and the ANNs matched the experimental data ($R^2 > 0.99$). Using Garson's equation to determine feature relevance, feature reduction was suggested. These investigations demonstrate the variety and efficacy of ML approaches in biomass gasification prediction, with ISVM-BGP showing promise for future applications.

5.2 Materials and Methods

ISVM-BGP, which stands for integrated SVMs for biomass gasification prediction, is a version proposed in this research as a method to enhance models that might be used for prediction related to biomass gasification. The ISVM-BGP machine addresses the complex challenges of using various biomass feedstock and operational conditions. This is accomplished through continuous tracking with adaptable manipulation, specific feature boom and improvement, and SVM layout. As a result of this technique's willpower to forecast reliability and thorough understanding of the gasification method, it's miles viable to obtain power answers that might be more environmentally friendly and efficient. It demonstrates the capability for utilization inside the manufacturing of bioenergy, the transformation of waste into power, and the improvement of sustainable electricity.

Forest waste changed into fuel in an air gas generator; lawn waste and sewage sludge become gasoline in a catalyzed CO_2co-gasifier; palm kernel shell (PKS) becomes fuel in a warmness generator; sorghum and lignite were turned into gas in a hydrothermal procedure gasifier; coconut shells and oil palm fronds were gas in a catalyzed air gasified; and gas and sawdust were gas in an air generator. To investigate the influence of several predictors, such as temperature, the ratio of moisture to biomass, the amount of oxygen present, the quantity of ash present, the speed of flow of the gasifying agent, the mass of the biomass, the weight of the ash, the size of the feedstock, the superficial acceleration, the ratio of adsorbent to biomass, the dimension of the biomass particles, the quantity of coal present, and the quantity of sorghum, had on the hydrogen-rich syngas, an experimental approach was utilized to investigate the effectiveness of these predictors. Figure 5.1, which represents the process, depicts the procedure of gasification and co-gasification. Both of these processes are described in detail. Upstream pre-processing, gasification, and downstream operation are the three distinct stages that may be distinguished from one another in the process. The entire technique can be divided into these three distinct phases. Various processes are incorporated in the upstream preparation of biomass and coal. These processes include the reduction of size, drying, and compaction, as well as the production of gasifying agents. In a stationary or fluidized gasifier, it is usual to carry out the hydrogenation or co-gasification procedure using coal and biomass. This is done to achieve the desired results. It is the responsibility of this precise kind of gasifier to catalyze the chemical process that is precipitated through warm electricity. As a result of

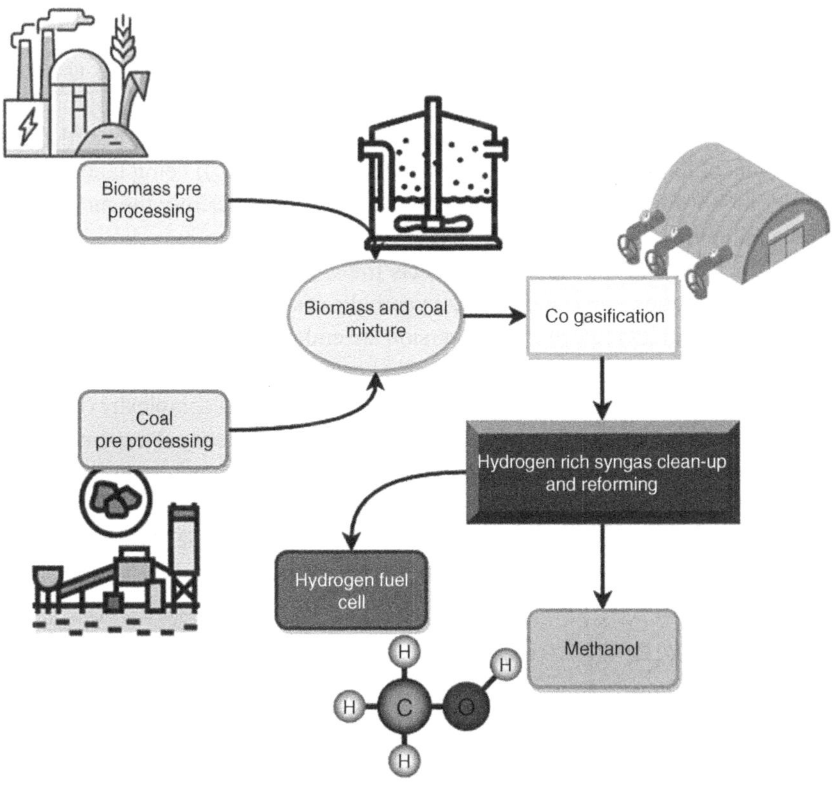

Figure 5.1 Gasification and mixed gasification of coal and biofuels.

the thermochemical procedure, which ends up in the formation of gaseous products, the most important is syngas, which is plentiful in hydrogen, by-products together with tar, ash, and particulate are created. These by-products are fashioned because of the manner.

The downstream processing technique includes the elimination of tar, ash, and particulate matter; in addition to reforming the syngas, this is created for several purposes. This syngas is then used for a whole lot of programs. Several packages can be particularly noteworthy, including the manufacture of methanol, the production of a hydrogen gasoline cellular, the production of Fischer-Tropsch liquid (FTL), and the development of blended heat and electricity (CHP). The gasification and co-gasification strategies have been depicted using a complete 124 datasets. These datasets were applied to demonstrate the procedures. Every unmarried one of these datasets was composed of a discrete collection of predictors, which are the parameters that might be used in the information entry process.

$$Q = \frac{\omega_1 \cdot \sin(B^2) + \sqrt{\omega_2 \cdot S} + \log\left(1 + f^{\omega_3 \cdot \tan i(F)^2}\right)}{\sqrt[3]{1 + \omega_4 \cdot \cos i(D)^2} \times \prod_{o=1}^{O}\left(1 + f^{-\omega_5 \cdot \cot i\left(U_o^2\right)}\right)} \tag{5.1}$$

Many elements are pivotal to the ISVM-BGP overall performance Q equation (5.1), which predicts the outcome of the biomass gasification approaches. In addition to the sine characteristic, the version's accuracy (B) is used to anticipate the results of biomass gasification, with the square root $((B^2))$ being taken under consideration. Efficiency (F) indicates how rapidly and efficiently the model supplies predictions, with its square root $((F)^2)$ impacting the hyperbolic tangent transformation, and robustness (S) reflects the version's consistency throughout varied eventualities. After undergoing hyperbolic cosine transformation, the computational resources required for the version are embodied with the aid of the computational cost $((D)^2)$. The computing time at each generation modified the usage of the hyperbolic cotangent, which is known as the temporal complexity (U_o^2). The importance of accuracy, durability, performance, computational expenses, and time complexity is decided with the aid of the weights that are used (ω_1, ω_2, ω_3, ω_4, and ω_5). Integrating iterative approaches and nested modifications, equation (5.1) assesses the ISVM-BGP's overall performance inside the complicated putting of biomass gasification predictions from each attitude.

$$ACC = \frac{1}{O}\sum_{j=1}^{O} \omega_j \left(1 - \frac{\left|\hat{Z}_j - Z_j\right|^{\delta}}{\left(\gamma + \max\left(\epsilon, \left|Z_j\right|\right)\right)^{\beta} + \sum_{k=1}^{N} \alpha_k \cdot \left|\hat{Y}_{jk} - Y_{jk}\right|^{\rho}}\right) \tag{5.2}$$

Various variables concern the complex evaluation of the biomass gasification forecasting version within the ACC accuracy equation. The O stands for the overall count number of data points used for the evaluation, and ω_j denotes the burden given to every fact point, deliberating variations in their significance. In the jth statistics factor, the projected gasification outcome is denoted as $\hat{Z}_j$, and the actual gasification outcome is Z_j. The numerical balance is assured by employing the modest wonderful constant ϵ, which prevents the department from zero aid. This degree of customization can be achieved by adjusting the sensitivity of the accuracy metric to discrepancies between the expected and actual outcomes using the parameters β, γ, and δ. The anticipated and actual amounts of the kth characteristic for the jth data point are represented by $\hat{Y}_{jk}$ and Y_{jk}, respectively, in equation (5.2) that includes N features. To account for feature changes in the evaluation and capture more nuanced interactions within the context of biomass gasification prediction, the complexity of equation (5.2) is increased by the addition of parameters α_k and ρ.

There is a type of algorithm known as a meta-algorithm, which combines several different soft computing methodologies into one forecasting model. Some examples of algorithms that fall within this category are ensemble approaches. Due to the numerous advantages of the ensemble technique in applications that involve automated decision-making, researchers, particularly those working in engineering and ML, have expressed an interest in the technique. The forecasts produced by installing various soft computing technologies were improved through an ensemble technique employed by the researchers who participated in this study. The objective of constructing a collective model is not to obtain a greater degree of prediction accuracy than the individual computation approach that is believed to be the most efficient separately. This is not the case when the ensemble model is developed. However, the accuracy of a combined model will be greater than the total precision of the individual procedures integrated into the model. This is because the ensemble model integrates multiple algorithms. A theoretically sound model may make inaccurate predictions when placed to testing and validation activities, even though it performed well during learning (generalization performance). This is because the model will be put through these activities. There is a possibility that soft computing approaches that perform equally in generalization might not be as successful in field predictions if the information set that is used to evaluate them is not comparable to the data that is gathered in the field. Considering this, it is logical to conclude that pooling the results of various diverse methodologies would lessen the likelihood of a scary prognosis being predicted.

The ensemble model can incorporate many data outcomes into its final product if the data gathering is sufficient. Employing a wide range of resampling-based methods, including jackknives and cross-validation, can train many soft computing methods on various datasets. This is accomplished in many different ways. For the sake of this investigation, the chosen resampling strategy includes a method known as K-fold cross-validation. Considering the pretty small size of the pattern, the selection was made rather. When a single model cannot decide the separation boundary due to the complexity of the scenario, it is feasible to execute an ensemble method through a divide-and-conquer method. This is practicable because it is feasible to use an ensemble approach. To set up suitable boundaries, this method requires splitting the dataset into portions that are easier to deal with, developing a single model on every one of those sections, and integrating the effects of the education process. This is accomplished to ensure that the boundary is applied as it should be. The system that was applied in the path of this investigation is illustrated in Figure 5.2.

$$E = \frac{\sum_{J=1}^{O} \omega_j \cdot \left(1 - \frac{|P_j - Q_j|}{\max(P_j, Q_j)}\right) + \beta \cdot \gamma \cdot \delta \cdot \alpha}{\sqrt{\sum_{j=1}^{O} \omega_j^2 + \beta \cdot \alpha + \frac{\epsilon}{\sqrt{\emptyset}}}} \tag{5.3}$$

The integrating SVMs for biomass gasification prediction (ISVM-BGP) version dependability is represented via E within equation (5.3). The cost of each dataset instance is captured by the ω_j, which stands for a weighting component. In the jth dataset example, P_j represents the discovered gasification final results, and Q_j represents the expected gasification outcome. The O represents the general depending on the dataset times utilized for the assessment. The model's performance over the years, its adaptability, the impact of external environmental factors, and the effect of factors β, γ, α, and δ are all affected. The baseline for the version's performance is represented by a regular ϵ, while the intricate nature of the gasification method is meditated with the aid of the aspect $\emptyset$. In the framework of biomass gasification predictions, the use of the ISVM-BGP model and the purpose of the overall equation (5.3) is to give a radical and distinctive reliability degree by employing the interaction of these elements.

Integrated SVMs for biomass gasification prediction are the model's title, which is depicted in Figure 5.3, a block diagram of the ISVM-BGP version. This modern gadget will help optimize diverse tactics and improve new electricity solutions by enhancing the precision and efficacy of the biomass combustion prediction. More specifically, it's going to improve the accuracy of the prediction. Among the key components contained within the input data, which is the primary part of the diagram, the 2 number one protected components are feedstock crafted from biomass and the parameters for operation. Once those inputs have been processed, the ISVM-BGP version will continue to the following level. From this point forward, support vector machines, known as SVMs, could be the driving force behind the model configuration section. The overall performance of the device may be optimized during the configuration technique by exceptionally tuning the parameters

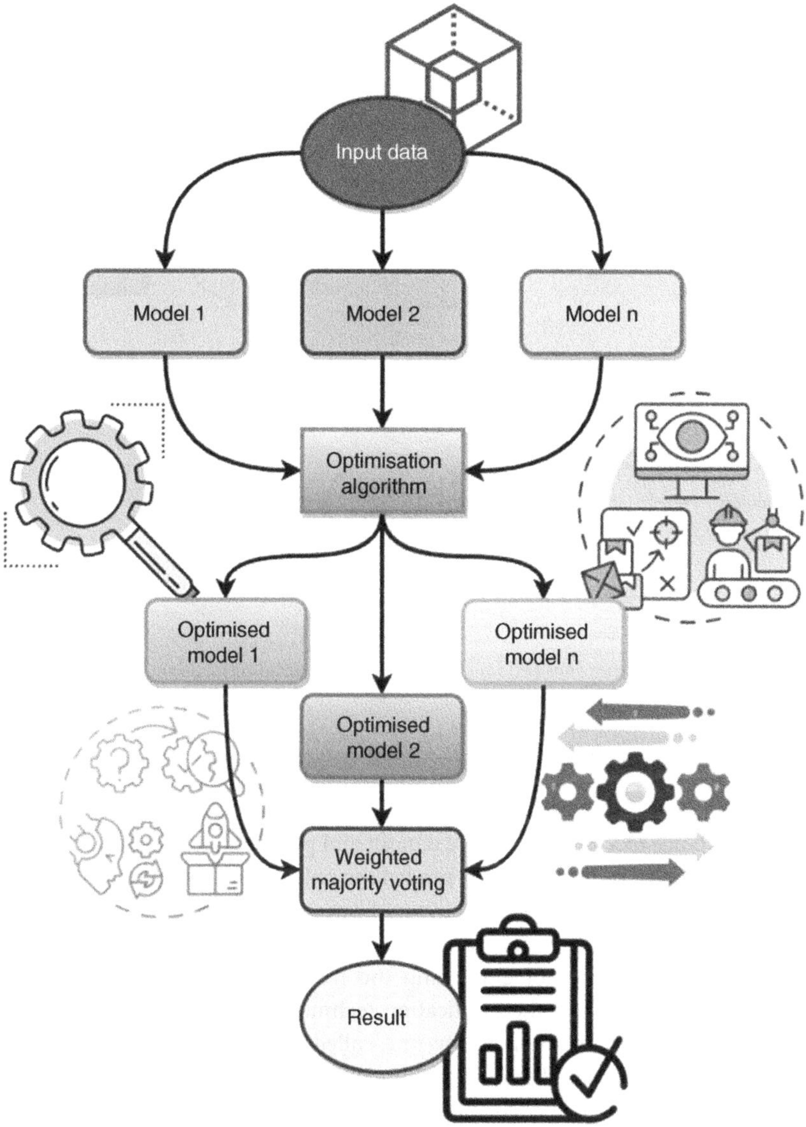

Figure 5.2 A comprehensive framework for the optimal ensemble model.

of the SVMs and choosing kernel capabilities that might be appropriate. The ISVM-BGP version, carried out inside the gasification prediction block, is the inspiration upon which the device is built. To implement this integrated approach, adaptive manipulation and real-time monitoring are applied. This is made possible

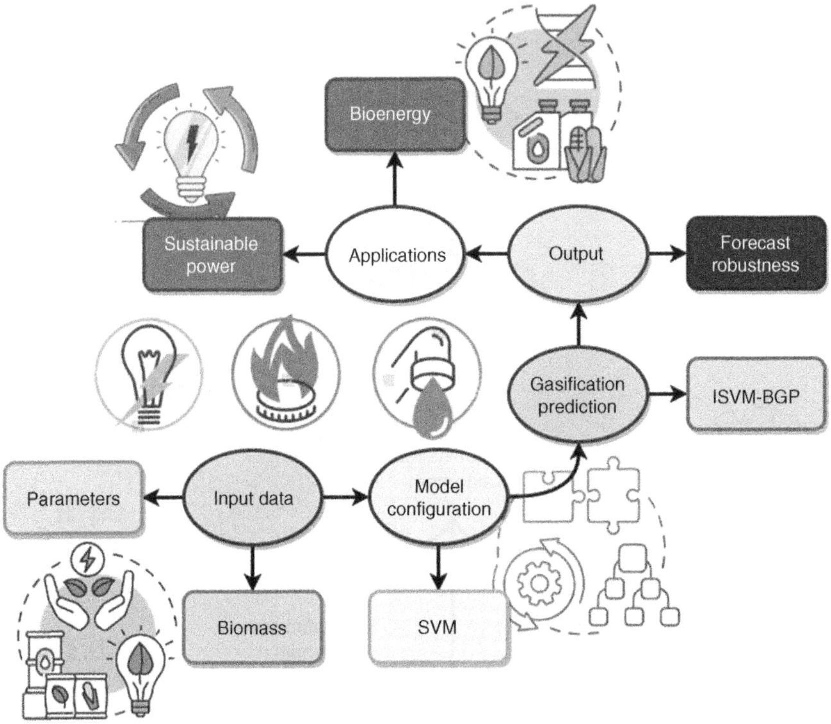

Figure 5.3 Using combined support vector machines to forecast biomass gasification.

with the guidance of a comprehensive dataset and precise characteristic engineering. The ISVM-BGP gadget is used to predict the outcomes of the complicated gasification tactics simultaneously. It is obvious from the output block that the ISVM-BGP model has successfully generated favorable outcomes. More specifically, it highlights the importance of enhancing the robustness of projections, which results in a visualization of the gasification technique that is extra accurate and trustworthy. The phase that follows this one, called applications, gives an instance of how the ISVM-BGP paradigm will be applied in several distinctive sectors. The idea has an extensive range of possible applications, some of the most promising of which consist of the manufacturing of bioenergy, the conversion of waste into electricity, and the technology of energy in an environmentally accountable way. An assessment of the coolest consequences that may be executed by deploying the ISVM-BGP paradigm is provided in the benefits block of the file. The technique that has been suggested may additionally have some distance-accomplishing ramifications and benefits, including advanced usage of resources, less effect on the environment, and elevated strength performance. It is feasible that these benefits may be realized. The integrated method for predicting biomass gasification that the ISVM-BGP model utilizes is depicted graphically in Figure 5.3, representing the technology. This approach considers all the significant components and stages involved in the process. It has been proven that this concept can encourage the development of alternative types of energy that are less harmful to the environment and positively impact the future.

$$\eta = \frac{\beta \cdot F_{\text{out}}}{\gamma \cdot F_{\text{in}}} \times \left(1 - \frac{\delta \cdot J_{\text{loss}}}{\alpha \cdot F_{\text{in}}}\right) \times \left(1 - \frac{\epsilon \cdot \cdot}{\vartheta \cdot F_{\text{in}}}\right) \times \left(1 - \frac{\eta \cdot F_{\text{emission}}}{\theta \cdot F_{\text{in}}}\right) \quad (5.4)$$

The entire process efficiency is represented by η in the provided efficiency equation (5.4) for biomass gasification. With β and γ serving as coefficients indicating the characteristics of the energy input and output, respectively, the terms $\beta \cdot F_{\text{out}}$ and $\gamma \cdot F_{\text{in}}$ represent the useful energy output and energy input, respectively. In gasification, the irreversibility and loss of impact are represented by $\delta \cdot J_{\text{loss}}$, the waste impact by $\epsilon \cdot F_{\text{waste}}$, and the emission impact by $\eta \cdot F_{\text{emission}}$. The influence of these factors is controlled by the coefficients that correspond to them, δ, ϵ, and η, while the denominators include α, ϑ, and θ, which normalize and assign weightage. To sum up, equation (5.4) captures an in-depth evaluation of the biomass gasification effectiveness, deliberating some parameters and how they interact during the strength conversion.

An effective approach for addressing the difficulties of biomass gasification prediction is the integrated SVMs for biomass gasification prediction, which is regularly called ISVM-BGP. Through the seamless integration of nonstop tracking, adaptive management, and advanced engineering of features, the ISVM-BGP improves prediction precision and gives a thorough comprehension of the gasification technique. Sustainable strength generation, bioenergy manufacturing, and waste-to-power conversion might all benefit from its versatility, which is demonstrated in many industries. Through a complete examination and simulation assessment, the ISVM-BGP confirms its effectiveness, presenting an exciting opportunity for a cleaner and more environmentally friendly energy solution.

5.3 Results and Discussion

Biomass gasification requires accurate predictive models due to its complexity, which includes several process units and biomass feedstock. The current research compares the performance, accuracy, and reliability of ISVM-BGP and SVM. ISVM-BGP uses adaptive control, better component engineering, and real-time analysis to predict biomass gasification.

The performance analysis shows that the ISVM-BGP model can predict the biomass gasification results even when the process is very complex. Real-time monitoring, adaptive management, and complex factor engineering help the model make accurate predictions. A key data center responsible for the different types of feedstock and process parameters allows ISVM-BGP to handle the complex interactions of the gasification process easily. By incorporating kernel functions into the SVM framework, the model can better handle nonlinear interactions, improving the prediction accuracy. Validation uses historical and synthetic data to test the model on multiple datasets. This ensures the integrity of the model. The ISVM-BGP has been successful in bioenergy production and waste-to-energy conversion, proving to be versatile. This shows the system's flexibility. The model's ability to improve energy efficiency, the environment, and resource use shows its practicality. Simulation studies have proven that the ISVM-BGP is accurate and dependable for decision-making and process optimization. In conclusion, the ISVM-BGP performance analysis is promising for biomass gasification optimization and cleaner, more sustainable energy research. As seen in Figure 5.4, a performance analysis comparing it to SVM reveals that it achieves an accuracy of 89.7%. As shown in Figure 5.5, a comparison between the performance analysis and the ISVM-BGP demonstrates a remarkable accuracy of 98.4%.

The ISVM-BGP accuracy analysis shows its ability to predict biomass gasification outcomes. Real-time monitoring lets the model adapt to the gasification's changing operational conditions. Adaptive control improves projections, enabling the model to learn and adapt from real-time feedback. A comprehensive dataset covering biomass feedstock and performance characteristics is essential for model

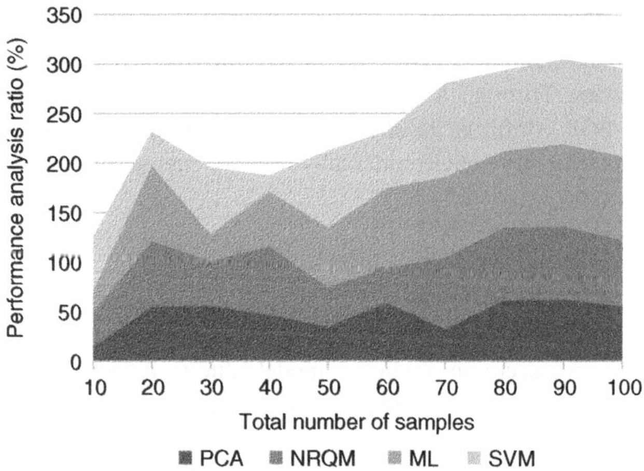

Figure 5.4 Performance analysis is compared with SVM.

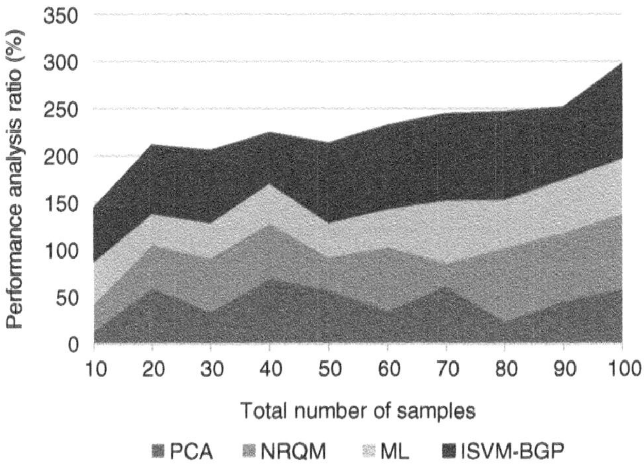

Figure 5.5 Performance analysis is compared with ISVM-BGP.

accuracy. Advanced engineering improves the version's capability to understand complex gasification processes and interactions. SVMs with optimized kernel features help seize the nonlinear relationships of biomass gasification dynamics. The model can be relied on through empirical validation across artificial and historical datasets. Validation research displays that the ISVM-BGP captures gasification degree records and covers a wide range of conditions effectively. The model optimizes resource utilization and efficiency by accurately predicting the gas results. The model's accuracy is particularly evident in bioenergy production and waste-to-energy conversion. The adaptability of the ISVM-BGP proves its durability and efficiency. Simulations confirmed the model's accuracy under many conditions, revealing real-world reliability and validity. The accuracy analysis shows that ISVM-BGP is a powerful tool for predicting biomass gasification results and creating greener, more sustainable options. The accuracy of this model affects decision-making, process optimization, and the evolution of green energy production. As shown in Figure 5.6, the accuracy analysis shows that the accuracy is 87.5% compared to the SVM process. Compared to ISVM-BGP, the validation test achieves a very high level of accuracy, 96.8%, which is shown in Figure 5.7.

The reliability study of the ISVM-BGP model evaluates the reliability, validity, and robustness of the biomass gasification model. The reliability of a model means that it can always produce results in different conditions. Real-time monitoring and adaptive management make ISVM-BGP flexible. These properties allow the model to respond to changes in the operating parameters during gasification, improving its reliability. The model is more reliable and works well in many situations because of its flexibility. A large dataset used for model building improves the

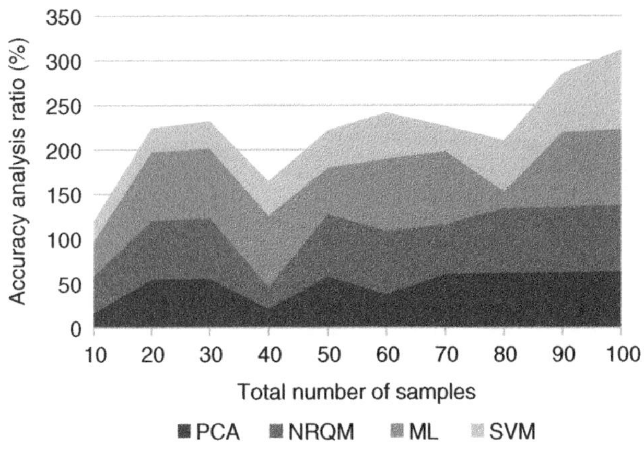

Figure 5.6 Accuracy analysis is compared with SVM.

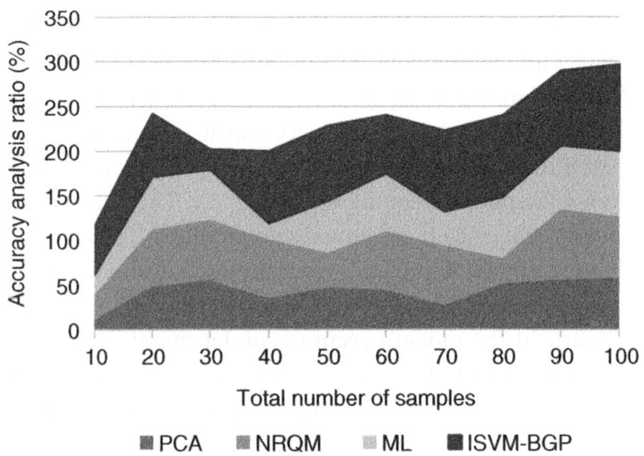

Figure 5.7 Accuracy analysis is compared with ISVM-BGP.

reliability. Using several biomass feedstock and working situations, the ISVM-BGP can manage hard fuel situations. Because it is comprehensive, this model is dependable and works in actual international situations with distinct biomass composition and electricity processing. SVM and stepped-forward kernel implementations make ISVM-BGP extra dependable. SVM captures the limitless interactions of biomass, making it dependable. Empirical validation guarantees that the version works on real and simulated data. The feedstock composition and system parameter variants can appreciably affect the gasoline yield in bioenergy production and waste-to-power conversion. ISVM-BGP robustness is essential. The simulation consequences show the regular and reliable overall performance of the ISVM-BGP under specific working conditions, which proves its reliability. Proper planning of the biomass gasification model promotes green and long-term energy and resource efficiency. Reliability analysis shows that the ISVM-BGP predicts biomass gasification results consistently and accurately. This benefits businesses looking for reliable options to improve gas processes and switch to healthier, more sustainable energy. Figure 5.8 shows that the comparison with SVM in dependency analysis results in an average accuracy of 88.4%. Figure 5.9 shows how the accuracy of the dependency analysis has increased to 94.2% compared to ISVM-BGP.

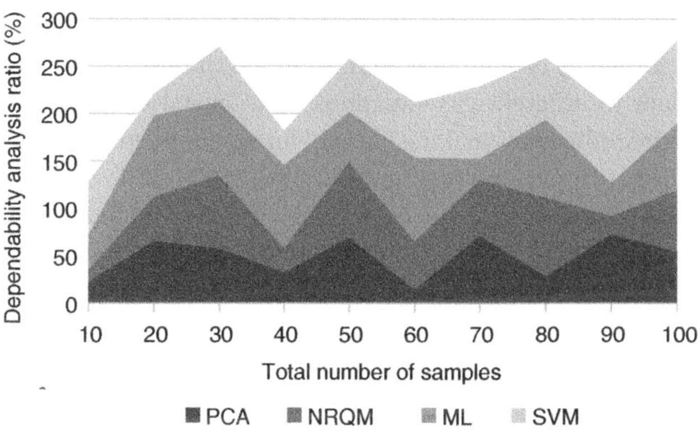

Figure 5.8 Dependability analysis is compared with SVM.

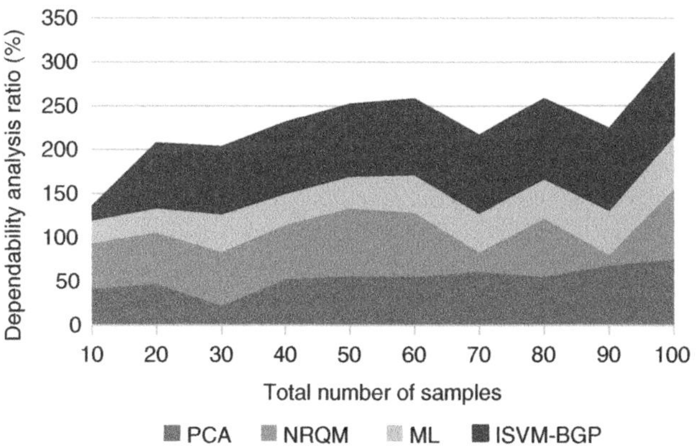

Figure 5.9 Dependability analysis is compared with ISVM-BGP.

The evaluation of the effectiveness of the ISVM-BGP model assesses its ability to provide accurate predictions quickly and efficiently to optimize the biomass gasification processes. Continuous control and variable control help make strong gasification changes, improving the model's performance. The flexibility of the ISVM-BGP lets it conform to operational parameter changes, improving real-time forecasting. This model works better with an immense fact series of biomass feedstock and method conditions to better apprehend the electricity generation potential of the gas. State-of-the-art engineering allows modeling and expertise of complicated gasification approaches. Comprehensive model validation guarantees its robust overall performance on artificial and historic datasets, enhancing its performance. The performance of ISVM-BGP helps in bioenergy manufacturing and waste-to-power conversion, which uses timely and accurate forecasts to maximize useful resource utilization and decrease environmental effects. Efficiency helps the model's aim of promoting sustainable and environmentally friendly power alternatives. The overall performance of the ISVM-BGP in several industries demonstrates its ability to adapt to different gas situations. This model can improve strength and useful resource efficiency, making it useful for development and decision-making. The performance evaluation shows that the ISVM-BGP correctly predicts the biomass gasification consequences. Due to its speedy adaptation to changing situations, computer-driven techniques, and large-scale use, this model can help us locate sustainable and green methods to supply biomass fuel. There is a development in the accuracy to 81.4% in the performance analysis compared to SVM, as shown in Figure 5.10. Compared with ISVM-BGP, the performance assessment reaches a completely excessive accuracy of 95.7%, as shown in Figure 5.11.

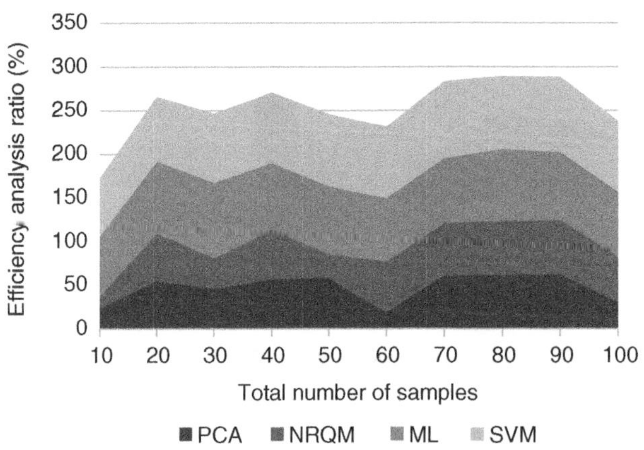

Figure 5.10 Efficiency analysis is compared with SVM.

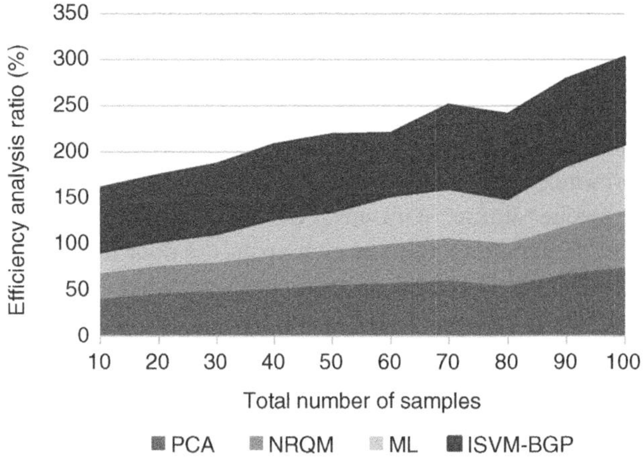

Figure 5.11 Efficiency analysis is compared with ISVM-BGP.

The effective, accurate, reliable, and efficient ISVM-BGP model predicts that biomass gasification results are higher than those of the SVM model. Real-time tracking, adaptive management, and superior issue engineering make the ISVM-BGP a promising tool for biomass gasification and sustainable power solutions.

5.4 Conclusion

The ISVM-BGP version has been advanced and applied to optimize the manner of biomass gasification. The research highlights the need for accurate prediction models for complicated gasification processes with a wide range of biomass feedstock, system parameters, and variables. Modern technologies, including improved kernel overall performance, real-time monitoring, adaptive control, and stepped-forward engineering, make the ISVM-BGP model a complete solution. The model's desires consist of enhancing the accuracy of the forecast and quickly imparting an entire image of the gasification method. The ISVM-BGP is flexible and appropriate for bioenergy manufacturing, waste-to-power, and sustainable electricity technology. The effects of gasification can be correctly expected through, for example, improving electricity efficiency, environmental impact, and resource utilization. A detailed simulation evaluation using artificial and historical databases confirms ISVM-BGP performance, accuracy, and reliability. Testing the version in one of the contexts indicates its flexibility and adaptability to actual global troubles. By combining advanced modeling techniques with real-world applications, ISVM-BGP improves the knowledge of gasification and permits answers to sustainable electricity. This version facilitates sustainable and green strength production by linking concepts and practices. The ISVM-BGP model nearly perfectly predicts the biomass gasification efficiency, accuracy, and flexibility. In many cases, efficient use shows a promising future in improving performance, reducing environmental impact, and developing alternative energy sources. As the world prioritizes sustainability, the ISVM-BGP model guides the energy sector toward a greener future. With advanced technology and a deep understanding of gasification dynamics, this research opens the door to groundbreaking energy sector developments and reviews a future where biomass gasification is more than a simple procedure; it generates environmentally friendly power.

References

1 Kim, J.Y., Kim, D., Li, Z.J. et al. (2023). Predicting and optimizing syngas production from fluidized bed biomass gasifiers: a machine learning approach. *Energy* 263: 125900.

2 Yang, Q., Zhang, J., Zhou, J. et al. (2023). A hybrid data-driven machine learning framework for predicting the performance of coal and biomass gasification processes. *Fuel* 346: 128338.

3 Bahadar, A., Kanthasamy, R., Sait, H.H. et al. (2022). Elucidating the effect of process parameters on the production of hydrogen-rich syngas by biomass and coal co-gasification techniques: a multi-criteria modeling approach. *Chemosphere* 287: 132052.

4 Kardani, N., Zhou, A., Nazem, M., and Lin, X. (2021). Modelling of municipal solid waste gasification using an optimized ensemble soft computing model. *Fuel* 289: 119903.

5 Zhao, S., Li, J., Chen, C. et al. (2021). Interpretable machine learning for predicting and evaluating hydrogen production via supercritical water gasification of biomass. *Journal of Cleaner Production* 316: 128244.

6 Xiaorui, L., Jiamin, Y., and Longji, Y. (2023). Predicting the high heating value and nitrogen content of torrefied biomass using a support vector machine optimized by a sparrow search algorithm. *RSC Advances* 13 (2): 802–807.

7 Tasneem, S., Ageeli, A.A., Alamier, W.M. et al. (2023). Development of machine learning-based models for describing processes in a continuous solar-driven biomass gasifier. *International Journal of Hydrogen Energy* 52: 718–738.

8 Kargbo, H.O., Zhang, J., and Phan, A.N. (2021). Optimization of two-stage biomass gasification for hydrogen production via artificial neural network. *Applied Energy* 302: 117567.

9 Khan, M.N.A., Haq, Z.U., Ullah, H. et al. (2023). Prediction of hydrogen yield from supercritical gasification process of sewage sludge using machine learning and

particle swarm hybrid strategy. *International Journal of Hydrogen Energy* 54: 512–525.

10 García Nieto, P.J., García-Gonzalo, E., Paredes-Sánchez, J.P. et al. (2019). Predictive modelling of the higher heating value in biomass torrefaction for the energy treatment process using machine-learning techniques. *Neural Computing and Applications* 31: 8823–8836.

11 Li, J., Li, L., Tong, Y.W., and Wang, X. (2023). Understanding and optimizing the gasification of biomass waste with machine learning. *Green Chemical Engineering* 4 (1): 123–133.

12 Elmaz, F., Yücel, Ö., and Mutlu, A.Y. (2019). Evaluating the effect of blending ratio on the co-gasification of high ash coal and biomass in a fluidized bed gasifier using machine learning. *Mugla Journal of Science and Technology* 5 (1): 1–12.

13 Umenweke, G.C., Afolabi, I.C., Epelle, E.I., and Okolie, J.A. (2022). Machine learning methods for modeling conventional and hydrothermal gasification of waste biomass: a review. *Bioresource Technology Reports* 17: 100976.

14 Gopirajan, P.V., Gopinath, K.P., Sivaranjani, G., and Arun, J. (2021). Optimization of hydrothermal gasification process through machine learning approach: experimental conditions, product yield and pollution. *Journal of Cleaner Production* 306: 127302.

15 Ozonoh, M., Oboirien, B.O., Higginson, A., and Daramola, M.O. (2020). Performance evaluation of gasification system efficiency using artificial neural network. *Renewable Energy* 145: 2253–2270.

16 Fang, Y., Ma, L., Yao, Z. et al. (2022). Process optimization of biomass gasification with a Monte Carlo approach and random forest algorithm. *Energy Conversion and Management* 264: 115734.

17 Elmaz, F., Yücel, Ö., and Mutlu, A.Y. (2020). Predictive modeling of biomass gasification with machine learning-based regression methods. *Energy* 191: 116541.

18 Ayodele, B.V., Mustapa, S.I., Kanthasamy, R. et al. (2022). Performance analysis of support vector machine, Gaussian Process Regression, sequential quadratic programming algorithms in modeling hydrogen-rich syngas production from catalyzed co-gasification of biomass wastes from oil palm. *International Journal of Hydrogen Energy* 47 (98): 41432–41443.

19 Ozbas, E.E., Aksu, D., Ongen, A. et al. (2019). Hydrogen production via biomass gasification, and modeling by supervised machine learning algorithms. *International Journal of Hydrogen Energy* 44 (32): 17260–17268.

20 Ascher, S., Watson, I., and You, S. (2022). Machine learning methods for modelling the gasification and pyrolysis of biomass and waste. *Renewable and Sustainable Energy Reviews* 155: 111902.

21 Yucel, O., Aydin, E.S., and Sadikoglu, H. (2019). Comparison of the different artificial neural networks in prediction of biomass gasification products. *International Journal of Energy Research* 43 (11): 5992–6003.

6

Role of Machine Learning Techniques in Modeling and Optimization of Biomass Gasification Parameters in a Downdraft Gasifier

Vikas Attri and Avdhesh Kr. Sharma

Mechanical Engineering Department, D.C.R. University of Science & Technology, Murthal (Sonepat), India

6.1 Introduction

The significance of distributed and decentralized power production is growing due to the rising energy demand and heightened environmental awareness. Presently, it is prudent to assert that the provision of energy for any nation is an unequivocal need. The evolution of all societies is significantly influenced by the accessibility and cost-effectiveness of energy. The degrees of access to energy and security have a substantial impact on several aspects of individuals' lives [1]. There are two main issues: the contamination of the environment caused by fossil fuels and a growing tendency to deplete natural resources [2, 3]. In order to address these grave concerns, it is imperative that the fossil fuel-based development model be abandoned. Additionally, new forms of renewable energy must be developed, such as bioenergy from bioresidues, solar energy, and wind energy. Among these options, bioresidues and solar energy stand out as the most promising renewable energy sources. However, the solar thermal power system faces significant challenges such as limited operation hours and fluctuating energy supply, which hinder its widespread adoption [4]. However, the characteristics provided by biomass, such as its abundant availability in tropical regions, render it a more promising renewable energy source in comparison to solar energy. The global waste biomass production is huge. The United States and Canada produce 72.5 Mt/year of forestry trash during harvesting and processing [5]. The use of biomass energy holds significant promise for developing nations as a promising alternative for small power production. The process of biomass gasification (BG) is the conversion of biomass feedstock into a gaseous fuel referred to as producer gas (PG) or syngas by a thermochemical process. The composition of this PG mainly comprises carbon monoxide (CO), hydrogen (H_2), carbon dioxide (CO_2), and methane (CH_4), with minor

Machine Learning for Sustainable Energy Solutions, First Edition. Edited by Zafar Said and Prabhakar Sharma.

amounts of other gaseous compounds. The system in which this occurs is called a gasifier, which is specifically designed to function under an oxygen-deficient environment [6]. For the purpose of the gasification process, several gasifier designs, such as fixed bed (downdraft and updraft), fluidized bed, and entrained bed, have been utilized for power production. For decentralized power production, PG from a downdraft gasifier is cleaner (it contains less tar contamination) and more convenient [7–9]. Without the requirement for gas cleaning procedures, the PG may be used for thermal applications such as gas burners. This eliminates the need for gas cleaning equipment. In order for gas to be used in gas engines, it is required to purify the gas. For utilizing this gas, the engine needs modifications to be operated in single-fuel mode [10]. BG has several benefits, including the production of renewable energy, the exploitation of waste, and the reduction of greenhouse gas emissions. However, it also poses some problems. The attainment of optimum gasification performance requires meticulous regulation of many factors, including the composition of the feedstock, the temperature of the gasifier, the duration of the reaction, and the equivalence ratio. Common methods for modeling and optimization often depend on empirical correlations or intricate simulations, which may be computationally demanding and lack precision.

Conventional modeling techniques can be difficult to apply when dealing with uncertain relationships between parameters. Thus, it is essential to develop a modeling approach that is both robust and efficient in order to accurately capture the intricate details of the gasification process. In recent times, there has been a significant focus on utilizing artificial neural networks (ANNs) and machine learning (ML) techniques to address these challenges in the field of gasification systems [11, 12]. Applying ML techniques to predict the operating input parameters for PG yield can be quite challenging. This is because multiple parameters may be influenced by a variety of input parameters simultaneously [13, 14]. There are various factors, including the steam-to-biomass ratio, biomass composition, and equivalence ratio, that can have a significant impact on the H_2 composition in PG. Therefore, it is crucial to employ effective strategies when developing ML models in order to achieve optimal accuracy. Thus, this study provides an overview of the fundamental principles of the BG and explores the different ML models that have been utilized in this area. ML allows for the deployment of more precise prediction models and the optimization of gasification processes, which in turn improve performance and make the process more economically viable.

6.2 Biomass Gasification

Biomass gasification is a thermochemical process that transforms solid biomass such as coal, wood biomass, and any solid waste into high-heat energy or PG. This technology is a very promising technique for producing an alternative fuel that is economically and environmentally benign and can be utilized for activities such as cooking, heating, generating electricity, producing biofuels, and synthesizing chemicals [15]. In the gasification technique, different types of biomass feedstock with different sizes can be utilized to generate PG, and biomass may be a more favorable choice for gasification than coal. Woody biomass consists of a larger proportion of volatile matter, which enables the generation of PG at a reduced temperature. PG is mostly composed of several gases, including H_2, CH_4, CO, and other hydrocarbons. Besides these gases, undesired by-products such as tars and other unwanted matter are commonly present in the PG [16, 17]. The complete gasification by-products are shown in Figure 6.1. The gasification takes place in a gasifier that can be categorized as a fixed-bed gasifier and a fluidized-bed gasifier. Fixed-bed gasifiers may be updraft, downdraft, or cross-draft, which are categorized on the basis of the design and flow rate of the PG [18]. Downdraft gasifiers are considered more suitable for small-scale power production. Mostly, downdraft gasifiers have power output ranging from 10 kW to 1 MW [19, 20]. In the downdraft gasifier, the biomass is supplied from the top, and this biomass undergoes various steps, which are explained in the next section. PG, which is generated by this gasifier, leaves from the bottom side.

6.2.1 Gasification Process

The process of BG in the gasifier consists of four parts, namely, drying, pyrolysis, combustion, and reduction, which are depicted in Figure 6.2.

1) **Drying:** The drying process of releasing water vapors occurs in a gasifier, where the temperatures are attained in the range of about 100–160 °C. Thus, the moisture removed from the biomass particles is mixed with the inflowing air and transported to the adjacent pyrolysis zone. In this process, heat is consumed

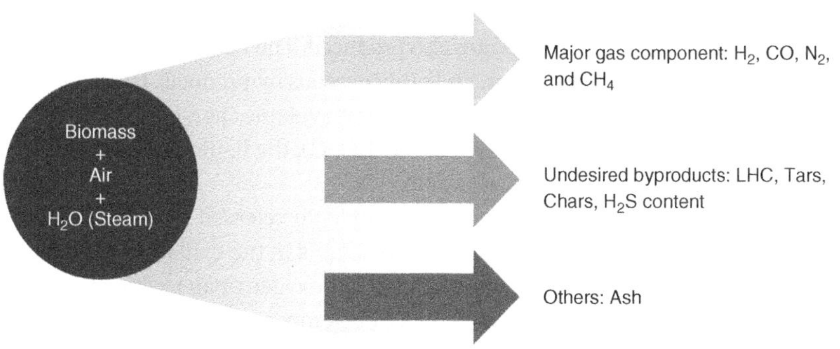

Figure 6.1 Gasification products.

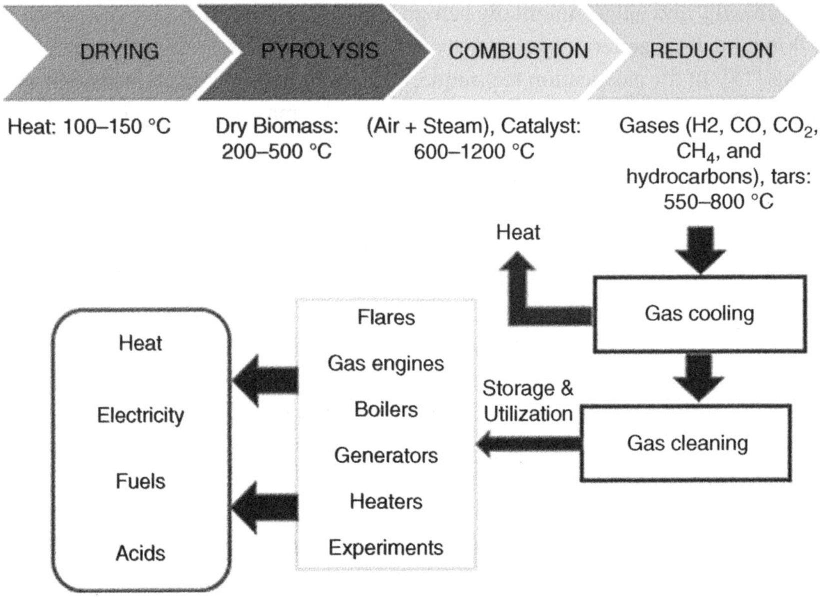

Figure 6.2 Steps of the gasification process.

to evaporate the moisture content. The drying process takes place in the drying zone or drying stage at a temperature of less than 160 °C.

2) **Pyrolysis:** The pyrolysis process is a thermal degradation process of biomass, which occurs when the derived biomass is heated to temperatures of 200–500 °C. During the thermal degradation process, the primary constituents of the biomasses (i.e., cellulose, hemicellulose, and lignin) break down into volatile matter and leave behind the solid substrate *char*. Volatiles include noncondensable heavy hydrocarbons that are transformed into *tar*. During the process, the formation of carboxyl and carbonyl group radicals triggers the reduction of molecular weight of the biomass constituents, generally amorphous cellulose. During the process, CO and CO_2 are also produced. The resulting cellulose (crystalline) decomposes into tar, char, and other gaseous by-products. Hemicellulose is broken down into a soluble polymer along with volatile char, fumes, and tar. At a higher temperature (between 300 °C and 500 °C), the lignin breaks down into acetic acid, methanol, water, and acetone [21, 22].
3) **Combustion:** The volatile matter, along with char released from the pyrolysis process, undergoes exothermic chemical reactions in the oxidation zone in the presence of limited exposure to the oxidizer (i.e., oxygen or air), which produces heat and energy, consequent of which the temperature rises in the range 1100–1500 °C. The volatiles and fraction of char substrate are consumed quickly to transform CO into CO_2 and H_2 into steam. It is claimed that oxygen

is a better gasifying medium than air for producing PG. Simultaneously, the part of heat released during the exothermic reaction zone or oxidation stage is utilized to provide the required heat or energy to sustain the pyrolysis and drying process [21, 22].

4) **Reduction:** When all the oxidant is consumed, the unconverted or remaining char and partially oxidized gaseous mixture, including unconverted tar, leaves the oxidation zone, thus entering the reduction zone to undergo a chemical reaction in a reduction environment. During this process, CO_2 is converted to two molecules of CO, while H_2O is converted to form H_2 and CO [23]. When all the oxidant is consumed, the unconverted or remaining char and partially oxidized gaseous mixture, including unconverted tar, leaves the oxidation zone, thus entering the reduction zone to obtain a chemical reaction in a reduction environment. Therefore, the CO_2 and water vapors from the oxidation zone react with char to produce CO, H_2, and some CH_4 components. The char is consumed almost completely, while some part of the tar leaves the gasifier if the temperature in the reduction zone is near 1000 °C [24]. The various chemical reactions that occur in the gasification process are shown in Table 6.1.

After these four processes in the gasifier, these mixtures of combustible gases require cooling and cleaning equipment so that the PG can be utilized in power production through IC engines. As a complicated process that follows gasification, cleaning the generated gas is an essential step in producing fuel with high efficiency [25]. The complete schematic diagram of the setup is shown in Figure 6.3 with auxiliary units.

6.2.2 Gasification Parameters

The quality of PG from the downdraft gasifier depends on various parameters such as the properties of the biomass feedstock, the design of the gasifier, the air-to-fuel ratio, the gasification temperature, the gasifying agent, the catalysts, and the operating conditions [26–28]. These factors significantly impact the composition of combustible gases in PG. These parameters are discussed later.

Table 6.1 Chemical reactions in the gasification process.

Oxidation reaction (exothermic)	Reduction reaction (endothermic)
R-I: $C + O_2 \rightarrow CO_2$	R-V:$C + CO_2 \rightarrow 2CO$
R-II: $C + 0.5O_2 \rightarrow CO$	R-VI: $C + 2H_2 \rightarrow CH_4$
R-III: $CO + 0.5O_2 \rightarrow CO_2$	R-VII: $CH_4 + H_2O \leftrightarrow CO + 3H_2$
R-IV: $H_2 + 0.5O_2 \rightarrow H_2O$	R-VIII: $CO + H_2O \leftrightarrow H_2 + CO_2$

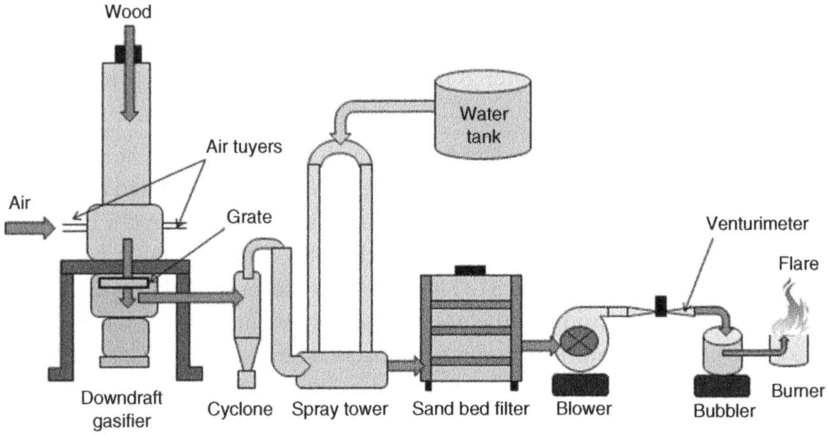

Figure 6.3 Layout of the downdraft gasifier with auxiliary units.

6.2.2.1 Biomass Characterization

The composition of combustible gases in PG is influenced by the composition (ultimate and proximate analysis), size, and moisture content of the biomass. In the downdraft gasifier, smaller-sized biomasses generate PG of superior quality as compared to large-sized biomass feedstock. PG derived from biomass feedstock with lower moisture content yields higher quality. In fixed-bed gasifiers, the feedstock size is limited [29]. The size of wood biomass should be 1–2 cm. The gasification process becomes much more energy efficient when the biomass is small [30]. The performance of the gasifier is also influenced by the uniformity of the biomass size. The efficiency of the gasifier improves with an increase in the homogeneity level of the biomass size [31]. The moisture content present in the biomass also affects the heating value of PG. Due to incomplete pyrolysis during gasification, biomass having more than 30% moisture content lowers the heating value of PG because heat is absorbed by the feedstock to dry the biomass [32].

6.2.2.2 Equivalence Ratio

Gasification may be facilitated by using air, steam, or oxygen as the agent. Air is primarily used as a gasification agent because of its abundant availability and cost-effectiveness. The equivalence ratio is a crucial process parameter in the gasification process. The equivalence ratio (ER) is defined as the ratio of the amount of air actually utilized in the gasification process to the amount of air required for complete combustion. The optimal equivalence ratio for efficient gasification typically falls between 0.2 and 0.4 [33]. Pyrolysis is the primary process in gasification when the equivalence ratio is below 0.2, whereas combustion becomes the dominant process when the equivalence ratio exceeds 0.4.

6.2.2.3 Gasification Temperature

The temperature rises as the ER increases, since this enhances the combustion process in the oxidation zone. The elevated temperature in the gasifier facilitated the decomposition of tar and increased the speed of the chemical reaction that converts char into gas, resulting in enhancing the heating value of PG [34]. The composition of the combustible gases in PG (H_2 and CO) is also enhanced with the increase in the gasification temperature. As anticipated, the oxidation zone often exhibits the highest temperature compared to other zones, but the temperatures of other zones fluctuate throughout the gasification process [35]. The variation arises from the inherent self-regulating characteristics of the downdraft gasifiers. The fluctuation in the airflow affects the heat generated during the oxidation process, resulting in oscillations in the gasification temperature [36].

6.2.2.4 Biomass Consumption Rate

The rate of biomass consumption is influenced by both the airflow rate and the size of the biomass. The rate at which biomass is consumed is directly proportional to the airflow rate and inversely proportional to the biomass size [37]. As the airflow rate increases, there is more oxygen in the combustion zone, which enhances the rate of the reactions. The oxidation process is enhanced by the smaller-sized biomass because it provides a higher total reaction area per unit volume of biomass.

6.2.2.5 Cold Gas Efficiency (CGE)

The cold gas efficiency (CGE) of a downdraft gasifier is defined as the ratio of the energy output to the energy input under atmospheric conditions. This parameter assesses how efficient the gasifier is to convert biomass energy into output PG energy. This can be calculated using equation (6.1).

$$\eta_{\text{CGE}} = \frac{\dot{m}_{\text{PG}}\text{LHV}_{\text{PG}}}{\dot{m}_{\text{biomass}}\text{LHV}_{\text{biomass}}} \tag{6.1}$$

6.2.2.6 Importance of Various Gasifying Agents in the Gasification Process

Air, steam, oxygen, and a combination of these may be used to increase the PG heating values during gasification. Because of its cost and availability, air is usually employed for gasification. Air gasification is simple in the reaction and reactor configuration [38]. When steam and oxygen gasifying agents are used, all the reactions are endothermic, requiring external heating. The reactions emit or absorb heat depending on air and oxygen [39]. By adjusting the flow rates of air or oxygen, the gasification process can be regulated as per the requirements. Increased gasification temperatures are necessary when there is a higher concentration of air or oxygen.

Table 6.2 Influence of the gasification parameters on the PG quality.

	Observations				
Parameters	**CO (% vol)**	**H_2 (% vol)**	**CH_4 (% vol)**	**LHV (MJ/kg)**	**CGE (%)**
Equivalence ratio ↑	↓	↓	↑	↓	↓
Temperature ↑	↑	↑	–	↑	↑
Moisture content ↑	↓	↑	↑	↑	↑
S/B ratio	↓	↑	↑	↑	↑
Air and oxygen ↑	↑	↑	↓	↑	↑

6.2.2.7 Effect of the Gasification Parameters on the Producer Gas

The production of gas in the gasification process is influenced by several factors, including the type of feedstock, the gasifying agent, the catalyst, and the operating conditions. These parameters have the potential to impact the composition of combustible gases in the PG [40]. Therefore, the reaction rate, the rate of conversion, and the yield of PG in this process are influenced by the following factors, as given in Table 6.2.

6.3 Machine Learning Techniques in Biomass Gasification

Artificial intelligence based on ML has been shown to be a useful supplement to conventional models used for the gasification process, contributing to a better understanding and control of its complexity. With the use of these methodologies, processes can be better designed to produce superior carbon conversion efficiency, which will reduce process time and costs for labor-intensive and complex procedures [41]. These methods have been widely employed in gasification processes to study the conversion reaction mechanisms and develop computer algorithms to achieve particular modelling goals [42]. Researchers have conducted numerous studies on the integration of ML in the gasification process, aiming to provide reliable outcomes for decision-making and forecasting. One can gain a deeper understanding of the methodology used to develop a strong and precise ML-based model. This approach is both flexible and cost-effective, making it highly regarded. In this section, we will focus on different ML-based modeling approaches for process development. The ML cycle for model selection is shown in Figure 6.4.

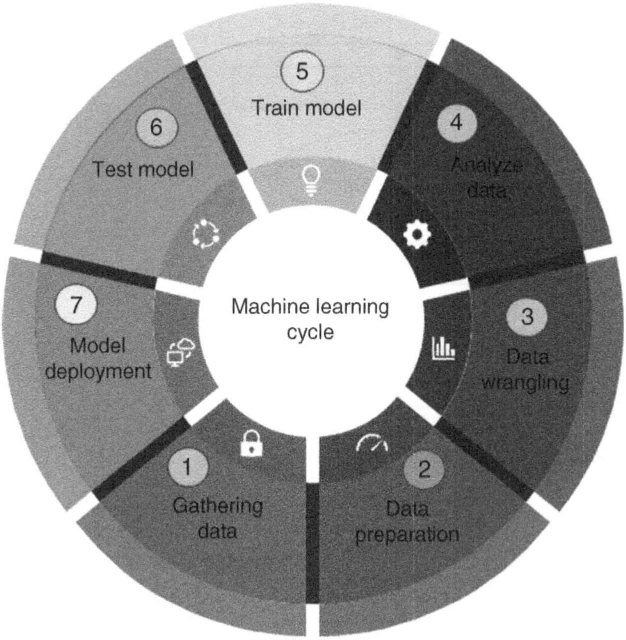

Figure 6.4 ML cycle for model selection.

6.3.1 Gaussian Process Regression

The Gaussian process regression (GPR) algorithm is a generic one that uses the Bayesian method to do regression in ML [43]. In the Bayesian method, all alternative values are assigned equal probability [44]. GPR has several advantages, including the capability to evaluate uncertainty in forecasts and the flexibility to manage limited datasets [45]. Constructing the predictive distribution with a GPR involves weighing feasible predictions based on their posterior distribution, allowing for predictions at unknown significant points [46]. The mean function and the covariance kernel functions of the GPR prior are selected and changed during the process of model selection. The mean function is often set to either 0 or the average value of the training data. Neither of these values is optimal. There is a kernel combination that is extensively utilized that captures the smoothness of functions, and that combination is the radial basis function kernel [47]. The algorithm for the model selection and configuration is shown in Figure 6.5.

6.3.2 Support Vector Machines

The support vector machine (SVM) is widely preferred for regression modeling due to its ability to produce reliable results with minimal computational resources [48].

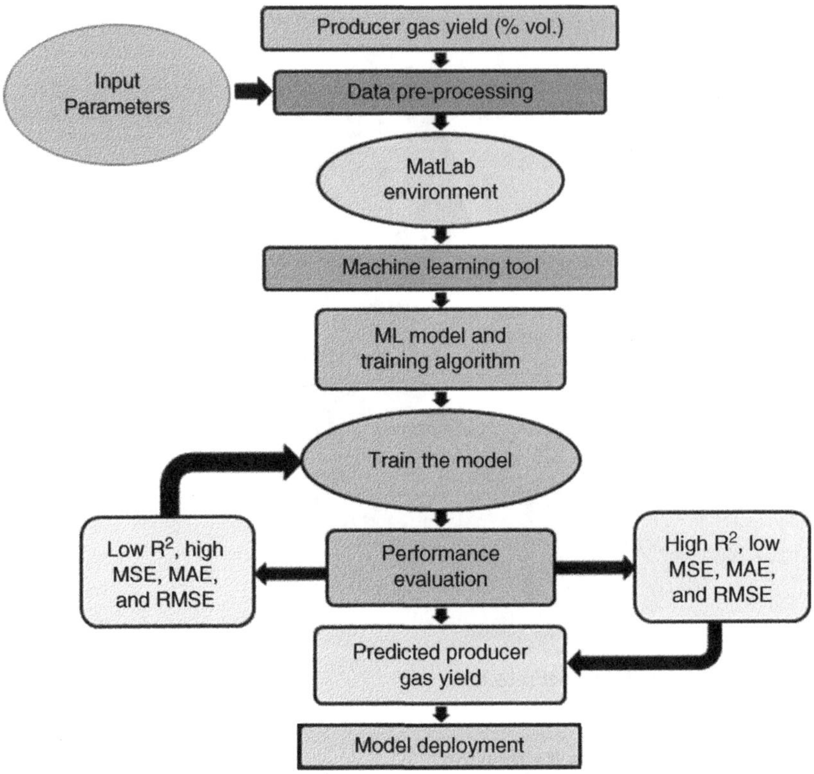

Figure 6.5 Flow diagram for the model configuration and modeling.

One of the goals of the SVM approach is to locate a hyperplane in the N-dimensional space that effectively divides the data points. It is possible to categorize the data with the assistance of hyperplanes, which act as decision boundaries. Data points on both sides of the hyperplane can be categorized using this feature. The number of characteristics that are included in the hyperplane is another factor that determines the size of the hyperplane. When doing the SVM modeling, the objective is to ensure that the data points are as close to the hyperplane as is practicable [49]. Hinge loss is employed to enhance the margin. When it comes to SVM, the regularization parameter is an extremely important factor in determining the optimal balance between maximizing the margins and minimizing the loss [50]. Empirical data indicates that the SVM exhibits strong performance in numerous practical scenarios. Research has demonstrated that they exhibit favorable outcomes in terms of predictive precision when applied to datasets of restricted magnitude. SVMs are particularly suitable for high-dimensional data sets. One additional benefit of SVMs is their reduced susceptibility to outliers, which can be attributed to the influence of SVM in generating the ideal hyperplane. There are several nonlinear mapping functions that may be utilized to solve these issues. Some examples of these functions are the sigmoid function, polynomials, and radial basis functions (RBFs). SVR, which stands for SVM regression, is a nonparametric regression model that stands out in comparison to traditional regression approaches such as polynomial fitting. Through the optimization of the distance between neighboring data points in order to calculate the regression hyperplane, it improves the performance of the SVM [51]. Support vector regression (SVR) is an extension of the SVM algorithm that allows it to be used for regression tasks instead of just classification. The function for regression is articulated in equation (6.2).

$$f(x, \alpha_i, \alpha_i^*) = \sum_{i=1}^{N} (\alpha_i^* - \alpha_i)(X_i^T X) + b \tag{6.2}$$

Here, $X_i^T X = \emptyset(x)$, $\emptyset(x)$ denotes the kernel function, α_i^* and α_i denote Lagrange multipliers, while N depicts the count of observations. SVM models forecast better without hidden layers than ML models. One can increase the model flexibility by using a variety of kernel functions and polynomial functions.

6.3.3 Artificial Neural Network

A very advanced method of ML, the ANN, is particularly effective in illuminating complex connections between the parameters that are input and the responses that are generated. In order to solve even the most difficult issues, this mathematical model takes its cues from the fundamental concept of biological neurons [51]. An ANN architecture comprises one input layer, one or more hidden layers, and an output layer. Once the data has been transmitted through the transfer function, it reaches the final layer known as the output layer. This layer consists of nodes that showcase the system's output. Because of its one-of-a-kind characteristics, the ANN is an ML technique that is extremely desirable for applications that require prediction. It is possible for ANN models to contain both nonlinear and linear input–output mappings, which gives them a great deal of flexibility. A typical neural network model is made up of three layers: the input layer, the hidden layer, and the output layer, as shown in Figure 6.6. These layers can have multiple neurons and are represented by equations (6.3) and (6.4).

$$H_{ij} = f\left(\sum_{i=1}^{I} X_{0i} W_{oi,j} + B_0\right) \tag{6.3}$$

$$Y_t = f\left(\sum_{i=1}^{J} H_{mj} W_{mj,t} + B_{Hm}\right) \tag{6.4}$$

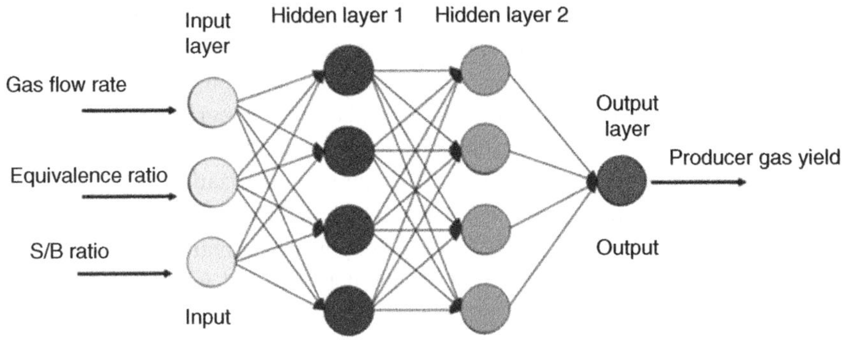

Figure 6.6 Block diagram of ANN.

In equations (6.3) and (6.4), *J*, *T*, and *I* correspondingly indicate the total number of neurons in the input, hidden, and output layers, respectively. *F* serves as the activation function; m denotes the hidden layer count. *B* and *W* stand in for the weight values and threshold, respectively.

An ANN model mandates that the input signal passes through numerous hidden layers from the input layer to the output layer. This transmission must be accomplished by using the activation function, bias vector, and weights throughout the process. In recent years, ANN technology has proven to be highly effective in a wide range of applications related to gasification [52]. These include predicting the parameters in gasifiers, analyzing the materials, evaluating the efficiency, and conducting economic and environmental analyses. When it comes to the gasification process, the input variables include biomass characteristics (i.e., proximate and ultimate analysis) [13, 53]. On the other hand, output variables encompass gas composition, heating values, emissions, wastes, and economics.

Various training procedures are employed to optimize the data for developing an ANN model. The MLFFBPNN, Gaussian Covariance Kernel, RBF, and the Levenberg–Marquardt algorithm are widely recognized as the four most popular algorithms in the field. One of the most commonly used neural networks in various fields is the multilayer feed-forward backpropagation neural network (MLFFBPNN). These networks utilize a backpropagation learning algorithm and are often employed in bioprocess optimization. The primary objective of this approach is to identify the optimal set of relation weights in order to minimize the expected error between the target and virtual values. The MLFFBPNN's ability to easily establish connections between inputs and generate outputs, coupled with its resilience and self-training capabilities, has made it a favored option for numerous applications.

6.3.4 Decision Trees

In geographic data mining, the decision tree (DT) approach is frequently utilized. This method is distinguished by its tree-like structure, which is possibly nonbinary or binary based on the analysis being performed. The DT model provides an illustration of the connection that exists between the values of objects and their qualities. Decision nodes, decision links, and decision leaves are the three main components that make up the dynamic programming (DT) approach. The storage of characteristics, the selection of criteria, and the performance of classifications are all significantly impacted by these key components. The branch points to the output of the feature attribute. In contrast, the leaf nodes are responsible for storing different types of solutions [54]. The path that leads from the root to the leaves is a representation of the input parameters, and the value of the targeted parameters is represented by each leaf independently. Constructing the DT model for the BG process depends on clearly stating the input as well as output vectors to be X and Y distributions, respectively. Equation (6.5) also shows a learnt dataset (D) in the DT. Following that, the DT model could be expressed as given in equation (6.6), and the input space may be divided into m units.

$$D = \{(x_1, y_1), (x_2, y_2), \ldots\ldots\ldots\ldots, (x_N, y_N)\} \tag{6.5}$$

$$f(x) = \sum_{m=1}^{M} C_m I(x \in R_m) \tag{6.6}$$

N represents the size of the sample data being used, while C_m represents the constant output in case of the mth unit.

Tree-based procedures are typically more straightforward to apply than other ML methods. Data often necessitates minimal preprocessing, and the techniques are well suited for handling missing predictors.

6.4 Model Performance Metrics

The data on the composition of the PG, the lower heating value, and the cold gas efficiency can be analyzed by many statistical metrics in order to analyze the performance of the downdraft gasifier. R^2 is a key measure that plays a part in determining how well the data fits the regression model. Utilizing the independent variables (inputs) allows for the prediction of a certain percentage of the variation in the dependent variable (output), which is measured by this method. When the value of R^2 is close to 1, it implies that the fit is more exact, and equation (6.7) may be used to calculate it.

$$R^2 = 1 - \frac{\text{SSR}}{\text{SST}} \tag{6.7}$$

where the sum of squared residuals, also known as SSR, is the sum of the squared differences between the predicted and observed values. The total sum of squares, also known as SST, corresponds to the sum of the squared differences between the actual values and their mean.

In order to evaluate the performance parameters, the mean squared error (MSE) is the parameter that is used. This value may be computed using equation (6.8). The mean squared variance between the expected and actual values is the definition of this term. The accuracy of the model is evaluated to determine its performance, with lower numbers indicating higher overall performance.

$$\text{MSE} = \frac{1}{n}\sum_{i=1}^{\text{n}} (Y_i - \hat{Y}_i)^2 \tag{6.8}$$

As seen in equation (6.9), the square root of the mean squared error (MSE) is the root mean squared error (RMSE). The error of the model is computed using the same units as the dependent variable throughout the process.

$$\text{RMSE} = \sqrt{\text{MSE}} \tag{6.9}$$

The MAE, which is provided in equation (6.10), is used to compute the average absolute difference between the predicted values and actual observed values. Both the MAE and MSE values that are lower suggest that the model is doing better.

$$\text{MAE} = \frac{1}{n}\sum_{i=1}^{n} |Y_i - \hat{Y}_i| \tag{6.10}$$

Herein, Y_i is the observed value, while $\hat{Y}_i$ is the forecasted value.

6.5 Application of the ML Model in Biomass Gasification

The ML approach has been extensively employed to accurately predict the performance parameters of gasification in various types of gasifiers, including downdraft, circulating fluidized-bed, and bubbling-bed gasifiers [55, 56]. The modeling and prediction of the gasification system performance have been the subject of a significant body of literature that has been produced over the course of time. A significant number of these studies have concentrated on examining the ways in which various operational factors influence performance. There are several benefits associated with the utilization of ML approaches when it comes to the

Table 6.3 Application of the ML models to optimize the gasification parameters.

Feedstock	ML technique	Inputs	Outputs	R^2 or RMSE	Ref
Coal	Random forest, Gaussian process regression	Operational conditions	PG composition	0.98–0.99	[57]
Different biomasses	ANN	Biomass composition and operational conditions	LHV of PG	0.99	[58]
Biomass (olive pit)	RF regression	Operating conditions	Composition and LHV of PG	0.813	[59]
Mixture of rubber and plastic	ANN	Operational conditions	H_2 gas composition	0.99	[60]
Sludge	SVM	Feedstock properties, reaction time, and gasification temperature	Syn gas yield, tar	0.99 $0.008 < MSE < 2.66$	[61]
Woody and herbaceous biomass	ANN	Operation conditions	PG yield	0.914	[62]
Woody biomass	SVM, RF	Operation conditions	Composition of PG	RMSE: 0.348–1.520	[41]
Biomass	SVM, ANN	Biomass properties	Product yield	0.8–0.932	[63]

prediction, evaluation, and optimization of gasification systems. Several more ML techniques, including RF, GBR, XGB, SVR/M, and NN, have garnered considerable amount of attention in the field of BG or waste gasification. There are many different applications of ML models that are covered in Table 6.3, which is used to predict the performance of BG systems.

6.6 Challenges and Prospects

There are numerous opportunities and obstacles when it comes to modeling downdraft gasifiers using ML techniques. ML has the potential to greatly improve the accuracy and efficiency of predictive models for gasifier performance. This can lead to real-time optimization and control in the operations. Through the utilization of extensive datasets that encompass a wide range of operating conditions and feedstock compositions, ML algorithms have the ability to uncover intricate relationships and patterns that may be overlooked by traditional models. Nevertheless, there are still obstacles to overcome in this field. These include the requirement for well-curated and varied datasets, as well as the ability to comprehend and explain the inner workings of ML models, which is of utmost importance in a domain that values transparency and understanding. In addition, ensuring the resilience and adaptability of the ML models across various gasifier designs and operating environments presents a major challenge. Addressing these challenges necessitates close collaboration between experts in gasification and ML. Together, they can create robust and transparent models that drive innovation in sustainable energy production.

6.7 Conclusion

The proposed research proficiently analyzed the role of ML techniques in the modeling and optimization of the gasification parameters of the downdraft gasifier. Biomass gasification is a promising technology for small-scale power production in rural areas due to a range of factors, including its alignment with sustainable goals. PG from the downdraft gasifier is influenced by various parameters and also by properties of the biomass feedstock. Several modeling approaches, like thermodynamic, CFD, and kinetic models, are commonly employed in the development of gasification systems. Conventional models can be impractical and often struggle to establish a reliable relationship between the input parameters and the output responses. Therefore, the incorporation of ML-based techniques is showing great potential in enhancing traditional modeling approaches with intelligent retrieval capabilities. Thus, this study thoroughly examined the utilization of the approach based on artificial intelligence in gasification technology. Harnessing the powerful predictive capabilities of ML models and integrating them into larger system models can help drive widespread adoption in industries. This approach reduces uncertainties for investors and policymakers, making it more attractive for large-scale deployment. To accomplish this, having enough training data is crucial. It's important to carefully consider the nature and availability of the training data when determining the most suitable ML method for a specific use case. Finally, with the increasing availability of data and computational power, new research opportunities have emerged. For example, more intricate models could be utilized for applications involving real-time process control and optimization.

References

1 Samadi, S.H., Ghobadian, B., and Nosrati, M. (2020). Prediction and estimation of biomass energy from agricultural residues using air gasification technology in Iran. *Renewable Energy* 149: 1077–1091. https://doi.org/10.1016/j.renene.2019.10.109.

2 Sharma, A.K. (2009). Experimental study on 75 kWth downdraft (biomass) gasifier system. *Renewable Energy* 34 (7): 1726–1733. https://doi.org/10.1016/j.renene.2008.12.030.

3 Raj, R., Singh, D.K., and Tirkey, J.V. (2023). Gasifier-engine performance analysis using co-gasification of mahua wood waste and saw-dust briquette blend: an experimental and optimization approach. *Biomass Conversion and Biorefinery* https://doi.org/10.1007/s13399-023-04285-8.

4 Sehrawat R, Sahdev R K &Tiwari S 2023 Heat storage material: a hope in solar thermal. *Environmental Science and Pollution Research*30: 11175–11198. https://doi.org/10.1007/s11356-022-24552-x

5 Cho, E.J., Trinh, L.T.P., Song, Y. et al. (2020). Biocon- version of biomass waste into high value chemicals. *Bioresource Technology* Elsevier Ltd https://doi.org/10.1016/j.biortech.2019.122386.

6 Sharma, A.K., Antil, S., and Sachdeva, G. (2023). Small-scale toroidal fluidized bed gasification system: development and experimental aspects. *Journal of Energy Resources Technology* 145 (5): https://doi.org/10.1115/1.4056447.

7 Kumar, A., Kumar, N., Baredar, P., and Shukla, A. (2015). A Review on Biomass Energy Resources, Potential, Conversion, and Policy in India. *Renewable and Sustainable Energy Reviews* Elsevier Ltd. 45: https://doi.org/10.1016/j.rser.2015.02.007.

8 Sikarwar, V.S., Zhao, M., Clough, P. et al. (2016). An overview of advances in biomass gasification. *Energy and Environmental Science* 9 (10): 2939–2977. https://doi.org/10.1039/C6EE00935B.

9 Sansaniwal, S.K., Pal, K., Rosen, M.A., and Tyagi, S.K. (2017). Recent advances in the development of biomass gasification technology: a comprehensive review. *Renewable and Sustainable Energy Reviews*. Elsevier Ltd 72: https://doi.org/10.1016/j.rser.2017.01.038.

10 Attri, V., Sharma, P., and Kr, S.A. (2020). Conversion of stationary diesel engine into producer gas-engine operated in spark ignited mode. *Advanced Science, Engineering and Medicine* 12 (11): 1353–1357. https://doi.org/10.1166/asem.2020.2602.

11 Ascher, S., Wang, X., Watson, I. et al. (2022). Interpretable machine learning to model biomass and waste gasification. *Bioresource Technology* 364: 128062.

12 Kushwah, A., Reina, T.R., and Short, M. (2022). Modelling approaches for biomass gasifiers: a comprehensive overview. *Science of the Total Environment* 834: 155243.

13 Attri, V. and Sharma, A.K. (2024). Modelling and optimization of gasification parameters of downdraft gasifier—an RSM approach. *Biomass Conversion and Biorefinery* 15: https://doi.org/10.1007/s13399-024-05368-w.

14 Jamei, M., Sharma, P., Ali, M. et al. (2024). Application of an explainable glass-box machine learning approach for prognostic analysis of a biogas-powered small agriculture engine. *Energy* 288: 129862.

15 Meena, M., Shubham, S., Paritosh, K. et al. (2021). Production of biofuels from biomass: predicting the energy employing artificial intelligence modelling. *Bioresource Technology* 340: 125642.

16 Zhang, W., Wu, Y., Huang, S. et al. (2022). Study on physicochemical characteristics, solidification and utilisation of tannery sludge gasification ash. *Journal of Environmental Management* 310: 114584.

17 Gautam, G., Adhikari, S., Thangalazhy-Gopakumar, S. et al. (2011). Tar analysis in syngas derived from pelletized biomass in a commercial stratified downdraft gasifier. *BioResources* 6 (4): 4652–4661. https://doi.org/10.15376/biores.6.4.4652-4661.

18 Susastriawan, A.A.P. and Saptoadi, H. (2017). Small-scale downdraft gasifiers for biomass gasification: a review. *Renewable and Sustainable Energy Reviews* 76: 989–1003.

19 Teixeira, G., Van de Steene, L., Martin, E. et al. (2012). Gasification of char from wood pellets and from wood chips: textural properties and thermochemical conversion along a continuous fixed bed. *Fuel* 102: 514–524.

20 Sheth, P.N. and Babu, B.V. (2009). Experimental studies on producer gas generation from wood waste in a downdraft biomass gasifier. *Bioresource Technology* 100 (12): 3127–3133.

21 Chawdhury, M.A. and Mahkamov, K. (2010). Development of a small downdraft biomass gasifier for developing countries. *Journal of Scientific Research* 3 (1): 51. https://doi.org/10.3329/jsr.v3i1.5613.

22 Bukar, A.A., Oumarou, M.B., Tela, B.M., and Eljummah, A.M. (2019). Assessment of biomass gasification: a review of basic design considerations. *American Journal of Energy Research* 7: 1–14.

23 Srivastava, T. (2013). Renewable energy (gasification). *Advance in Electronic and Electric Engineering* 3 (9): 1243–1250. http://www.ripublication.com/aeee.htm.

24 Samiran, N.A., Ng, J.-H., Jaafar, M.N.M. et al. (2016). H2-rich syngas strategy to reduce NO_x and CO emissions and improve stability limits under premixed swirl combustion mode. *International Journal of Hydrogen Energy* 41: 19243–19255.

25 Tang, F., Chi, Y., Jin, Y. et al. (2022). Gasification characteristics of a simulated waste under separate and mixed atmospheres of steam and CO_2. *Fuel* 317: 123527.

26 Attri, V. and Sharma, A.K. (2024). Real-time experimental investigation of a multi-flow biomass gasifier by using various locally available wood biomass. *Engineering Research Express* 6 (1): https://doi.org/10.1088/2631-8695/ad2e52.

27 Wongsiriamnuay, T., Kannang, N., and Tippayawong, N. (2013). Effect of operating conditions on catalytic gasification of bamboo in a fluidized bed. *International Journal of Chemical Engineering* 2013: 297941.

28 Ren, J., Cao, J.P., Zhao, X.Y. et al. (2019). Recent advances in syngas production from biomass catalytic gasification: a critical review on reactors, catalysts, catalytic mechanisms and mathematical models. *Renewable and Sustainable Energy Reviews* 116: 109426.

29 Yin, R., Liu, R., Wu, J. et al. (2012). Influence of particle size on performance of a pilot-scale fixed-bed gasification system. *Bioresource Technology* 119: 15–21.

30 Alauddin, Z.A.B.Z., Lahijani, P., Mohammadi, M., and Mohamed, A.R. (2010). Gasification of lignocellulosic biomass in fluidized beds for renewable energy development: a review. *Renewable and Sustainable Energy Reviews* 14 (9): 2852–2862.

31 Patel, V.R., Upadhyay, D.S., and Patel, R.N. (2014). Gasification of lignite in a fixed bed reactor: influence of particle size on performance of downdraft gasifier. *Energy* 78: 323–332.

32 McKendry, P. (2002). Energy production from biomass (part 3): gasification technologies. *Bioresour Technol* 83 (1): 55–63.

33 Olgun, H., Ozdogan, S., and Yinesor, G. (2011). Results with a bench scale downdraft biomass gasifier for agricultural and forestry residues. *Biomass and Bioenergy* 35 (1): 572–580.

34 Guo, F., Dong, Y., Dong, L., and Guo, C. (2014). Effect of design and operating parameters on the gasification process of biomass in a downdraft fixed bed: an experimental study. *International Journal of Hydrogen Energy* 39 (11): 5625–5633.

35 Liu, H., Hu, J., Wang, H. et al. (2012). Experimental studies of biomass gasification with air. *Journal of Natural Gas Chemistry* 21 (4): 374–380.

36 Balu, E. and Chung, J.N. (2012). System characteristics and performance evaluation of a trailer-scale downdraft gasifier with different feedstock. *Bioresource Technology* 108: 264–273.

37 Tinaut, F.V., Melgar, A., Perez, J.F., and Horrillo, A. (2008). Effect of biomass particle size and air superficial velocity on the gasification process in a downdraft fixed bed gasifier. An experimental and modelling study. *Fuel Processing Technology* 89 (11): 1076–1089.

38 Zhang, Y., Li, B., Li, H., and Liu, H. (2011). Thermodynamic evaluation of biomass gasification with air in autothermal gasifiers. *Thermochimica Acta* 519 (1–2): 65–71.

39 Lv, P., Wang, J., Bai, Y. et al. (2022). CO_2 gasification of petroleum coke with use of iron-based waste catalyst from FT synthesis. *Thermochimica Acta* 711: 179205.

40 Awais, M., Li, W., Munir, A. et al. (2021). Experimental investigation of downdraft biomass gasifier fed by sugarcane bagasse and coconut shells. *Biomass Conversion and Biorefinery* 11 (2): 429–444. https://doi.org/10.1007/s13399-020-00690-5.

41 Mutlu, A.Y. and Yucel, O. (2018). An artificial intelligence based approach to predicting syngas composition for downdraft biomass gasification. *Energy* 165: 895–901.

42 Hasanzadeh, R., Mojaver, P., Azdast, T. et al. (2023). Developing gasification process of polyethylene waste by utilization of response surface methodology as a machine learning technique and multi-objective optimizer approach. *International Journal of Hydrogen Energy* 48 (15): 5873–5886.

43 Gao, W., Karbasi, M., Hasanipanah, M. et al. (2018). Developing GPR model for forecasting the rock fragmentation in surface mines. *Engineering with Computers* 34: 339–345.

44 Hu, D., Su, X., Liu, X. et al. (2020). Bayesian-based probabilistic fatigue crack growth evaluation combined with machine-learning-assisted GPR. *Engineering Fracture Mechanics* 229: 106933.

45 Mahmoodzadeh, A., Mohammadi, M., Noori, K.M.G. et al. (2021). Presenting the best prediction model of water inflow into drill and blast tunnels among several machine learning techniques. *Automation in Construction* 127: 103719.

46 Taki, M., Rohani, A., Soheili-Fard, F., and Abdeshahi, A. (2018). Assessment of energy consumption and modeling of output energy for wheat production by neural network (MLP and RBF) and Gaussian process regression (GPR) models. *Journal of Cleaner Production* 172: 3028–3041.

47 Koschwitz, D., Frisch, J., and Van Treeck, C. (2018). Data-driven heating and cooling load predictions for non-residential buildings based on support vector machine regression and NARX Recurrent Neural Network: a comparative study on district scale. *Energy* 165: 134–142.

48 Zendehboudi, A., Baseer, M.A., and Saidur, R. (2018). Application of support vector machine models for forecasting solar and wind energy resources: a review. *Journal of Cleaner Production* 199: 272–285.

49 Mohammadi, M., Khorrami, M.K., Vatani, A. et al. (2021). Genetic algorithm based support vector machine regression for prediction of SARA analysis in crude oil samples using ATR-FTIR spectroscopy. *Spectrochimica Acta Part A: Molecular and Biomolecular Spectroscopy* 245: 118945.

50 Wu, J., Wang, Y.G., Tian, Y.C. et al. (2021). Support vector regression with asymmetric loss for optimal electric load forecasting. *Energy* 223: 119969.

51 Alfarra, F., Özcan, H. K., Cihan, P., Ongen, A., Guvenc, S. Y., and Ciner, M. N. (2024, March 1). Artificial intelligence methods for modeling gasification of waste biomass: a review. Environmental Monitoring and Assessment. *Springer Science and Business Media Deutschland GmbH.* https://doi.org/10.1007/s10661-024-12443-2.

52 Sun, C., Ai, L., and Liu, T. (2024). The PSO-ANN modeling study of highly valuable material and energy production by gasification of solid waste: an artificial intelligence algorithm approach. *Biomass Conversion and Biorefinery* 14 (2): 2173–2184.

53 Hai, T., Alsharif, S., Dhahad, H.A. et al. (2022). The evolutionary artificial intelligence-based algorithm to find the minimum GHG emission via the integrated energy system using the MSW as fuel in a waste heat recovery plant. *Sustainable Energy Technologies and Assessments* 53: 102531.

54 Ghosh, A. and Maiti, R. (2021). Soil erosion susceptibility assessment using logistic regression, decision tree and random forest: study on the Mayurakshi river basin of Eastern India. *Environmental Earth Sciences* 80 (8): 328.

55 Yang, Y., Shahbeik, H., Shafizadeh, A. et al. (2023). Predicting municipal solid waste gasification using machine learning: a step toward sustainable regional planning. *Energy* 278: 127881.

56 Sezer, S., Kartal, F., and Özveren, U. (2021). Prediction of chemical exergy of syngas from downdraft gasifier by means of machine learning. *Thermal Science and Engineering Progress* 26: 101031.

57 Ceylan, Z. and Ceylan, S. (2021). Application of machine learning algorithms to predict the performance of coal gasification process. In: *Applications of Artificial Intelligence in Process Systems Engineering*, 165–186. Elsevier. T. Nejat Veziroglu, International Association for Hydrogen Energy, Miami, Florida, United States of America.

58 Sezer, S. and Özveren, U. (2021). Investigation of syngas exergy value and hydrogen concentration in syngas from biomass gasification in a bubbling fluidized bed gasifier by using machine learning. *International Journal of Hydrogen Energy* 46 (39): 20377–20396.

59 Ozbas, E.E., Aksu, D., Ongen, A. et al. (2019). Hydrogen production via biomass gasification, and modeling by supervised machine learning algorithms. *International Journal of Hydrogen Energy* 44 (32): 17260–17268.

60 Ayodele, B.V., Mustapa, S.I., Kanthasamy, R. et al. (2021). Modeling the prediction of hydrogen production by co-gasification of plastic and rubber wastes using machine learning algorithms. *International Journal of Energy Research* 45 (6): 9580–9594.

61 Shahbeik, H., Rafiee, S., Shafizadeh, A. et al. (2022 Nov). Characterizing sludge pyrolysis by machine learning: towards sustainable bioenergy production from wastes. *Renewable Energy* 1 (199): 1078–1092.

62 George, J., Arun, P., and Muraleedharan, C. (2018). Assessment of producer gas composition in air gasification of biomass using artificial neural network model. *International Journal of Hydrogen Energy* 43 (20): 9558–9568.

63 Chen, X., Zhang, H., Song, Y., and Xiao, R. (2018). Prediction of product distribution and bio-oil heating value of biomass fast pyrolysis. *Chemical Engineering and Processing Process Intensification* 130: 36–42.

7

Response Surface Methodology-Based Multiattribute Optimization of a Hydrogen-Powered Dual-Fuel Engines

Sanjeev Kumar[1], Prabhu Paramasivam[2], and Abdul Razak[3]

[1] *Department of Mechanical Engineering, Delhi Skill and Entrepreneurship University, Delhi, India*
[2] *Department of Research and Innovation, Saveetha School of Engineering, SIMATS, Chennai, Tamil Nadu, India*
[3] *Department of Mechanical Engineering, P. A. College of Engineering (Affiliated to Visvesvaraya Technological University, Belagavi), Mangaluru, India*

7.1 Introduction

The global energy shortage has become a critical challenge, intensifying the urgency for nations to reevaluate their energy production and consumption patterns. Driven by population growth, rapid industrialization, and rising urban infrastructure, the worldwide demand for energy is surpassing supply capacity, raising questions about energy security and economic stability [1, 2]. Historically dominant in the global energy grid, traditional fossil fuels are limited and more sensitive to geopolitical concerns and supply chain interruptions. Furthermore, aggravating the issue of climate change is the great reliance on coal, oil, and natural gas, which has greatly contributed to accumulating greenhouse gases (GHGs) in the atmosphere [3]. Aiming to balance the amount of emitted GHGs with corresponding offsets or removal from the atmosphere by mid-century, nations all around have committed ambitious net-zero targets in response to the environmental crises linked with excessive carbon dioxide (CO_2), methane (CH_4), and nitrous oxide (N_2O) [4]. As the Paris Agreement describes, achieving net-zero emissions is essential to keeping the global temperature increase to 1.5 °C over preindustrial levels under control. With yearly CO_2 emissions returning quickly following pandemic-induced decreases, the present trend of global emissions implies an increasing disparity between objectives and actual performance, notwithstanding these pledges [5, 6]. Large-scale adoption of renewable energy technology, improvements in energy efficiency, and major investments in carbon capture and storage

Machine Learning for Sustainable Energy Solutions, First Edition. Edited by Zafar Said and Prabhakar Sharma.

(CCS) solutions are needed for the extraordinary change in the global energy scene that the shift toward net-zero demands. Moreover, reaching net zero is not simply a technical one but also a socioeconomic and political one that calls for global collaboration, fair transitions for underdeveloped nations, and coordinated policy approaches endorsing sustainable development [7, 8]. Maintaining environmental health, economic resilience, and social well-being depends on addressing the worldwide energy shortfall and concurrently lowering greenhouse gas emissions. Scientific studies stressing the need for urgency highlight how delayed action may cause lasting damage disproportionately, impacting disadvantaged communities. Therefore, the worldwide quest for net-zero emissions marks a basic change toward reconsidering energy generation, distribution, and use practices, thus preparing the ground for a more sustainable and resilient future [9–11].

Diesel engines have long been the backbone of the transportation sector, widely adopted for their robustness, fuel efficiency, and high torque characteristics. Because diesel propulsion can provide continuous power over long distances, heavy-duty vehicles, commercial trucks, maritime vessels, and rail systems have especially relied on it [12, 13]. Diesel engines can, however, present major environmental problems even with their technical benefits. Harmful pollutants like nitrogen oxides (NO_x), particulate matter (PM), carbon monoxide (CO), and unburned hydrocarbons (HC) are emitted when diesel fuels burn, all of which seriously compromise air quality and cause great health hazards [14, 15]. Moreover, diesel combustion is a main contributor to greenhouse gas emissions, mostly carbon dioxide (CO_2), therefore aggravating the worldwide problem of climate change. Globally, regulatory systems are tightening emission criteria like the Euro VI and US EPA regulations, advocating greener substitutes and therefore exerting major pressure on the diesel engine industry. In response to these growing environmental concerns and the need for sustainable solutions, dual-fuel engines have emerged as a promising transitional technology [16]. Usually, a traditional fuel like diesel and an alternative, lower-carbon fuel such as natural gas, biogas, or hydrogen-based carriers, dual-fuel engines run concurrently by using two distinct fuels. Dual-fuel systems greatly lower the total carbon footprint and other regulatory pollutants by partly substituting cleaner fuels for diesel. This method addresses the urgent requirement for pollution reductions while preserving the benefits of diesel engines—like great efficiency and dependability. Moreover, dual-fuel technology provides flexibility as the engine may switch to complete diesel operation in case the secondary fuel is not available, therefore providing a sensible option for areas lacking developed alternative fuel infrastructure [17]. Notwithstanding the possible advantages, there are still difficulties, including building strong control systems to govern the complicated dynamics of dual-fuel operations, optimizing combustion techniques, and guaranteeing fuel compatibility. Still, continuous research and development initiatives seek to improve the performance, stability, and environmental advantages of dual-fuel engines, therefore establishing them as a significant first step toward better transportation systems and significantly helping to meet global decarbonization targets [18, 19].

Optimization plays a critical and transformative role in enhancing the performance, efficiency, and environmental sustainability of dual-fuel engines. Achieving perfect operating conditions becomes intrinsically difficult in dual-fuel systems, in which two separate fuels with differing physical and chemical characteristics are concurrently burned [20]. By means of optimization methods, researchers and engineers can methodically find the ideal mix of engine parameters—that is, injection timing, injection pressure, pilot fuel quantity, air–fuel ratio, ignition delay, and combustion phasing—that maximizes efficiency while minimizing emissions [21]. Optimizing helps initially mostly by increasing the combustion efficiency. Particularly when running on a large percentage of gaseous fuels, dual-fuel engines often suffer from issues like incomplete combustion, banging, and uneven ignition [22]. Higher brake thermal efficiency (BTE) and lower fuel consumption follow from the fine-tune of important parameters made possible by optimization under the most favorable thermodynamic circumstances. Second, the control of emissions is much aided by optimization. By using sophisticated statistical approaches, machine learning algorithms, and multiobjective optimization strategies, researchers can find operating windows wherein detrimental emissions like NO_x, CO, HC, particulate matter, and CO_2 are reduced without compromising engine performance [23, 24]. The complex engine behavior and outcome of different parameter combinations are modeled and predicted using the response surface methodology (RSM), genetic algorithms (GA), simulated annealing (SA), and hybrid machine learning models like XGBoost regression. These prediction instruments let researchers effectively explore a large design space and choose the best settings that strike performance with regulatory compliance [24, 25]. Third, optimization improves the dual-fuel engines' flexibility and durability throughout different running environments. Engine behavior may be much influenced by factors like variations in load fluctuations, fuel mix variance, and ambient temperature. The dependability and commercial feasibility of dual-fuel systems may be raised by an improved control method dynamically adjusting engine setting in real time to preserve stability and optimal performance under varied situations. Furthermore, optimization helps to maximize economic gains by lessening the pressures on engine components resulting from less-than-ideal combustion and by reducing fuel costs via effective use. All things considered, optimization is an essential instrument for the evolution and improvement of dual-fuel engines, therefore enabling the worldwide shift toward more environmentally friendly transportation technologies [26–28].

Despite extensive research on hydrogen-diesel dual-fuel engines, significant literature gaps persist in fully understanding and optimizing their complex combustion dynamics. Many earlier studies have mostly concentrated on single-variable experiments or partial parametric investigations, sometimes overlooking the complex interdependencies among many operational factors. Moreover, little research has been conducted on applying strong statistical optimization models like response surface methodology (RSM) to concurrently simulate and maximize important performance and emission properties. Most published studies either treat performance or emissions alone, lacking a comprehensive, multiobjective optimization method that efficiently balances both elements. The uniqueness of the present work is in using RSM to methodically create predictive models for important engine outputs—including brake thermal efficiency, peak pressure, and major emissions like NO_x, CO, and unburned hydrocarbons—across a broad range of compression ratios, injection timings, and hydrogen substitution ratios. This chapter presents a scientifically based approach to find ideal operating settings that improve combustion efficiency while greatly lowering environmental impact by building statistically proven links between input factors and engine responses. The main goal of this study is to close the present research gap by providing an integrated modeling and optimization framework utilizing RSM for hydrogen-diesel dual-fuel engines, hence increasing their economic and environmental viability. This all-encompassing approach not only clarifies dual-fuel combustion events but also offers useful recommendations for engine design and control with the aim of creating more environmentally friendly and effectively running transportation systems.

7.2 Materials and Methods

7.2.1 Test Engine Setup and Fuel

In the present experimental study, a 5.2 kW Kirloskar make, single-cylinder, water-cooled, four-stroke diesel engine equipped with a mechanical governor for maintaining constant speed operation was employed. Originally intended for conventional diesel running, the engine was changed to run under a hydrogen-diesel dual-fuel system. To exactly apply and quantify various engine loads during the testing, an eddy current dynamometer was linked to the engine. While diesel fuel came from an approved Indian Oil Corporation pumping station to ensure consistency, hydrogen gas, the major energy source, was commercially purchased from a recognized provider in Delhi, ensuring high purity levels. Hydrogen was added to the intake manifold in the dual-fuel mode, and a small quantity of diesel served as a pilot fuel to start combustion. Under the controlled laboratory settings, this arrangement allowed the methodical assessment of the performance, combustion behavior, and emissions. Based on the standard ASTM criteria, the

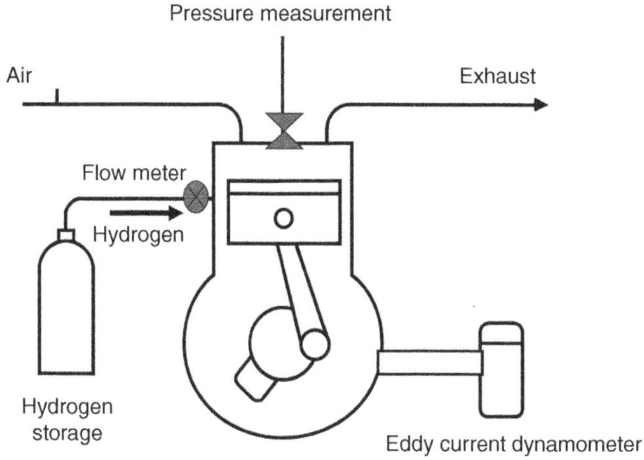

Figure 7.1 Test setup.

Table 7.1 Properties of the hydrogen gas.

Property	Value	Unit
Chemical formula	H_2	–
Molecular weight	2.016	g/mol
Lower heating value (LHV)	120	MJ/kg
Higher heating value (HHV)	141.8	MJ/kg
Autoignition temperature	∼585	°C
Flammability range in air	4–75	% (by volume)
Flame temperature	∼2045	°C
Density at 0 °C, 1 atm	0.0899	kg/m^3

physicochemical parameters of hydrogen gas and diesel fuel were assembled, which are essential for the knowledge of the combustion characteristics in the dual-fuel operation. The specifics are listed below (Figure 7.1, Tables 7.1 and 7.2).

7.2.2 Analysis of Variance

Analysis of variance (ANOVA) is a powerful statistical tool used to determine whether there are statistically significant differences between the means of three or more independent groups. Developed by Ronald Fisher, ANOVA compares the means of several samples to assist in evaluating the effect of one or more variables.

Table 7.2 Properties of the diesel fuel.

Property	Value	Unit
Chemical composition	C_{10}–C_{20} hydrocarbons	—
Lower heating value (LHV)	42.4	MJ/kg
Cetane number	45–55	—
Density at 15 °C	820–845	kg/m^3
Flash point	52.5	°C
Kinematic viscosity at 40 °C	3.45	cSt

It runs under the presumption that the populations from which the samples come from follow a normal distribution and have equal variances [29–31]. ANOVA becomes essential in studies that require an experimental design to find which input factors dramatically influence the result. It breaks down the overall variance in the data into elements ascribed to several sources of variation, including both inside and between groups. Central to ANOVA, the F-test measures group variance against group variance to obtain an F-statistic indicating whether observed variations are probably the result of actual effects or random chance. Reversing the null hypothesis with a significant F-statistic indicates that not all group means are equal. Advanced optimization methods such as response surface methodology (RSM), which helps create prediction models and determine the impact of several factors, also derive from ANOVA. Its exact mathematical background guarantees the dependability and statistical validity of the experimental results [32, 33].

7.2.3 Response Surface Methodology

Response surface methodology (RSM) is a statistical and mathematical technique employed to model and analyze problems in which a response of interest is influenced by several variables, with the objective of optimizing this response. Originally proposed by Box and Wilson in 1951, RSM is especially helpful when several input factors could affect the performance criteria or quality standards of a process [34, 35]. The approach consists of conducting a set of planned investigations to effectively investigate the interactions between response and explanatory factors. Usually of the second order, fitting polynomial equations to the experimental data helps to find the ideal circumstances for the desired reactions. RSM is fundamentally based on its capacity to assess how the interactions of several factors affect one or more response variables. Design of experiment (DoE) methods such as Box–Behnken design and central composite design (CCD) methodically alter the input variables to obtain useful results. Response surfaces—graphical representations showing the connections between the inputs and the response—can then be produced using the resultant models [36, 37]. These surfaces help one to see the ideal areas and grasp the type of reaction over the experimental field. RSM finds extensive use in the physical and biological sciences, industry, and engineering, among other domains. Its uses include manufacturing techniques and chemical reaction optimization to enhance product compositions and system performance. The method's strength is its efficiency; it offers a thorough understanding of the system under investigation and lowers the experimental trial count relative to conventional techniques. RSM is a useful instrument for researchers and engineers trying to improve process performance and product quality, as it helps to identify ideal operating conditions and clarifies the difficult relationships among variables.

7.3 Results and Discussion

7.3.1 Correlation Analysis

The data collected from engine testing was employed for modeling and optimization. The correlation heatmap in Figure 7.2 displays the relationships between the load, H_2 concentration, BSFC, BTE, HC, CO, and NO_x. With BTE, HC, CO, and NO_x, load has high positive relationship; with BSFC, it exhibits strong negative correlations. Showing relatively weak links across the matrix, H_2 concentration seems to be essentially uncorrelated with other variables. Strongly inversely associated with load, BTE, and NO_x, BSFC suggests that these variables rise as BSFC lowers. Reflecting how engine efficiency increases with increasing load and emissions, BTE is highly positively associated with load and NO_x and somewhat with HC and CO. The perfect correlation between HC and CO indicates that under the investigated conditions, they rise and decrease simultaneously. Maintaining a high negative association with BSFC, NO_x shows substantial positive correlations with load and BTE and modest correlations with HC and CO. The graphic generally shows that the load greatly affects both the performance and emissions criteria and that the H_2 concentration has a little effect inside the system [38, 39].

7.3.2 Analysis of Variance

The outcomes of the ANOVA are listed in Table 7.3. The brake-specific fuel consumption (BSFC) analysis shows that the model is highly significant with an F-value of 145.62 and a p-value less than 0.0001. The load has the largest effect; its sum of squares is 0.25, and its very high F-value of 537.27 indicates that BSFC greatly drops as the load rises. Higher loads increase the engine efficiency, which

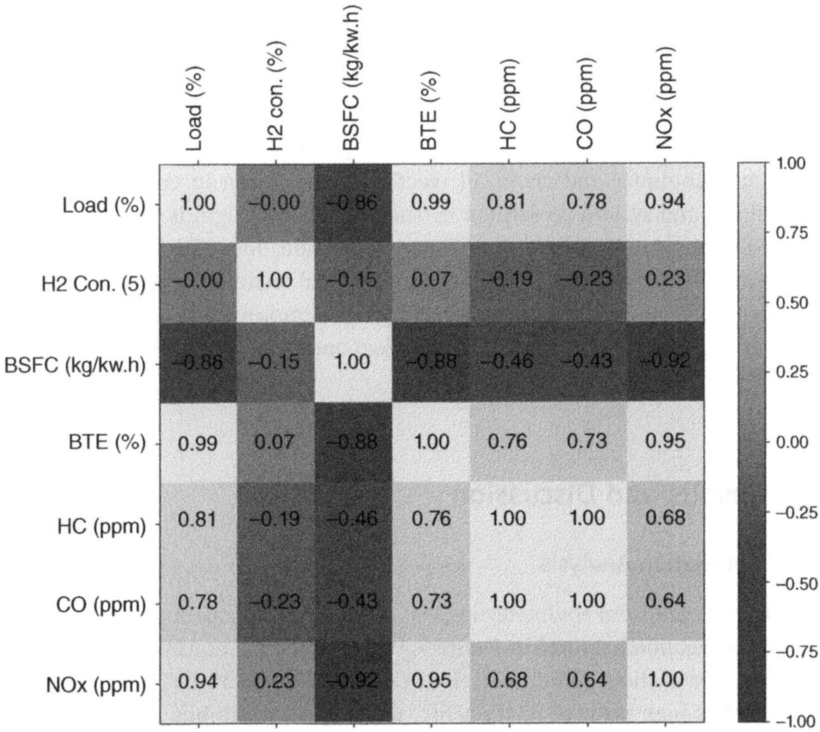

Figure 7.2 Correlation heatmap.

Table 7.3 ANOVA outcomes.

Response	Source	Sum of the squares	d_f	Mean square	F value	p-value
BSFC	Model	0.34	5	0.068	145.62	<0.0001
	A-Load	0.25	1	0.25	537.27	<0.0001
	B-HSR	0.009002	1	0.009002	19.16	0.0006
	AB	4.46E-05	1	4.46E-05	0.095	0.7624
	A^2	0.071	1	0.071	150.5	<0.0001
	B^2	0.002509	1	0.002509	5.34	0.0366
	Residual	0.006578	14	0.00047		
	Cor Total	0.35	19			
BTE	Model	1246.02	5	249.2	261.26	<0.0001
	A-Load	1210.85	1	1210.85	1269.42	<0.0001

Table 7.3 (Continued)

Response	Source	Sum of the squares	d_f	Mean square	F value	p-value
	B-HSR	6.84	1	6.84	7.17	0.018
	AB	0.62	1	0.62	0.65	0.4352
	A^2	2.44	1	2.44	2.56	0.1317
	B^2	0.2	1	0.2	0.21	0.6574
	Residual	13.35	14	0.95		
	Cor Total	1259.37	19			
HC	Model	13 275.26	5	2655.05	76.52	<0.0001
	A-Load	9015.15	1	9015.15	259.84	<0.0001
	B-HSR	540.56	1	540.56	15.58	0.0015
	AB	50.18	1	50.18	1.45	0.2491
	A^2	3626.53	1	3626.53	104.52	<0.0001
	B^2	42.58	1	42.58	1.23	0.2866
	Residual	485.74	14	34.7		
	Cor Total	13 761	19			
CO	Model	12 254.03	5	2450.81	56.49	<0.0001
	A-Load	8019.42	1	8019.42	184.85	<0.0001
	B-HSR	689.33	1	689.33	15.89	0.0014
	AB	121.95	1	121.95	2.81	0.1158
	A^2	3529.49	1	3529.49	81.36	<0.0001
	B^2	13.21	1	13.21	0.3	0.5897
	Residual	607.35	14	43.38		
	Cor Total	12 861.38	19			
NO_x	Model	1 488 000	5	297 600	66.62	<0.0001
	A-Load	1 283 000	1	1 283 000	287.22	<0.0001
	B-HSR	84 664.01	1	84 664.01	18.96	0.0007
	AB	16 102.71	1	16 102.71	3.61	0.0784
	A^2	16 174.82	1	16 174.82	3.62	0.0778
	B^2	5658.86	1	5658.86	1.27	0.2792
	Residual	62 526.86	14	4466.2		
	Cor Total	1 550 000	19			

lowers the fuel consumption per unit of power output by a logical means of inverse relationship. With an F-value of 19.16 and a p-value of 0.0006, the hydrogen substitution ratio (HSR) likewise dramatically influences the BSFC, implying that a somewhat higher hydrogen enrichment somewhat increases the combustion efficiency. The relatively high p-value (0.7624) of the interaction term AB, however, indicates that load and HSR have very little joint influence. With A^2 displaying an F-value of 150.5, stressing the nonlinear effects of load on BSFC, the quadratic factors A^2 and B^2 have a major influence. A good model fit is shown by a small residual value of 0.006578. The study generally shows that adjusting the load and hydrogen concentration may greatly lower the BSFC, thus improving the fuel economy.

The brake thermal efficiency (BTE) response demonstrates a highly significant model with an F-value of 261.26 and a p-value of less than 0.0001. With a sum of squares of 1210.85 and an amazing F-value of 1269.42, the load parameter rules supremely and indicates that the BTE increases dramatically as the load rises. Higher loads are predicted as they lower heat losses and improve the engine's general energy conversion efficiency. With an F-value of 7.17 and a p-value of 0.018, which indicates that hydrogen enrichment somewhat increases the thermal efficiency, HSR has a much lesser but still statistically significant effect on BTE. The low p-values of the interaction term AB and the quadratic terms A^2 and B^2 imply that their combined or nonlinear impacts are small. With a 13.35 residual error, the model fit is really decent. In terms of science, increasing loads cause the combustion temperature to rise, hence lowering specific losses and thus boosting the thermal efficiency. Hydrogen encourages full combustion by virtue of its increased flame speed and wider flammability limits, hence supporting the noted increase in BTE. For the best engine efficiency, then load and hydrogen concentration must be adjusted.

The hydrocarbon (HC) emissions model is statistically highly significant, with an F-value of 76.52 and a p-value of less than 0.0001. With a sum of squares of 9015.15 and an amazing F-value of 259.84, load is the most important element, emphasizing that HC emissions are quite load dependent. Richer mixes cause HC emissions to first increase as the load increases; yet, improved combustion helps to stabilize or reduce at extremely high loads. With an F-value of 15.58 and a p-value of 0.0015, HSR also contributes significantly as hydrogen addition improves cleaner combustion, hence lowering HC emissions. The statistically negligible interaction term AB and the quadratic terms point to either very small combined or nonlinear effects. The residual variance of 485.74 exhibits a decent match. Hydrocarbons are essentially generated scientifically by incomplete combustion. Hydrogen lowers unburnt hydrocarbons since it increases flame propagation and lowers the quenching effects. Thus, the successful reduction of HC emissions in internal combustion engines depends on the optimal load and suitable hydrogen mixing.

The carbon monoxide (CO) emission response shows a highly significant model with an F-value of 56.49 and a p-value of less than 0.0001. Load is the most influential factor; it has an F-value of 184.85 and a sum of squares of 8019.42, suggesting that CO emissions rise most definitely with load. Greater loads require richer mixes, which encourage incomplete combustion and therefore higher CO generation. With an F-value of 15.89 and a p-value of 0.0014, HSR also greatly influences CO emissions, implying that hydrogen addition promotes leaner and more complete combustion, hence lowering CO levels. Less important are the interaction and quadratic terms, which suggest that their impacts are small in relation to the main determinants. The residual error of 607.35 points to a decent fit for the model. Scientifically, CO is directly the result of incomplete combustion. The greater flame speed of hydrogen guarantees a more thorough oxidation of carbon during combustion, therefore reducing CO emissions. Thus, two good ways to reduce CO emissions while yet keeping effective engine running are load regulation and hydrogen enrichment.

The nitrogen oxide (NO_x) emissions model reveals very strong significance, with an F-value of 66.62 and a p-value of less than 0.0001. Load is the main contributor in the case of the NO_x model; the F-value is 287.22, and its sum of squares is quite big, 1,283,000. This suggests that NO_x generation increases dramatically with increasing load, mainly because larger loads enhance combustion temperatures, which promote NO_x creation via thermal processes. With an F-value of 18.96 and a p-value of 0.0007, HSR has a significant but smaller impact. Because of its high energy content, hydrogen raises combustion temperatures, therefore encouraging the generation of NO_x. The lack of a statistically significant interaction or curvature effects is suggested by interaction AB and quadratic terms A^2 and B^2. With a moderate residual error of 62,526.86, the fit is really well. Fundamentally, NO_x generation follows the Zeldowicz process and is temperature dependent [40]. While hydrogen improves efficiency and reduces other pollutants, its introduction could aggravate NO_x generation because of the higher in-cylinder temperatures. Controlling NO_x emissions thus depends on a precise balance between the hydrogen concentration and load, which benefits from the good impact of hydrogen on other performance criteria.

The following models were developed using ANOVA for each response in terms of predictors:

$$\begin{aligned} \text{BSFC} = {} & 0.57530 - 8.85143 \times 10^{-3} \times \text{Load} + 6.12792 \times 10^{-3} \\ & \times \text{HSR} - 1.42857 \times 10^{-5} \times \text{Load} \times \text{HSR} + 5.68571 \\ & \times 10^{-5} \times \text{Load}^2 - 1.57955 \times 10^{-3} \times \text{HSR}^2 \end{aligned} \tag{7.1}$$

$$\text{BTE} = 7.67051 + 0.26334 \times \text{Load} + 0.17132 \times \text{HSR} - 1.67771 \times 10^{-3} \times \text{Load} \times HSR - 3.34286 \times 10^{-4} \times \text{Load}^2 + 0.013955 \times \text{HSR}^2 \tag{7.2}$$

$$\text{HC} = 38.51828 - 0.61797 \times \text{Load} + 0.63350 \times \text{HSR} - 0.015146 \times \text{Load} \times \text{HSR} + 0.012876 \times \text{Load}^2 - 0.20577 \times \text{HSR}^2 \tag{7.3}$$

$$\text{CO} = 97.62829 - 0.60138 \times \text{Load} + 0.098821 \times \text{HSR} - 0.023611 \times \text{Load} \times \text{HSR} + 0.012702 \times \text{Load}^2 - 0.11463 \times \text{HSR}^2 \tag{7.4}$$

$$\text{NO}_x = 287.19226 + 8.89872 \times \text{Load} - 10.39135 \times \text{HSR} + 0.27132 \times \text{Load} \times \text{HSR} - 0.027192 \times \text{Load}^2 + 2.37207 \times \text{HSR}^2 \tag{7.5}$$

7.3.3 Surface Diagrams and Predictions

Figure 7.3 illustrates the brake-specific fuel consumption (BSFC) model using a 3D surface plot and an actual vs predicted plot, focusing on the effects of the engine load and the hydrogen supply rate (HSR). Investigating the 3D surface map in Figure 7.3a, the BSFC shows a nonlinear pattern, whereby at low engine loads of around 20% and a minimum HSR of 2 LPM, BSFC values are very high, around 0.59 kg/kWh. The BSFC falls drastically to around 0.35 kg/kWh when the engine load approaches 80% and the HSR rises to 12 LPM. This shows that because of better combustion efficiency and lower energy losses, higher load operation together with more hydrogen replenishment greatly lowers the fuel consumption per unit of energy production. Data points in the actual against anticipated plot (Figure 7.3b) form a close cluster around the 45° reference line, therefore supporting the predictive power of the model. Further encapsulating the improved tendency is the color gradient from blue at higher BSFC values to red at lower BSFC values. Notably, the surface plot curvature at mid-load and mid-HSR values points to a more sensitive interaction effect, whereby little hydrogen supply at medium loads maximizes the BSFC decrease most effectively. Consequently, the model not only shows important operating zones where hydrogen enrichment offers the greatest fuel efficiency gains but also matches the data very well.

Figure 7.4 presents the brake thermal efficiency (BTE) model through a three-dimensional surface plot (Figure 7.4a) and an actual vs predicted plot (Figure 7.4b). BTE is displayed in the surface plot as a function of the hydrogen supply rate (HSR) between 2 and 12 l per minute (LPM) and the engine load between 20% and 80%. Numerically, the BTE is about 8% with a lower load of 20% and an HSR of 2 LPM; it climbs dramatically to about 28% as the load rises to 80% and the HSR to 12 LPM. The surface plot shows a smooth and virtually flat gradient, demonstrating a strong positive linear connection, whereby both

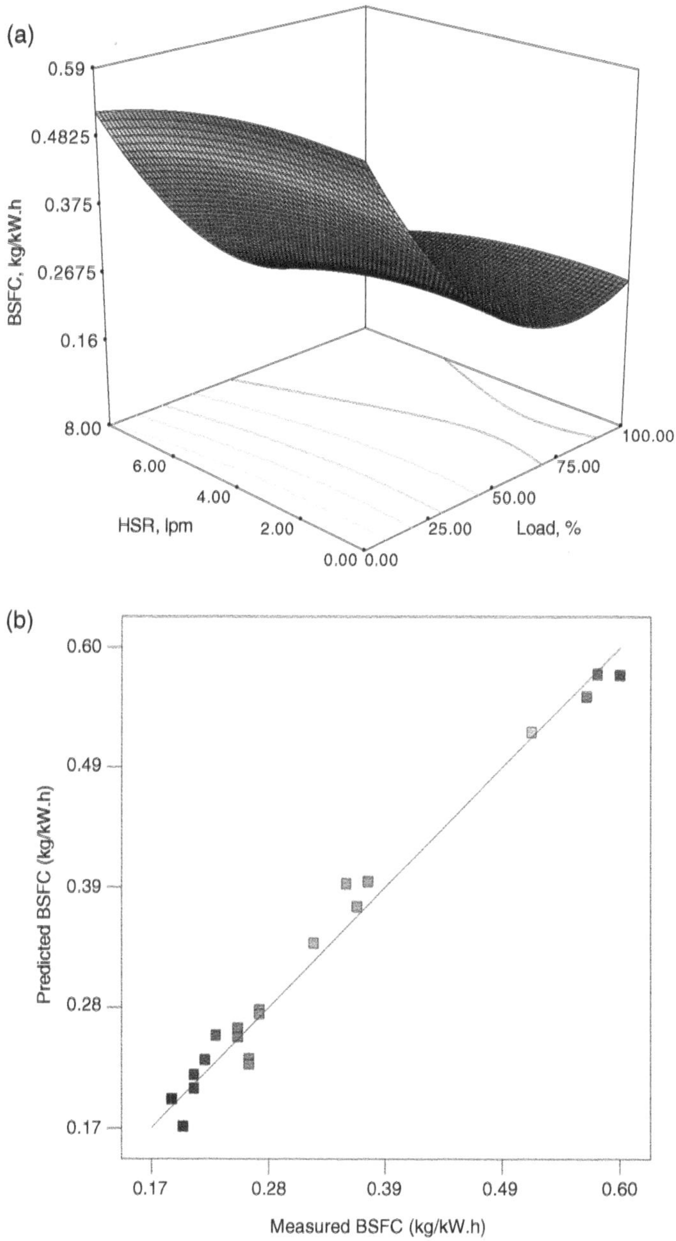

Figure 7.3 BSFC model: (a) surface plot and (b) predicted vs measured values.

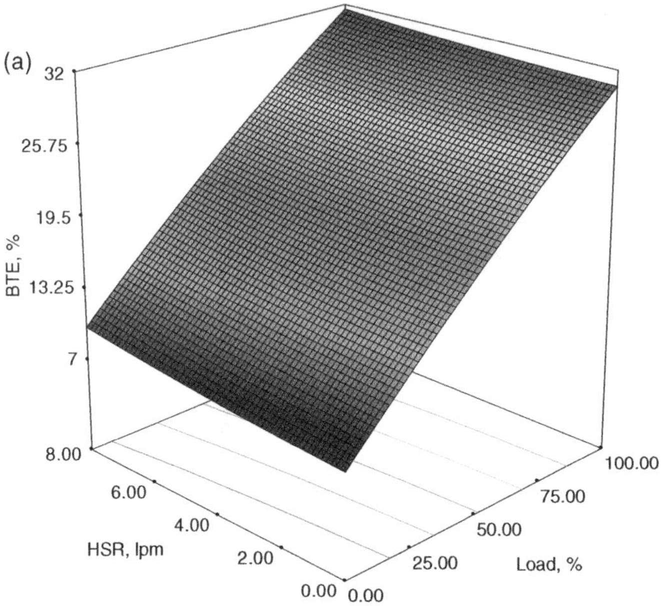

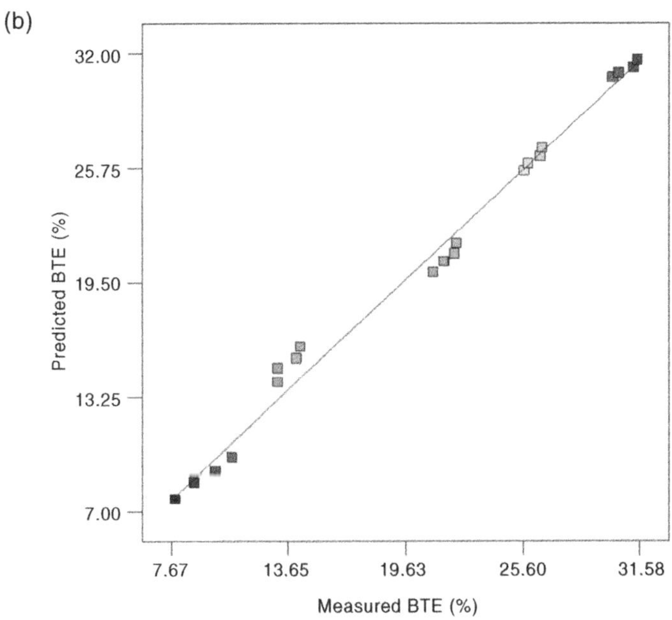

Figure 7.4 BTE model: (a) surface plot and (b) predicted vs measured values.

increased load and hydrogen supplementation greatly improve the BTE. The data points in the actual against anticipated plot strongly match along the diagonal line with BTE values ranging from about 8% to 28%, therefore indicating great model accuracy. The color evolution from blue (lower BTE values) to red (higher BTE values) shows even more consistent improvement in BTE with operational improvements. The quadratic regression model with the minimum variance between the anticipated and actual values is validated by the overall numerical confirmation that higher hydrogen flow rates and more engine loads synergistically increase the thermal efficiency.

Figure 7.5 presents the hydrocarbon (HC) emission model using two visualizations: a surface plot (Figure 7.5a) on the left and a scatter plot (Figure 7.5b) on the right. The surface map shows the change of HC emissions with respect to engine load in percentage and hydrogen supply rate (HSR) in liters per minute (LPM). The engine load runs from 0% to 100% on the y-axis; the x-axis displays the hydrogen supply rate ranging from 0 to around 6 LPM. Indicating the HC emissions is the vertical z-axis. With emission values starting from around 10 parts per million (ppm) up to roughly 50 ppm, the surface is colored from blue at the lower emissions to red at the higher emissions. The HC emissions first drop at lower engine loads but then start to increase once more beyond certain points when the hydrogen supply rate rises from 0 to 6 LPM. For low engine loads around 20% and low HSR close to 1 LPM, for instance, HC emissions are very high, around 45 ppm. The HC emissions can be seen first falling initially but then somewhat increasing again when the engine load approaches 80% and HSR is around 5 LPM, suggesting a nonlinear connection. Positively, the scatter plot gives a better perspective on the experimental data point distribution. Using the same color map as the surface plot, each point reflects a measurement at particular HSR and engine load circumstances, colored based on HC emission levels. Particularly with low hydrogen supply rates and moderate to high engine load, the data points cluster more densely at lower emission levels (blue zone). Points progressively move into yellow and red zones when the hydrogen flow rate and load both rise, suggesting higher HC emissions exceeding 40 ppm at high HSR and near 100% full engine loads. This research trend shows that although hydrogen supplementation first lowers HC emissions because of improved combustion, excessive hydrogen supply, especially at high engine loads, leads to combustion instability or rich circumstances, which increases hydrocarbon emissions once again.

Figure 7.6 shows the carbon monoxide (CO) emission model, represented by two plots: a surface plot (Figure 7.6a) on the left and a scatter plot (Figure 7.6b) on the right. Whereas the engine load is displayed along the y-axis from 0% to 100%, the hydrogen supply rate (HSR) in liters per minute (LPM) is plotted along the x-axis, roughly from 0 to 6 LPM, in the surface plot. Measuring in parts per million (ppm), the z-axis shows the CO emissions. The surface shows a color gradient, from blue

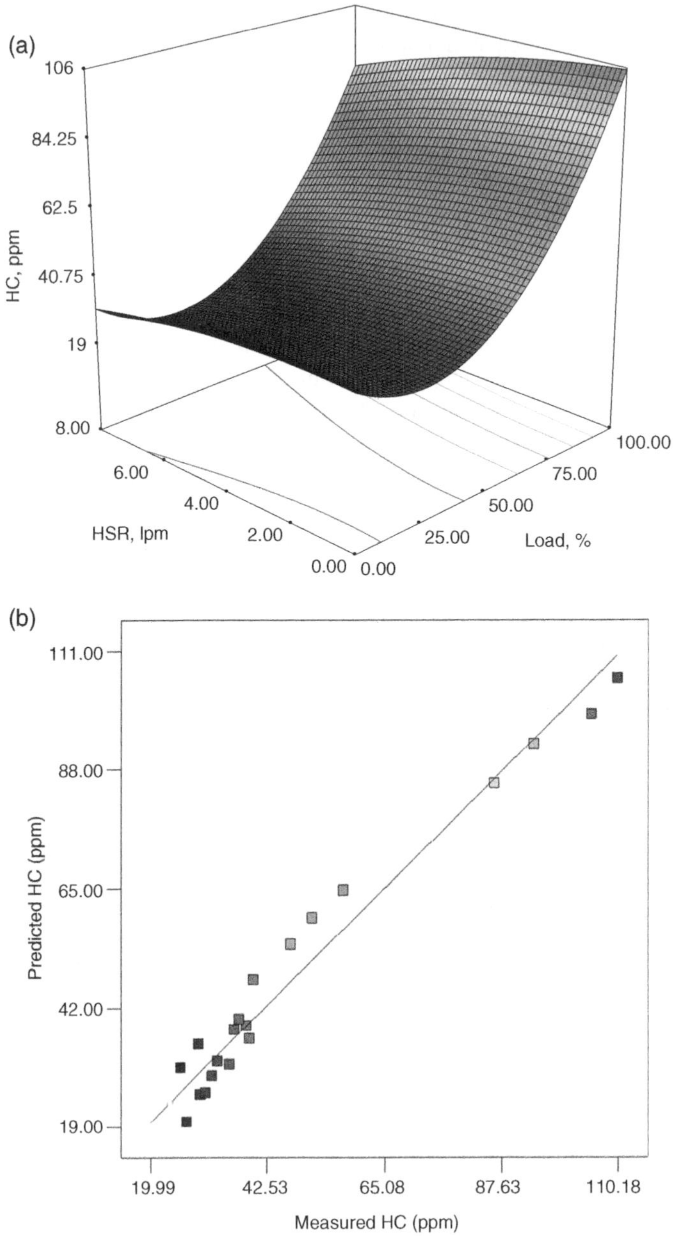

Figure 7.5 HC model: (a) surface plot and (b) predicted vs measured values.

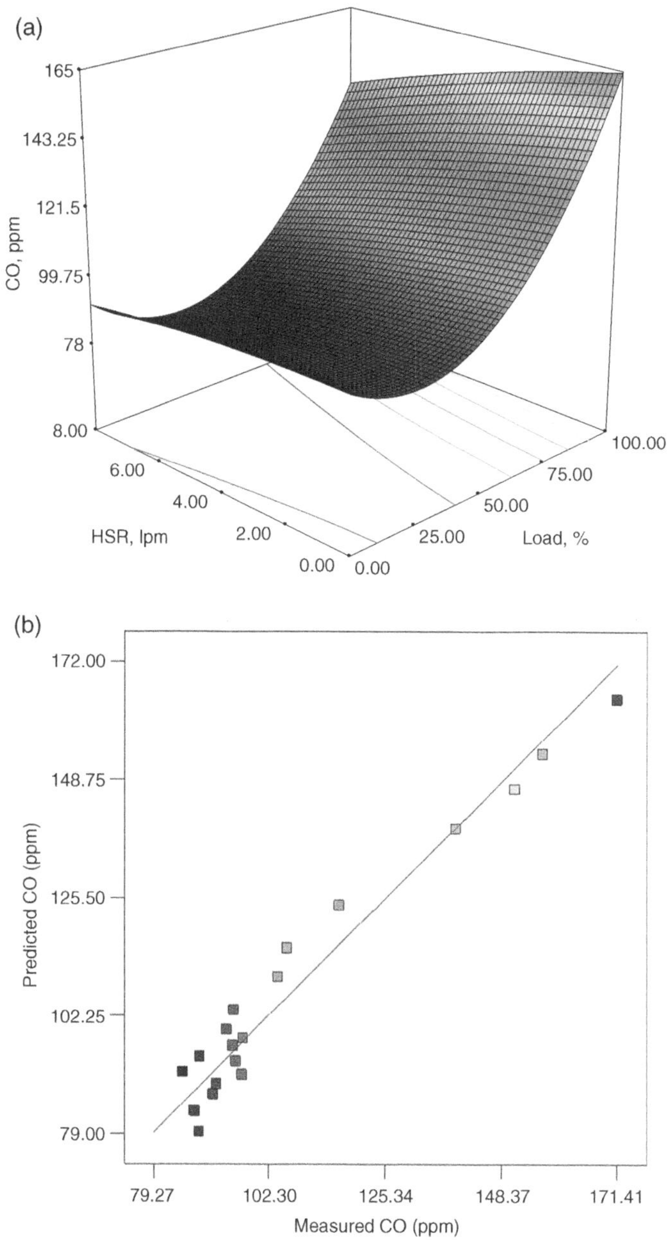

Figure 7.6 CO model: (a) surface plot and (b) predicted vs measured values.

at low CO levels to red at higher CO levels. The lowest CO emissions, about 0.5% at low hydrogen supply rates and moderate engine loads, are found. Particularly at higher engine loads over 60%, the CO emissions first drop but then progressively climb again, reaching values around 2.5% at HSR close to 6 LPM and engine load around 100% when the hydrogen supply grows above 2 LPM. Like the surface plot, the scatter plot on the right shows the experimental data points in color-coded fashion. This indicates that when the HSR is less than 3 LPM and the engine load ranges from 40% to 80%, most points are focused in the lower CO emission band. When the hydrogen supply reaches 4 LPM and the engine load approaches 100%, the data points become more dispersed and tinted yellow to red, indicating more CO emissions. One can ascribe this scientific behavior to the effect of hydrogen on combustion. At first, mild hydrogen enrichment increases more complete combustion, hence lowering the CO generation. But at very high hydrogen flow rates and strong engine loads, the air-fuel combination may become too rich or the combustion temperatures may change, resulting in incomplete combustion and a rise in CO emissions.

Figure 7.7 depicts the NO_x emission model using a surface plot (left) and a scatter plot (right). In the surface plot (Figure 7.7a), the HSR (LPM) is along the x-axis, spanning from 0 to about 6, while the engine load (%) is on the y-axis, ranging from 0 to 100 (Figure 7.7b). From blue at low NO_x levels (~100 ppm), the surface changes to crimson at high NO_x levels (~1000 ppm). Low load (~0–40%) and low HSR (~0–2 LPM) help maintain rather low NO_x emissions (~100–300 ppm). NO_x emissions significantly increase when both HSR and load increase outside of 3 LPM and 60%, respectively, peak at 1000 ppm at ~6 LPM and 100% load. Using data points moving from blue (low NO_x) to red (high NO_x), the scatterplot validates this tendency as the HSR and load increase. While yellow, orange, and red points rule at greater HSR and near-full load, most blue points cluster at lower HSR and loads. This scientific behavior corresponds with the combustion characteristics of hydrogen: adding H_2 raises flame temperature and promotes thermal NO_x generation by the Zeldovich mechanism. Low HSR first enhances combustion without significant NO_x spikes; then, high loads cause significant thermal NO_x from increased peak temperatures by using excessive H_2.

Regarding HC emissions (Figure 7.5), an increase in HSR initially decreases HC levels due to enhanced combustion; but, at around 4–5 LPM and at elevated engine loads (exceeding 80%), HC levels begin to rise again, ultimately reaching around 50 ppm. Regarding CO emissions (Figure 7.6), a comparable pattern is seen in which low HSR (up to around 3 LPM) decreases CO emissions, but, at elevated HSR and loads, CO increases significantly, reaching a peak near 2.5%. Conversely, NO_x emissions (Figure 7.7) consistently escalate with increased HSR and load, with NO_x levels advancing from around 100 ppm at low HSR and load to almost 1000 ppm at 6 LPM and full load. H_2 enrichment enhances combustion efficiency

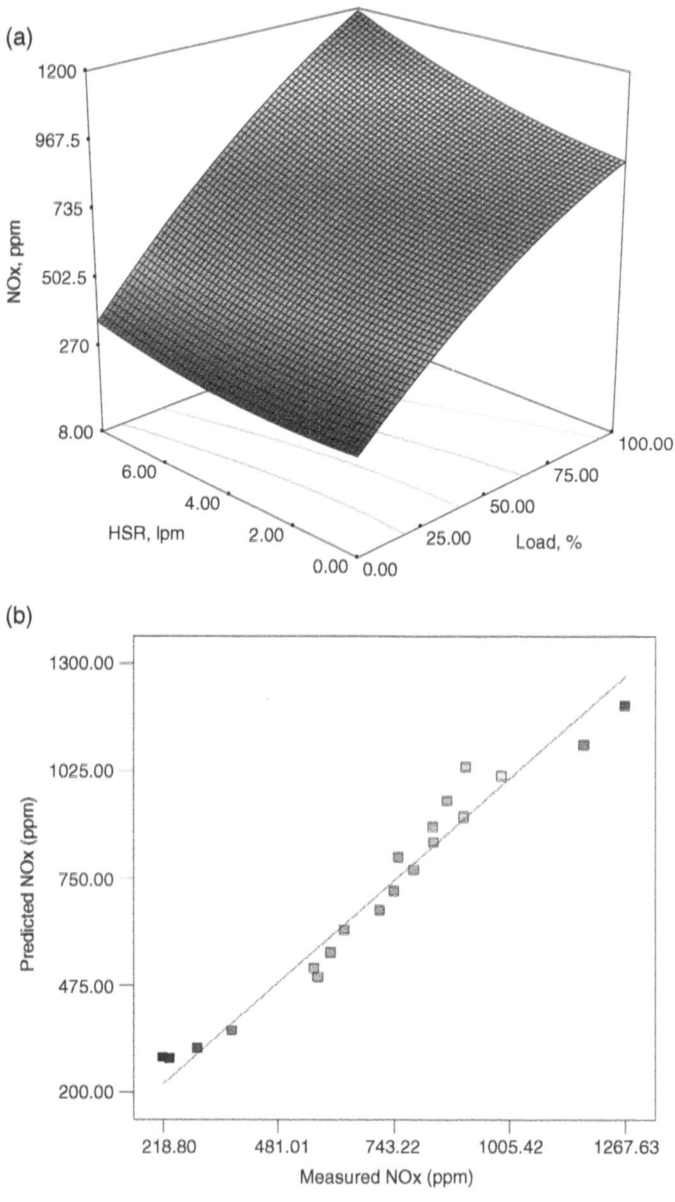

Figure 7.7 NO_x model: (a) surface plot and (b) predicted vs measured values.

by reducing HC and CO emissions; however, excessive H_2 at elevated loads results in incomplete combustion of HC and CO, significantly increasing NO_x emissions due to elevated flame temperatures. Each emission responds uniquely based on the influence of H_2 on combustion temperature and the equilibrium of the air–fuel ratio.

7.3.4 Parametric Optimization

The data presented is the result of desirability-based optimization using the response surface methodology (RSM) for optimizing multiple performance parameters. The optimization includes several goals with different constraints and weights, aiming to balance multiple objectives such as minimizing fuel consumption and emissions while maximizing efficiency. The examined parameters are load, HSR, BSFC, BTE (brake thermal efficiency), HC (hydrocarbon emissions), CO (carbon monoxide emissions), and NO_x (nitrogen oxide emissions). The optimization procedure designates the unique significance to each parameter, with BTE and load receiving the greatest priorities (5 and 3, respectively). The findings present the ideal answer for each parameter: Load is 70.31, HSR is 8.00, BSFC is 0.173939, BTE is 25.8525, HC is 42.0975, CO is 98.3108, and NO_x is 999.687. The desirability index for this solution is 0.708, signifying a positive result that meets the optimization objectives. The values fall between the established lower and higher bounds, attaining the optimal balance between competing objectives (Table 7.4).

Table 7.4 Optimized results.

Parameter	Goal	Lower limit	Upper limit	Weight	Importance	Optimized results
Load (%)	is in range	0	100	1	3	70.31
HSR, (lpm)	is in range	0	8	1	3	8
BSFC (kg/kWh)	minimize	0.19	0.6	1	1	0.17394
BTE (%)	maximize	7.88	31.51	1	5	25.8525
HC (ppm)	minimize	25.99	110.18	1	1	42.0975
CO (ppm)	minimize	85.13	171.41	1	1	98.3108
NO_x (ppm)	minimize	218.8	1267.63	1	1	999.687
Desirability						0.708

7.4 Conclusion

This study presents a thorough RSM-based multiobjective optimization framework for hydrogen-diesel dual-fuel engines, concurrently addressing key performance and emission characteristics. The principal conclusions are as follows:

- Moderate hydrogen supply rates substantially improve brake-specific fuel consumption (BSFC) and brake thermal efficiency (BTE), concurrently reducing hydrocarbon (HC) and carbon monoxide (CO) emissions.
- At elevated engine loads and hydrogen enrichment levels, nitrogen oxide (NO_x) emissions markedly escalate due to increased flame temperatures.
- The predictive models exhibit remarkable concordance with the experimental data, corroborated by the actual vs predicted plots.
- The ideal operational point attained a load of 70.31%, an HSR of 8.00 LPM, a BTE of 25.85%, and an attractiveness index of 0.708.

This work elucidates the intricate combustion interactions in dual-fuel engines and offers practical suggestions for the construction of cleaner and more efficient hydrogen-fueled transportation systems. The proposed technique may be applied to different alternative fuel systems, facilitating the advancement of sustainable internal combustion technology.

References

1 Ishaq, H. and Dincer, I. (2021). Comparative assessment of renewable energy-based hydrogen production methods. *Renewable and Sustainable Energy Reviews* 135: https://doi.org/10.1016/j.rser.2020.110192.

2 R, Now (2019). Renewables 2019 Global Status Report. *Renewable Energy Policy Network for the 21st Century.*

3 Mehanovic, D., Al-Haiek, A., Leclerc, P. et al. (2023). Energetic, GHG, and economic analyses of electrified steam methane reforming using conventional reformer tubes. *Energy Conversion and Management* 276: https://doi.org/10.1016/j.enconman.2022.116549.

4 Algayyim, S.J.M., Saleh, K., Wandel, A.P. et al. (2024). Influence of natural gas and hydrogen properties on internal combustion engine performance, combustion, and emissions: a review. *Fuel* 362: 130844. https://doi.org/10.1016/J.FUEL.2023.130844.

5 Jacoby, H. (2021). The Paris agreement. Https://ClimateMitEdu/Explainers/Paris-Agreement (accessed 17 January 2025).

6 Vinuales, J.E. (2016). The Paris agreement on climate change. *German Yearbook of International Law* 59 Paris.

7 Gangadhari, R.K., Karadayi-Usta, S., and Lim, W.M. (2023). Breaking barriers toward a net-zero economy. *Natural Resources Forum* 49: https://doi.org/10.1111/1477-8947.12378.

8 Tan, X. and Tan, T. (2022). Biofuels from biomass toward a net-zero carbon and sustainable world. *Joule* 6: https://doi.org/10.1016/j.joule.2022.06.005.

9 Patra, S., Verma, J., Mishra, Y.K. et al. (2023). The positioning of biofuel cells-based biobatteries for net-zero energy future. *Journal of Energy Storage* 72: https://doi.org/10.1016/j.est.2023.107919.

10 Kouchaki-Penchah, H., Bahn, O., Bashiri, H. et al. (2024). The role of hydrogen in a net-zero emission economy under alternative policy scenarios. *International Journal of Hydrogen Energy* 49: https://doi.org/10.1016/j.ijhydene.2023.07.196.

11 Ganter, A., Gabrielli, P., and Sansavini, G. (2024). Near-term infrastructure rollout and investment strategies for net-zero hydrogen supply chains. *Renewable and Sustainable Energy Reviews* 194: https://doi.org/10.1016/j.rser.2024.114314.

12 Hebbar, G.S. (2014). NOx from diesel engine emission and control strategies—a review. *International Journal of Mechanical Engineering and Robotics Research* 3: 471–482.

13 Kumar, S., Cho, J.H., Park, J., and Moon, I. (2013). Advances in diesel-alcohol blends and their effects on the performance and emissions of diesel engines. *Renewable and Sustainable Energy Reviews* 22: https://doi.org/10.1016/j.rser.2013.01.017.

14 Zaharin, M.S.M., Abdullah, N.R., Najafi, G. et al. (2017). Effects of physicochemical properties of biodiesel fuel blends with alcohol on diesel engine performance and exhaust emissions: a review. *Renewable and Sustainable Energy Reviews* 79: https://doi.org/10.1016/j.rser.2017.05.035.

15 Geng, P., Cao, E., Tan, Q., and Wei, L. (2017). Effects of alternative fuels on the combustion characteristics and emission products from diesel engines: a review. *Renewable and Sustainable Energy Reviews* 71: 523–534. https://doi.org/10.1016/j.rser.2016.12.080.

16 Ni, P., Wang, X., and Li, H. (2020). A review on regulations, current status, effects and reduction strategies of emissions for marine diesel engines. *Fuel* 279: https://doi.org/10.1016/j.fuel.2020.118477.

17 Wagemakers, A.M.L.M. and Leermakers, C.A.J. (2012). Review on the effects of dual-fuel operation, using diesel and gaseous fuels, on emissions and performance. *SAE International Journal of Engines* 2012010869 https://doi.org/10.4271/2012-01-0869.

18 Hung, N.T., Trang, N.T., and Thang, N.T. (2022). Quantile relationship between globalization, financial development, economic growth, and carbon emissions: evidence from Vietnam. *Environmental Science and Pollution Research* 29: https://doi.org/10.1007/s11356-022-20126-z.

19 Zheng, Y., Yu, S., and Caporin, M. (2024). Spatial effect of biomass energy consumption on carbon emissions reduction: the role of globalization. *Environmental Science and Pollution Research* 31. https://doi.org/10.1007/s11356-024-32849-2.

20 Veza, I., Spraggon, M., Fattah, I.M.R., and Idris, M. (2023). Response surface methodology (RSM) for optimizing engine performance and emissions fueled with biofuel: review of RSM for sustainability energy transition. *Results in Engineering* 18: https://doi.org/10.1016/j.rineng.2023.101213.

21 Alruqi, M., Sharma, P., Bora, B.J., and Ghosh, A. (2023). Multi-objective optimization of ternary blends of Algal biodiesel–diesel–1-decanol to mitigate environmental pollution in powering a diesel engine using RSM, ANOVA, and artificial bee colony. *Environmental Science and Pollution Research* https://doi.org/10.1007/s11356-023-30948-0.

22 Singh, A., Sinha, S., Choudhary, A.K. et al. (2021). An experimental investigation of emission performance of heterogenous catalyst jatropha biodiesel using RSM. *Case Studies in Thermal Engineering* 25: 100876. https://doi.org/10.1016/j.csite.2021.100876.

23 Sharma, P. and Sahoo, B.B. (2022). An ANFIS-RSM based modeling and multi-objective optimization of syngas powered dual-fuel engine. *International Journal of Hydrogen Energy* 47: 19298–19318. https://doi.org/10.1016/j.ijhydene.2022.04.093.

24 Sunil Kumar, K., Surakasi, R., Patro, S.G.K. et al. (2024). Performance, combustion, and emission analysis of diesel engine fuelled with pyrolysis oil blends and *n*-propyl alcohol-RSM optimization and ML modelling. *Journal of Cleaner Production* 434. https://doi.org/10.1016/j.jclepro.2023.140354.

25 Tuan Le, T., Sharma, P., Bora, B.J. et al. (2024). Co-gasification of waste biomass-low grade coal mix using downdraft gasifier coupled with dual-fuel engine system: multi-objective optimization with hybrid approach using RSM and Grey Wolf Optimizer. *Process Safety and Environmental Protection* 191: 234–248. https://doi.org/10.1016/j.psep.2024.08.066.

26 Djimtoingar, S.S., Derkyi, N.S.A., Kuranchie, F.A., and Yankyera, J.K. (2022). A review of response surface methodology for biogas process optimization. *Cogent Engineering* 9: https://doi.org/10.1080/23311916.2022.2115283.

27 Weremfo, A., Abassah-Oppong, S., Adulley, F. et al. (2023). Response surface methodology as a tool to optimize the extraction of bioactive compounds from plant sources. *Journal of the Science of Food and Agriculture* 103: https://doi.org/10.1002/jsfa.12121.

28 Susaimanickam, A., Manickam, P., and Joseph, A.A. (2023). A Comprehensive review on RSM-coupled optimization techniques and its applications. *Archives of Computational Methods in Engineering* 30: https://doi.org/10.1007/s11831-023-09963-4.

29 Emerson, R.W. (2022). ANOVA assumptions. *Journal of Visual Impairment and Blindness* 116: https://doi.org/10.1177/0145482X221124187.

30 Rouder, J.N., Engelhardt, C.R., McCabe, S., and Morey, R.D. (2016). Model comparison in ANOVA. *Psychonomic Bulletin and Review* 23: https://doi.org/10.3758/s13423-016-1026-5.

31 Rouder, J.N., Schnuerch, M., Haaf, J.M., and Morey, R.D. (2023). Principles of model specification in ANOVA designs. *Computational Brain & Behavior* 6: https://doi.org/10.1007/s42113-022-00132-7.

32 Liu, Q. and Wang, L. (2021). t-Test and ANOVA for data with ceiling and/or floor effects. *Behavior Research Methods* 53: https://doi.org/10.3758/s13428-020-01407-2.

33 P, Dimitriou, Z, Peng, D, Lemon, et al. (2013). Diesel Engine Combustion Optimization for Bio-diesel Blends Using Taguchi and ANOVA Statistical Methods. *SAE Technical Papers, vol. 6, SAE International.* 10.4271/2013-24-0011.

34 Sarabia, L.A. and Ortiz, M.C. (2009). Response surface methodology. *Comprehensive Chemometrics* 1: 345–390. https://doi.org/10.1016/B978-044452701-1.00083-1.

35 Hadiyat, M.A., Sopha, B.M., and Wibowo, B.S. (2022). Response surface methodology using observational data: a systematic literature review. *Applied Sciences (Switzerland)* 12: https://doi.org/10.3390/app122010663.

36 Nazarpour, M., Taghizadeh-Alisaraei, A., Asghari, A. et al. (2022). Optimization of biohydrogen production from microalgae by response surface methodology (RSM). *Energy* 253: 124059. https://doi.org/10.1016/j.energy.2022.124059.

37 Pereira, L.M.S., Milan, T.M., and Tapia-Blácido, D.R. (2021). Using response surface methodology (RSM) to optimize 2G bioethanol production: a review. *Biomass and Bioenergy* 151: 106166. https://doi.org/10.1016/J.BIOMBIOE.2021.106166.

38 Dodo, U.A., Ashigwuike, E.C., and Abba, S.I. (2022). Machine learning models for biomass energy content prediction: a correlation-based optimal feature selection approach. *Bioresource Technology Reports* 19: 101167. https://doi.org/10.1016/J.BITEB.2022.101167.

39 Il, P.S. and Oh, T.H. (2010). Correlation and simple linear regression. *Journal of Veterinary Clinics* 27: https://doi.org/10.1007/978-3-319-89993-0_6.

40 Rao, A., Liu, Y., and Ma, F. (2022). Study of NOx emission for hydrogen enriched compressed natural along with exhaust gas recirculation in spark ignition engine by Zeldovich' mechanism, support vector machine and regression correlation. *Fuel* 318: 123577. https://doi.org/10.1016/J.FUEL.2022.123577.

8

Addition of Nanoparticles to Biodiesel–Diesel Blends to Improve Engine Efficiency and Reduce Tailpipe Emission

Mudasar Zafar[1], Abida Hussain[2], Tauseef Ahmed[3], Ahmed Daabo[4], and Farman Ullah[5]

[1] School of Mathematics, Actuarial and Quantitative Studies (SOMAQS), Asia Pacific University of Technology & Innovation (APU), Bukit Jalil, Kuala Lumpur, Malaysia
[2] Department of Fundamental and Applied Sciences, Universiti Teknologi PETRONAS, Bandar Seri Iskandar, Malaysia
[3] Department of Mechanical Engineering, Universiti Teknologi PETRONAS, Bandar Seri Iskandar, Malaysia
[4] Mining Engineering Department, College of Petroleum and Mining Engineering, The University of Mosul, Mosul, Iraq
[5] Department of Physics, University of Science and Technology, Bannu, Khyber Pakhtunkhwa, Pakistan

8.1 Introduction

Energy provision is vital to economic and social development and to raising the standard of living. Most currently used world energy sources originate from fossil fuels, and one day, they will be exhausted if we do not invent technologies that can apply the additional fuels for energy consumption. Worldwide energy demand is increasing more rapidly than the population. This growth forecast is expected to expand to as much as 1.5% in 2030, according to excessive energy use in developing nations, especially Asia [1]. Fuel energy export is an industry that is growing throughout the world, and as mentioned in other related studies, it is in direct competition with the fastest growing alternative fuels to fossil fuel-based products, which these countries seek not to depend on. The lack of development of new and renewable energy sources led by the energy sector is primarily due to sustainability problems; so many nongovernmental organizations must use new and renewable energy sources [2].

One of the most promising alternative fuels to address this issue is biodiesel. Biodiesel is nontoxic, biodegradable, and renewable and has properties closely related to diesel fuel [3]. It is from vegetable oils, as well as animal [4]. However, these oils have high viscosity and cannot be directly scaled up for fuel. For instance, these raw oils must be processed through trans esterification to separate glycerin as a by-product and esters, resulting in biodiesel [5]. Like petroleum diesel, biodiesel

Machine Learning for Sustainable Energy Solutions, First Edition. Edited by Zafar Said and Prabhakar Sharma.

can run in compression-ignition engines with little or no modification [6]. In addition, biodiesel has more benefits regarding engine wear, cost, and accessibility. Biodiesel, when burned, creates less harmful pollutants to human health [7]. Moreover, it offers improved lubricity over diesel fuel. The need to lower greenhouse gas emission levels and the threats posed by global warming have sparked a renewed interest in alternative fuels and emission control technology within the transportation sector [8]. Biodiesel is a renewable, environmentally friendly diesel-like fuel produced from multiple sources of vegetable oil or animal fats that have become a promising alternative to regular fossil diesel in recent years due to the capability of biodiesel to mitigate harmful greenhouse gas emissions and dependence on fossil fuel resources [9].

However, like any other ICEs, they are pollutants that will generate environmental and health problems. Understanding the origin of biodiesel engine emissions is required to develop methods to control them [10, 11]. However, biodiesel produces harmful air and human health pollutants like regular diesel. The specific emissions from biodiesel engines are also classified as nitrogen oxides (NO*x*), carbon monoxide (CO), unburned hydrocarbons, and particulate matter [12, 13]. These pollutants can create air pollution, build smog, and result in public health effects. The low sulfur content of biodiesel has led to additional particle emissions, which require soot emission control [14, 15]. Biodiesel as a feedstock for alternative diesel fuel can mitigate some environmental burdens of conventionally produced fuels [16]. However, determining the aggregate environmental cost of the emissions released while burning biodiesel is important for understanding the technology's advantages and disadvantages [17, 18].

Nanoparticles have unique surface and chemical properties, making them promising candidates for improving combustion and reducing emissions [19]. Nanoparticles could also improve the catalytic performance in, and pollutant capture from, exhaust streams by their small size and high surface area. The addition of nanoparticles to biodiesel formulations has been a hot topic of research and consideration because of its apparent capacity to enhance several fuel properties and engine efficiency [20]. It is observed that addition of nanoparticles alters the kinematic viscosity, density, flash point, cetane number, calorific values, brake thermal efficiency (BTE), brake-specific fuel consumption (BSFC), combustion characteristics, cylinder pressure, heat release rate (HRR), and emissions, which comprise carbon monoxide (CO), hydrocarbons (HC), nitrogen oxides (NO_x), and smoke [21–23].

This comprehensive review aims at a complete review of the use of nanoparticles as additives to curb the emissions of biodiesel engines. They will cover various dimensions of the use of nanoparticles from their development and characterization, emission control practices, potential environmental and health impacts, and experimental research results. The emerging demand for cleaner transportation fuels is driving technical developments; the facts and insights outlined in this review may assist policymakers, researchers, and the industry in expediting environmental, economic, and efficient biodiesel engine technology. It will also offer guidance regarding the issues and potential regarding the use of routine nanoparticles in biodiesel engine emissions and formulate the prospects for future research and developments in this crucial area of knowledge.

This work has shown that the engine performance and emissions characteristics depend highly on the fuel blend, nanoparticle additive type, and composition. The literature indicates a positive effect of nanoparticles on the biodiesel properties (BTE, BSFC, and emissions reduction), but a very sharp variation in the nanostructure biodiesel effects can be observed, which is highly dependent on the physicochemical properties of the nanoparticles (i.e., type, size, and/or concentration) and the biodiesel blend type. Thus, additional exploration and optimization must be performed to maximize the beneficial effects of the nanoparticle additives and the complex nature of the biodiesel engine performance and emissions. Such a comprehensive review provides valuable information on the current state of research on the impact of nanoparticles in biodiesel engines. It further emphasizes the requirement for more research to optimize the use of this technology to run the engine sustainably and efficiently.

8.2 Background and Performance of Biodiesel Blends in Engine Efficiency

While diesel engines are widely used and highly efficient, they have significant limitations and environmental issues. Alternative fuels, however, come with significant disadvantages, including the high output of harmful pollutants (primarily nitrogen oxides (NO_x) and particulate matter (PM)) during combustion that contribute to air pollution and pose health risks [24]. Even worse, diesel engines rely greatly on fossil fuels, which are nonrenewable resources and lead to greenhouse gas emissions, especially carbon dioxide (CO_2) [25]. With diesel fuel becoming increasingly expensive and the global drive to ensure energy security and environmental sustainability becoming more pressing, there is a need for fuel that is both cleaner and can be sourced from renewable sources [26]. Long-term use of traditional diesel, on the other hand, can give rise to wear and tear in the engine, injector clogging, and rising maintenance bills in the future [27]. To address these challenges, researchers and industries have turned their focus to alternative fuels, such as biodiesel mixtures, as they have the potential to mitigate emissions and enhance combustion efficiency and the use of sustainable energy sources [28, 29].

8.2.1 Properties of the Biodiesel

An engine's efficiency is primarily influenced by the properties of the fuel it uses, including the flash point, pour point, cetane number, density, viscosity, heating value, and energy value (Figure 8.1) [30]. The properties of these blends should adhere to the limits established by organizations such as ASTM, EN, and ISO to ensure good performance. The recent trend is to blend two or three biofuels with diesel and a few additives to enhance the fuel quality [31, 32]. These biodiesel blends have different physical and chemical properties from one another and may influence the operation of the engine and the emissions in the exhaust gases [33].

Biodiesel has been shown to improve specific important fuel properties that directly affect an engine's performance and efficiency when blended with conventional diesel fuel [34]. The calorific value is used to understand how much energy is contained by the fuel under combustion. A higher calorific value typically translates into better power output from the engine [35]. The calorific value of pure biodiesel is lower than that of diesel fuel but is increased by blending biodiesel with diesel. For example, the calorific value of jatropha oil is approximately 39.8 MJ/kg, whereas palm oil and coconut oil have about 40 MJ/kg and 38.5 MJ/kg, respectively [36]. For context, regular diesel has an energy density of 46 MJ/kg. The blending of biodiesel with diesel helps to enhance the longevity and thus the energy content to power the engine, resulting in more power and efficiency of the engine [37, 38].

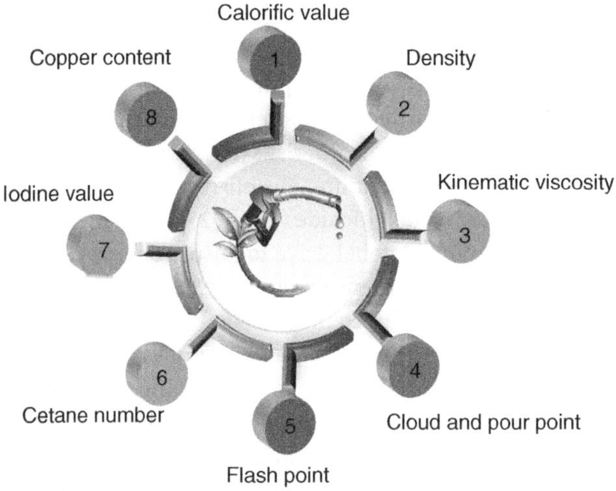

Figure 8.1 Properties of the engine efficiency.

Another important aspect is fuel density or the amount of fuel injected and how it burns inside the engine. With biodiesel generally having a higher density than diesel, this may cause the formation of larger fuel droplets after injection [39]. This could affect how well the fuel mixes with air and burns [40]. For instance, soybean biodiesel is (931 kg/m^3) denser than coconut biodiesel (807.3 kg/m^3). What makes biodiesel significant is the proper blend to get the density right, so managing how the diesel gets combusted and sent to be combusted benefits the engine operating strategy and reduces unburnt diesel emissions [41]. Fuel viscosity, or how thick the substance is, is a major contributor to the combustion quality [42]. Fuels that are too lean can cause leakage and wear in engine parts, while thicker fuels create big drops that burn ineffectively and generate increased degrees of smoke [43]. Biodiesel is even thicker and viscous at low temperatures, resulting in incomplete combustion. However, when mixed with diesel, its viscosity is reduced to an acceptable threshold. For instance, the kinematic viscosity of the sunflower oil and its blends varies from 4.5 to 5.8 mm^2/s while those of palm oil and its blends range from 3.2 to 4.5 mm^2/s. Hence, a spray of the fuel becomes finer and mixes well with air, which ultimately improves combustion and leads to better fuel economy [44, 45].

The cetane number is another important aspect that indicates how quickly the fuel ignites under pressure. Higher cetane numbers translate to more rapid ignition, smoother engine operation, and reduced noise and emissions [46]. Biodiesel naturally has a higher cetane number than regular diesel. The cetane number for palm biodiesel is 54.6 and jatropha 51.6, and sunflower biodiesel is only slightly lower at 37.9. The cetane number is improved with a standard diesel-biodiesel blend, which reduces the ignition delay and enhances combustion, leading to better combustion performance and lower fuel leftovers [47]. The flash point of a fuel is the minimum temperature at which it can produce sufficient vapor to burn. This means that the fuel is safer to store and transport because it has a higher flash point. Biodiesel has a substantially higher flash point than diesel, usually between 92 and 120 °C. Biodiesel blends are, therefore, safer to save and deal with, particularly throughout bulk storage [48].

The cold flow properties (cloud and pour points) are also paramount in cold climates. These values are the temperatures at which the fuel begins to coalesce or ceases to flow [49]. Saturated fatty acids cause biodiesel to have higher cloud and pour points than diesel. The cloud point and pour point of the palm oil biodiesel are 16 and 15 °C, while those of soybean biodiesel are 9 and 2 °C, respectively [50]. To reduce these values and allow for better performance and less chance of fuel line blockages in colder climates, biodiesel is blended with diesel in some areas [51]. One of the most important characteristics for long-term storage and engine application is the iodine value (IV), which indicates the degree of unsaturation in biodiesel [52]. A higher iodine value means more double bonds, which cause the fuel to oxidize and deposit. Soybean biodiesel, with an iodine value of 128–143, is grouped with coconut biodiesel, whose iodine value is low, about 19. Biodiesel with a lower iodine value or blending with diesel is recommended to improve fuel stability and limit engine deposit formation [53].

Finally, metals like copper in biodiesel can also impact its quality during storage and use. Copper can adversely react with biodiesel, which will change the properties of the biodiesel and decrease its performance. The metal content was also detected and controlled using techniques like stripping voltammetry. Minimizing copper will improve the quality of biodiesel blends and the diesel engine operability with no surprises. Biodiesel blends improve many important fuel properties for engine performance. Such blends are more sustainable and practical alternatives to conventional diesel, and they provide better energy content, ignition quality, fuel flow, and safety while enabling cleaner combustion and lower emissions.

8.2.2 Performance

The power output of an internal combustion engine is determined by how efficiently the fuel turns into usable power. All the previous studies of engine performance using biodiesel–diesel blends evaluated the performance using brake thermal efficiency (BTE), brake-specific fuel consumption (BSFC), and engine power output [54]. Observing the behavior of these parameters reveals how differences in biodiesel blends affect engine behavior vs diesel. Brake thermal efficiency is the ratio of the amount of practical work from combustion that the engine converts via the chemical energy of the fuel into mechanical energy. Research has found that the biodiesel blend thermal efficiency (BTE) is marginally lower than that of pure diesel [55]. By way of example, work by researchers found that acceptable levels of efficiency were obtained using a B25 blend (25/75 biodiesel/diesel), requiring no modifications to the engine [56]. Other research discovered that the lower efficiency of biodiesel was primarily due to its higher viscosity and lower heating value [57]. One study found that 100% beef tallow biodiesel (B100) had lower efficiency than B5 through B78 biodiesel due to lower calorific values [58]. However, in another study, Moringa oil methyl esters blends exhibited lower thermal efficiency than diesel in all tested load conditions due to the higher viscosity and density values [59].

Likewise, palm oil biodiesel produces marginally lower thermal efficiency than diesel, requiring approximately 10% more fuel to maintain the same output [60]. Interestingly, the studies also revealed that the presence of oxygen in the biodiesel was working to improve the combustion quality, even as the efficiency reduced just a bit [61]. The researchers noted that by adding alcohol to the biodiesel blends, the thermal efficiency improved (0.56), particularly with 25% alcohol in the blend, on par with some chemicals as performance additives, capable of optimizing plant performance [62]. An additional parameter used to measure fuel performance is the small specific fuel consumption [63]. It is the fuel consumed in the generation of one unit of power. The brake-specific fuel consumption (BSFC) usually increases with the fraction of biodiesel in the blend. This is due to biodiesel containing less energy than diesel [64]. On the other hand, BSFC was reduced with increasing engine load for various biodiesel contents. For example, jatropha biodiesel blends exhibited better performance and lower BSFC than vegetable oils because of their lower viscosity and improved flow [65]. Even the B20 and B30 blends appeared to show similar characteristics as diesel fuel, particularly at full load. However, it was known that 100% biodiesel (e.g., palm oil fatty acid methyl esters) increased BSFC by 4–5% and fell in brake power and torque slightly. This was due to the low heating value of the biodiesel [66]. This can lead to reduced combustion efficiency, high temperature, and high fuel consumption. Researchers show that BSFC increases with the addition of oxygen-rich additives such as triacetin, and there is a direct correlation between lower fuel heating value and higher BSFC [55].

Engine power output is another factor that may differ when using biodiesel blends. Most studies show a slight drop in power with increasing percentages of biodiesel in the fuel mix [67]. For example, a study utilizing 10%, 30%, and 50% biodiesel blends in a single-cylinder engine found that increasing the biodiesel content reduced power [68]. A study with fixed compression and equivalence ratios found that the power-driven increased slightly with a load but eventually decreased with higher loads and biodiesel percentages [69]. Some unique biodiesel feedstocks, such as hazelnut soap stock and waste sunflower oil, were said to keep or improve the power of multicylinder engines. This variation is often associated with the fuel's oxygen content, viscosity, and flow rate. These higher-viscosity fuels also tend to produce fewer internal fuel pump leaks, which can marginally improve the available power in certain situations. We often experience a slight but noticeable drop in the power output as biodiesel has a lower heating value than petroleum diesel [70].

In conclusion, a biodiesel–diesel mixture has several effects on the engine performance. The presence of high levels of biodiesel in the blend may, however, lead to conditions where slight reductions in brake thermal efficiency and power can be observed due to the nature of the fuel, i.e., higher viscosity and lower calorific value, but careful blending and additives may help balance this out. The oxygen content of the biodiesel can also improve combustion, offsetting some efficiency losses. Understanding the bioscience of these performance changes is essential for optimizing biodiesel utilization in diesel engines.

8.2.3 Performance of Biodiesel Blends in Emission Characteristics

Emission and combustion characteristics are substantially affected by the blends of biodiesel used in compression ignition (CI) engines. CI engines are responsible for the emissions of oxides of nitrogen (NO_x), hydrocarbons (HC), carbon monoxide

(CO), and particulate matter (PM) [71]. The emissions of these materials will vary depending on the engine type, fuel type, conditions under which the engine operates, and combustion parameters. Blends of biodiesel in diesel have been shown in several studies to lower the corresponding emissions of CO, HC, and PM relative to pure diesel [22, 72]. In some cases, NO_x emissions may show a slight increase. It is likely to decrease with the blending of biodiesel into diesel fuel. The high oxygen content in biodiesel contributes to more thorough combustion and is mainly responsible for this effect [73]. It was found that lower HC emissions were observed per unit of energy for fuel blends B10 and B20 compared to pure diesel. These result in better combustion and less unburnt hydrocarbons due to the shorter ignition delay and higher cetane number of biodiesel. However, with higher alcohols, such as butane, higher HC emissions could be possible because of longer ignition delays and partial combustion, particularly at higher engine loads [74].

CO emissions are also typically lower in biodiesel blends than in diesel fuel. Once again, this is attributed to the oxygen in biodiesel, which aids the more complete fuel combustion [75]. The studies showed that nonethanol blending, for example, B20 and B7E5, reduces CO emissions, especially under high load conditions. However, under some conditions, especially low load or poor air-fuel mixing, CO emissions may increase somewhat, more so with alcohols included in the blend [76]. The fuel-air mixture and the timing of combustion heavily influence the CO levels. In comparison, higher percentages of biodiesel tend to positively affect NO_x emissions [77]. The reason is that the biodiesel's higher combustion temperature and oxygen content encouraged the formation of nitrogen oxides. However, small quantities of alcohols like propanol or butanol can lower NO_x under certain conditions. It has been proven that the application of 4–8% alcohol blends results in decreasing NO_x emissions by as much as 20%. However, the increase in engine speed and temperature increases NO_x emissions due to more complete combustion and in-cylinder temperatures. Alcohol blends can have a cigarette lighter effect that causes delayed ignition and may also play a role in forming more NO_x [78].

Biodiesel replacement for diesel substantially reduces particulate matter (PM) emissions. The reason is that biodiesel has oxygen, lowering the soot formation. Biodiesel blends also affect the size, shape, and structure of the particles. Studies conducted with advanced equipment, including transmission electron microscopy (TEM), show that biodiesel leads to the formation of smaller and more reactive particles [79]. This, in turn, means that the emissions from biodiesel are less harmful than diesel emissions. The number and type of particles emitted under various engine conditions could also be affected by adding waste lubricating oil or alcohol to biodiesel blends [80].

Biodiesel blends also affect the combustion process in CI engines. The ignition delay, heat release rate, and cylinder pressure are important combustion characteristics. Biodiesel blends have a shorter ignition delay, the time between fuel injection and ignition, because the blends have a higher cetane number and improved oxygen content. This promotes quicker and more effective combustion [81]. Increasing the percentage of jatropha methyl ester biodiesel in the blend decreased the ignition delay under all load conditions, for instance [82]. Another vital combustion parameter is the heat release rate. At low engine loads, biodiesel blends have a higher heat release rate than diesel, and the opposite can be true at high engine loads. This shift is linked to the burning phase of the fuel and the availability of oxygen. Under full-load conditions, diesel has a more excellent peak heat release due to better fuel-air mixing and longer ignition delay, while biodiesel is more stable because of its oxygenated nature, resulting in smoother combustion [83].

To summarize, while biodiesel–diesel blends reduce harmful emissions, including CO, HC, and PM, they marginally increase NO_x emissions under some operating conditions. They also affect the combustion properties, such as the ignition delay time and soot emissions. This indicates the potential of biodiesel as a cleaner substitute for fossil diesel, specifically when engine conditions are optimized or in addition to appropriate additives.

8.3 Mechanisms of Nanoparticles in Combustion Improvement

From both economic and environmental perspectives, biofuels have emerged as an attractive alternative fuel with the potential to replace nonrenewable conventional fuels [84]. To be efficient, all alternative biofuels must be simple to rejuvenate, using higher speeds and be environmentally friendly [85]. One of the principal problems for the alternative fuels industry has been emissions; it is claimed that the most efficient solution to reduce emissions from vehicles is to develop automobiles powered by alternative fuels such as electricity, biofuels, and hydrogen [86]. Concerns over the sustainability of the use of these alternative fuels, risk, and what will happen to them when their resources are exhausted mean a detailed assessment of the factors influencing their potential automotive filtration is necessary [87]. Even if fuel cell and battery-powered vehicles are locally emission free, they are limited by resource scarcity (high availability), infrastructure constraints, and cost, resulting in low market penetration and customer acceptance [88]. Different types of nanoparticles (see Figure 8.2) have been extensively researched due to their ability to improve biodiesel–diesel fuel mix performance in internal combustion engines owing to their nanometric size and high surface area-to-volume ratio [89]. Due to these attributes, nanoparticles can function as combustion catalysts, including improving the igniting quality of fuels, accelerating fuel oxidation, and promoting overall combustion efficiency [90]. Metal nanoparticles like iron (Fe), copper (Cu), silver (Ag), and zinc

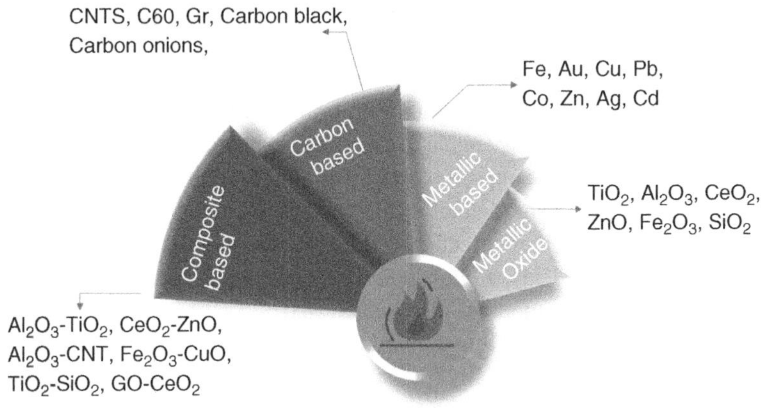

Figure 8.2 Use of different types of nanoparticles in the combustion improvement.

(Zn), synthesized using top-down or bottom-up methods, greatly encourage uniform and comprehensive combustion owing to their high reactivity and surface energy [91]. Metal oxide nanoparticles like TiO_2, Al_2O_3, and CeO_2 offer enhanced thermal stability and serve as oxygen donors throughout the combustion process, lowering the emissions of CO, unburned HC, and PM [20]. They are more catalytic; therefore, they can release heat quicker and enhance the brake thermal efficiency (BTE). Similarly, carbon-based nanoparticles, like carbon nanotube (CNT), graphene, and carbon black, have high thermal conductivities and structural strength, which can improve fuel atomization, heat transfer, and the combustion rate [92]. This leads to improved combustion chamber performance, resulting in low ignition delay and combustion duration, ultimately contributing to better engine smoothness and reduced brake-specific fuel consumption. These composite-type nanoparticles, derived from metals, oxides, and carbon nanostructures, present multifunctional benefits, such as enhanced fuel dispersibility, more stable combustion, and lower tendencies for knocking [93].

Biodiesel–diesel blends with nanoparticles showed notable enhancements in brake thermal efficiency, decreases in brake-specific fuel consumption, and stark reductions in tailpipe emissions, making them a good way for a cleaner and more efficient functioning engine.

8.4 Biodiesel–Diesel Blends Nanoparticle Method

In addition to the type of nanoparticle used, the method of synthesis and dispersion during synthesis has also been found to be effective in integrating nanoparticles into biodiesel–diesel blends (refer to Figure 8.3) [94]. The preparation and

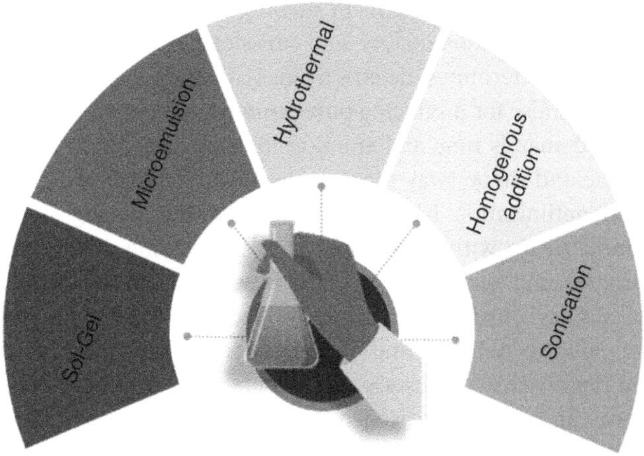

Figure 8.3 Preparation techniques of the biodiesel–diesel blends.

mixing technique of nanoparticles into fuel plays an important role in affecting their stability and surface activity and, finally, influencing engine performance and emission characteristics [95]. The sol–gel method is one of the most widely used techniques for preparing high-purity metal and metal oxide nanoparticles by converting a liquid sol into a solid gel, followed by separation steps using centrifugation or filtration [96]. No engine modification is required for its applications in engines, as the cerium oxide nanoparticles synthesized by this method and dispersed into the *Nerium oleander* biofuel emulsions have successfully reduced NO_x and smoke opacity emissions in compression ignition engines [97]. The microemulsion principle creates nanoparticles in nanoscale droplets formed by surfactant and cosurfactant molecules, thus having the ability to precise control over particle size and shape; hematite-phase Fe_2O_3 nanoparticles synthesized by microemulsion have shown promise for improving combustion [94]. High-liquid–solid reactivity associates the hydrothermal method with high-temperature and high-pressure conditions in sealed reactors in vacuum and nonvacuum environments, providing high-stability nanoparticles with uniform morphology and dopants [98]. When used on cerium oxide-based nanoadditives, this approach gave rise to enhanced combustion, improved brake thermal efficiency (BTE), and lower NO_x and hydrocarbon emissions. The homogeneous adhesion technique will result in the systematic dispersion of the nanoparticles within your fuel because it dissolves them in a solvent before mixing. Improved secondary atomization, complete combustion, and enhanced BTE at optimum injection timings were reported using nickel oxide nanoparticles synthesized by this method [99]. Finally, the sonication method, by using ultrasound, destroys agglomerates of the particles, allowing them to be homogeneously distributed in the fuel without surfactants, improving the fuel homogeneity and leading to more stable combustion [100].

Each fuel synthesis technique proffers merits in shaping nanoparticle behavior to impact engine efficiency, combustion quality, and emissions. How particles are prepared and mingled deeply determines the effectiveness of biodiesel–diesel use. The sol–gel pathway earns praise for producing pure, uniform nanoparticles with morphology mastery but demands time and emotion for moisture. The microemulsion path permits size and shape sway through steady surfactants, yet scalability stays limited and optimization is critical. Hydrothermal crafting yields highly crystallized nanoparticles with slim dispersions and doping ability, yet necessitates high pressures and temperatures using costly, specialized instruments. Homogeneous addition ensures even dispersion, sustaining strength and combustion, although solvent removal and dispersion durability risks curb practicality. Sonication rapidly cleaves clusters and scatters nanoparticles evenly; nevertheless, permanence and energy are needed when magnified and restricted. Grasping each technique's strengths and bounds is key to choosing the most suitable based on the desired particle traits, fuel compatibility, and engine application [101].

8.4.1 Limitations

While the benefits are great, there are also some limitations of biodiesel–diesel mixtures that should be considered. A significant concern is the rise in brake-specific fuel consumption because of the lower energy content of biodiesel compared to diesel. Therefore, additional fuel must be sent through the system to create equivalent power, particularly at higher blend ratios or when the engine is operated under full load. Biodiesel has a slightly higher viscosity and density than petroleum, which may impact fuel spray and atomization, particularly in cold climates where the flow properties deteriorate.

Despite the improved fuel properties and combustion efficiency after adding nanoparticles to biodiesel–diesel show, there seem to be some limitations. A big problem is the stability of the nanoparticles in the fuel mixture because agglomeration or sedimentation after a while can affect the uniform distribution and thus result in unstable engine operation. In addition, the ideal dose and type of nanoparticles are still under investigation, and high concentrations can lead to injector choking, wear due to abrasives, or unintentional effects on engine parts. The nanoparticle generation and dispersion methods must be inexpensive to facilitate large-scale commercialization. On the other hand, the emissions of nanoparticles during combustion can have unknown environmental and health effects, particularly the long-term inhalation of nano-sized particles. These hurdles must be dissolved via extra research before nanoparticle-enhanced biodiesel can be taken into consideration as an alternative gasoline that is effective and safe to use.

8.4.2 Future Work

Future studies must focus on customizing biofuel mixes and engine parameters to overcome the present limitations and progress engine productivity and emission oversight. Analyzing additives like alcohols (for example, ethanol, butanol) or nanoparticles could enhance ignition and diminish nitrogen oxide emanations. Engine configuration techniques, such as exhausted gas reusing (EGR), variable infusion timing, and post-treatment frameworks, should be investigated in biofuel consolidation. Long-term investigations should zero in on investigating the controlled utilization of nanoparticles to enhance the burning and outflow attributes of biodiesel–diesel blends. Nanoparticles, such as cerium oxide, aluminum oxide, zinc oxide, and titanium dioxide, have improved ignition quality, diminished particulate emanations, and expanded warm productivity because of their catalytic properties and high surface-to-volume ratio. Scientists ought to research ideal fixations and dispersion strategies to guarantee a stable suspension of nanoparticles inside the fuel, forestalling blocking, or deposit issues. Moreover, engine adjustment procedures, such as changing the infusion timing or weight, ought to be created explicitly for nanoparticle-blended fuels to ultimately utilize their advantages. Life cycle appraisals and toxicological considerations are fundamental to comprehend the natural and well-being ramifications of utilizing nanoparticles in burning frameworks. Finally, consolidating progressed fuel additives with keen engine control procedures may offer a guaranteed way to cleaner and more productive biodiesel-based transportation arrangements.

8.5 Conclusion

This study investigates the effects of biodiesel–diesel blends on internal combustion engine sustainability to enhance sustainability and reduce dependence on fossil fuels. At lower blend ratios (B10, B20, or B25), the experimental findings confirm that biodiesel blends may perform close to conventional diesel regarding brake thermal efficiency and power output. Although the brake-specific fuel consumption is somewhat high because of the biodiesel's lower calorific value, the performance is still satisfactory, and larger-scale engine modifications are not required. In terms of emissions, biodiesel blends have shown significant decreases in carbon monoxide (CO), hydrocarbons (HC), and particulate matter (PM), which can be primarily attributed to the more excellent oxygen content of biodiesel that encourages more complete combustion. However, nitrogen oxide (NO_x) emissions will likely increase slightly due to the increase in-cylinder temperatures and fuel-bound oxygen. From the perspective of combustion characteristics, biodiesel, compared to petroleum diesel, enhances the ignition behavior and decreases the ignition delay, which improves fuel burning and engine smoothness. In general, biodiesel blends can be used as alternative fuels, with the advantage of performing cleaner in CI engines.

Author Contributions

Mudasar Zafar, writing review paper, formal analysis, and literature review; Abida Hussain, writing review and editing and introduction; Tauseef Ahmed, graphics, validation, literature review; Ahamed Daabo, methodology; and Farman Ullah, results and discussion.

Statement of Interest

All authors contributed equally to the paper, and there are no conflicts of interest.

Acknowledgment

Authors would like to thank Asia Pacific University of Technology & Innovation (APU) for supporting this work to be published.

References

1 Gren, L., Malmborg, V.B., Falk, J. et al. (2021). Effects of renewable fuel and exhaust aftertreatment on primary and secondary emissions from a modern heavy-duty diesel engine. *Journal of Aerosol Science* 156: 105781.
2 Vargün, M., Özsezen, A.N., Botsalı, H., and Sayın, C. (2023). A study on the impact of fuel injection parameters and boost pressure on combustion characteristics in a diesel engine using alcohol/diesel blends. *Process Safety and Environmental Protection* 177: 29–41.
3 Jin, C. and Wei, J. (2023). The combined effect of water and nanoparticles on diesel engine powered by biodiesel and its blends with diesel: a review. *Fuel* 343: 127940.
4 Doppalapudi, A.T., Azad, A., and Khan, M. (2023). Advanced strategies to reduce harmful nitrogen-oxide emissions from biodiesel fueled engine. *Renewable and Sustainable Energy Reviews* 174: 113123.
5 Soudagar, M.E.M., Mujtaba, M.A., Safaei, M.R. et al. (2021). Effect of Sr@ ZnO nanoparticles and ricinus communis biodiesel-diesel fuel blends on modified CRDI diesel engine characteristics. *Energy* 215: 119094.

6 Hosseinzadeh-Bandbafha, H., Panahi, H.K., Dehhaghi, M. et al. (2023). Applications of nanotechnology in biodiesel combustion and post-combustion stages. *Renewable and Sustainable Energy Reviews* 182: 113414.

7 Manimaran, R., Mohanraj, T., and Ashwin, R. (2023). Green synthesized nano-additive dosed biodiesel-diesel-water emulsion blends for CI engine application: performance, combustion, emission, and exergy analysis. *Journal of Cleaner Production* 413: 137497.

8 Vellaiyan, S. (2024). Exploring composite particles in ammonia decomposition for algae biodiesel integrated fuel and assessing energy and ecology metrics. *Process Safety and Environmental Protection* 190: 714–728.

9 Gad, M., Ağbulut, Ü., Afzal, A. et al. (2023). A comprehensive review on the usage of the nano-sized particles along with diesel/biofuel blends and their impacts on engine behaviors. *Fuel* 339: 127364.

10 Ramalingam, K., Vellaiyan, S., Venkatesan, E.P. et al. (2023). Challenges and opportunities of low viscous biofuel—a prospective review. *ACS Omega* 8 (19): 16545–16560.

11 Magara-Gomez, K.T., Olson, M.R., Okuda, T. et al. (2012). Sensitivity of hazardous air pollutant emissions to the combustion of blends of petroleum diesel and biodiesel fuel. *Atmospheric Environment* 50: 307–313.

12 Ogunkunle, O. and Ahmed, N.A. (2021). Overview of biodiesel combustion in mitigating the adverse impacts of engine emissions on the sustainable human–environment scenario. *Sustainability* 13 (10): 5465.

13 Yang, H.-H., Chien, S.-M., Lo, M.-Y. et al. (2007). Effects of biodiesel on emissions of regulated air pollutants and polycyclic aromatic hydrocarbons under engine durability testing. *Atmospheric Environment* 41 (34): 7232–7240.

14 Aljaafari, A., Fattah, I.M., Jahirul, M.I. et al. (2022). Biodiesel emissions: a state-of-the-art review on health and environmental impacts. *Energies* 15 (18): 6854.

15 Kularathne, I., Gunathilake, C., Rathneweera, A. et al. (2019). The effect of use of biofuels on environmental pollution—a review. *International Journal of Renewable Energy Research* 9 (3): 1355–1367.

16 Smith, K.R. (2013). *Biofuels, Air Pollution, and Health: A Global Review.* Plenum Press.

17 Živković, S. and Veljković, M. (2018). Environmental impacts the of production and use of biodiesel. *Environmental Science and Pollution Research* 25: 191–199.

18 McCormick, R.L. (2007). The impact of biodiesel on pollutant emissions and public health. *Inhalation Toxicology* 19 (12): 1033–1039.

19 Modi, V., Rampure, P.B., Babbar, A. et al. (2024). Nanoparticle-enhanced biodiesel blends: a comprehensive review on improving engine performance and emissions. *Materials Science for Energy Technologies* 7: 257–273.

20 Sekoai, P.T., Ouma, C.N., Du Preez, S.P. et al. (2019). Application of nanoparticles in biofuels: an overview. *Fuel* 237: 380–397.

21 Elkelawy, M., Mohamad, H.A., Abo-Samra, S., and Elshennawy, I.A.-E. (2023). Nanoparticles additives for diesel/biodiesel fuel blends as a performance and emissions enhancer in the applications of direct injection diesel engines: a comparative review. *Journal of Engineering Research* 7 (1): 112–121.

22 Mahgoub, B.K. (2023). Effect of nano-biodiesel blends on CI engine performance, emissions and combustion characteristics–review. *Heliyon* 9 (11): e21367.

23 Kumar, S. and Goga, G. (2023). Review analysis on the performance & emission characteristics of a diesel engine fuelled with various gaseous & bio fuels. *Materials Today: Proceedings* 80: 285–292.

24 Riayatsyah, T., Sebayang, A.H., Silitonga, A.S. et al. (2022). Current progress of *jatropha curcas* commoditisation as biodiesel feedstock: a comprehensive review. *Frontiers in Energy Research* 9: 815416.

25 Singh, D., Sharma, D., Soni, S.L. et al. (2021). A comprehensive review of physicochemical properties, production process, performance and emissions characteristics of 2nd generation biodiesel feedstock: *Jatropha curcas*. *Fuel* 285: 119110.

26 Kumar, N. and Raheman, H. (2021). Combustion, performance, and emission characteristics of diesel engine with nanocera added water emulsified Mahua biodiesel blend. *Environmental Progress and Sustainable Energy* 40 (4): e13572.

27 A, Sholihah, S, Murti, A, Juwita, et al. (2021). Characterization of mixed biodiesel and petrodiesel as transportation fuel." *International Conference on Sustainable Biomass (ICSB 2019)*, 15–17 October 2019, Atlantis Press.

28 Parvez, M., Lal, S., Khan, O. et al. (2024). Sustainable regulatory framework for ethanol and biodiesel blending with petro-fuels in India to meet fuel demand: a review of biofuel policies. *Biofuels* 16: 1–14.

29 Animon, C.R., George, S.E., Alex, N.E. et al. (2023). Advancing the creation of biodiesel from Jatropha curcas L.: a promising source of renewable energy. *International Journal of Research and Analytical Reviews (IJRAR)* 10: 176–181.

30 Saxena, P., Jawale, S., and Joshipura, M.H. (2013). A review on prediction of properties of biodiesel and blends of biodiesel. *Procedia Engineering* 51: 395–402.

31 Wakil, M., Kalam, M., Masjuki, H. et al. (2015). Influence of biodiesel blending on physicochemical properties and importance of mathematical model for predicting the properties of biodiesel blend. *Energy Conversion and Management* 94: 51–67.

32 Hoekman, S.K., Broch, A., Robbins, C. et al. (2012). Review of biodiesel composition, properties, and specifications. *Renewable and Sustainable Energy Reviews* 16 (1): 143–169.

33 Yuan, W., Hansen, A., and Zhang, Q. (2003). Predicting the physical properties of biodiesel for combustion modeling. *Transactions of ASAE* 46 (6): 1487–1493.

34 Bhale, P.V., Deshpande, N.V., and Thombre, S.B. (2009). Improving the low temperature properties of biodiesel fuel. *Renewable Energy* 34 (3): 794–800.

35 Pawar, S., Hole, J., Bankar, M. et al. (2023). Use of fatty acid chemical composition for predicting higher calorific value of biodiesel. *Materials Today: Proceedings.*

36 Sivaramakrishnan, K. and Ravikumar, P. (2012). Determination of cetane number of biodiesel and its influence on physical properties. *ARPN Journal of engineering and Applied Sciences* 7 (2): 205–211.

37 Kaisan, M., Anafi, F., Nuszkowski, J. et al. (2020). Calorific value, flash point and cetane number of biodiesel from cotton, jatropha and neem binary and multi-blends with diesel. *Biofuels* 11: 321–327.

38 Amin, A., Gadallah, A., El Morsi, A. et al. (2016). Experimental and empirical study of diesel and castor biodiesel blending effect, on kinematic viscosity, density and calorific value. *Egyptian Journal of Petroleum* 25 (4): 509–514.

39 Pratas, M.J., Freitas, S.V., Oliveira, M.B. et al. (2011). Biodiesel density: experimental measurements and prediction models. *Energy and Fuels* 25 (5): 2333–2340.

40 Hoang, A.T. (2021). Prediction of the density and viscosity of biodiesel and the influence of biodiesel properties on a diesel engine fuel supply system. *Journal of Marine Engineering & Technology* 20 (5): 299–311.

41 Verduzco, L.F.R. (2013). Density and viscosity of biodiesel as a function of temperature: empirical models. *Renewable and Sustainable Energy Reviews* 19: 652–665.

42 Yoon, S.H., Park, S.H., and Lee, C.S. (2008). Experimental investigation on the fuel properties of biodiesel and its blends at various temperatures. *Energy and Fuels* 22 (1): 652–656.

43 Joshi, R.M. and Pegg, M.J. (2007). Flow properties of biodiesel fuel blends at low temperatures. *Fuel* 86 (1–2): 143–151.

44 Knothe, G. and Steidley, K.R. (2005). Kinematic viscosity of biodiesel fuel components and related compounds. Influence of compound structure and comparison to petrodiesel fuel components. *Fuel* 84 (9): 1059–1065.

45 Eremeeva, A., Khasanov, A., Oleynik, I. et al. (2025). Development of biofuel as marine low-viscosity fuels with environmentally friendly components. *International Journal of Engineering* 38: 273–279.

46 Mekonnen, K.D., Endris, Y.A., and Abdu, K.Y. (2024). Alternative methods for biodiesel cetane number valuation: a technical note. *ACS Omega* 9 (6): 6296–6304.

47 Suardi, S., Mahmuddin, F., Klara, S. et al. (2024). Performance and emission characteristics of diesel engines using biodiesel from waste cooking oil with cetane number improver. *International Journal of Marine Engineering Innovation and Research* 9 (3): 528.

48 Bukkarapu, K.R. and Krishnasamy, A. (2022). A critical review on available models to predict engine fuel properties of biodiesel. *Renewable and Sustainable Energy Reviews* 155: 111925.

49 Santos, S.M., Wolf-Maciel, M.R., and Fregolente, L.V. (2023). Cold flow properties: applying exploratory analyses and assessing predictive methods for biodiesel and diesel-biodiesel blends. *Sustainable Energy Technologies and Assessments* 57: 103220.

50 Gomez-Rodriguez, K.A., Chavarria-Hernandez, J.C., and Martinez-Tapia, G.E. (2021). Estimation of cold flow properties of biodiesel from fatty acid composition. *Energy and Fuels* 35 (2): 1442–1448.

51 Su, B., Wang, L., Xue, Y. et al. (2021). Effect of pour point depressants combined with dispersants on the cold flow properties of biodiesel-diesel blends. *Journal of the American Oil Chemists' Society* 98 (2): 163–172.

52 Pradana, Y.S., Makertihartha, I.G.B., Indarto, A. et al. (2024). A review of biodiesel cold flow properties and its improvement methods: towards sustainable biodiesel application. *Energies* 17 (18): 4543.

53 Umeuzuegbu, J.C. (2021). Empirical modeling of kinetics of transesterification and cold flow properties of used soya oil biodiesel. *International Journal of Innovative Research in Engineering and Management (IJIREM)* 8: 21–41.

54 Baste, S.V. and Umale, S.S. (2024). Effect of injection pressure on brake thermal efficiency and specific fuel consumption when fueled with methyl ester. *AIP Conference Proceedings* 2985 (1): 020011. AIP Publishing.

55 Mohite, S.2022).Importance of performance and emission characteristics in biodiesel. *Biennial International Conference on Future Learning Aspects of Mechanical Engineering,* 3–5 August, Springer.

56 Ogunsola, A.D., Durowoju, M.O., Ogunkunle, O. et al. (2023). Shea butter oil biodiesel synthesized using snail shell heterogeneous catalyst: performance and environmental impact analysis in diesel engine applications. *Sustainability* 15 (11): 8913.

57 Wahome, A., Ngunjiri, G., Shitanda, D., and Ogola, W. (2013). Performance characteristics of blended rice bran biodiesel in a diesel engine. *International Journal of Engineering Science Invention* 2: 35–41.

58 Hasan, M. and Rahman, M.M. (2017). Performance and emission characteristics of biodiesel–diesel blend and environmental and economic impacts of biodiesel production: a review. *Renewable and Sustainable Energy Reviews* 74: 938–948.

59 Rahman, M.M., Hassan, M.H., Kalam, M.A. et al. (2014). Performance and emission analysis of *Jatropha curcas* and *Moringa oleifera methyl* ester fuel blends in a multi-cylinder diesel engine. *Journal of Cleaner Production* 65: 304–310.

60 Ndayishimiye, P. and Tazerout, M. (2011). Use of palm oil-based biofuel in the internal combustion engines: performance and emissions characteristics. *Energy* 36 (3): 1790–1796.

61 Papong, S., Chom-In, T., Noksa-nga, S., and Malakul, P. (2010). Life cycle energy efficiency and potentials of biodiesel production from palm oil in Thailand. *Energy Policy* 38 (1): 226–233.

62 Khiraiya, K., Ramana, P., Panchal, H. et al. (2021). Diesel-fired boiler performance and emissions measurements using a combination of diesel and palm biodiesel. *Case Studies in Thermal Engineering* 27: 101324.

63 Kadam, A., Patil, A., Kulkarni, M. et al. (2022). Mathematical model for brake-specific fuel consumption for neem biodiesel on single cylinder diesel engine. *Industrial Engineering Journal* 15 (10): 905–918.

64 Gülüm, M. (2022). Effects of compression ratio, blending ratio and engine speed on fuel cost, performance and exhaust emissions of a diesel engine fueled with bio-derived alternative fuels. *Sustainable Energy Technologies and Assessments* 53: 102464.

65 Ansari, A.M., Memon, L.A., Ghannam, M.T., and Selim, M.Y. (2023). Impact of biodiesel blended fuel with nanoparticles on performance and noise emission in compression ignition engine. *International Journal of Thermofluids* 19: 100390.

66 Muhaisen, N.M. and Graisa, M.M. (2024). The effect of biodiesel on the diesel engine. *International Journal of Electrical Engineering and Sustainability* 36–45.

67 Palani, Y., Devarajan, C., Manickam, D., and Thanikodi, S. (2022). Performance and emission characteristics of biodiesel-blend in diesel engine: a review. *Environmental Engineering Research* 27 (1).

68 Selvan, B.K., Das, S., Chandrasekar, M. et al. (2022). Utilization of biodiesel blended fuel in a diesel engine–combustion engine performance and emission characteristics study. *Fuel* 311: 122621.

69 Chaudhary, A., Panchal, S.H., Surana, A. et al. (2022). Performance, emission and combustion characteristics of various biodiesel blends. *Journal of Thermal Analysis and Calorimetry* 1–25.

70 Emaish, H., Abualnaja, K.M., Kandil, E.E., and Abdelsalam, N.R. (2021). Evaluation of the performance and gas emissions of a tractor diesel engine using blended fuel diesel and biodiesel to determine the best loading stages. *Scientific Reports* 11 (1): 9811.

71 Nayak, S.K., Nižetić, S., Huang, Z. et al. (2022). Influence of injection timing on performance and combustion characteristics of compression ignition engine working on quaternary blends of diesel fuel, mixed biodiesel, and *t*-butyl peroxide. *Journal of Cleaner Production* 333: 130160.

72 Alenezi, R.A., Norkhizan, A., Mamat, R. et al. (2021). Investigating the contribution of carbon nanotubes and diesel-biodiesel blends to emission and combustion characteristics of diesel engine. *Fuel* 285: 119046.

73 Elkelawy, M., Mohamad, H.A.E.D., Elhamid, E.A., and El-Gamal, M.A. (2023). Experimental Investigation of the biodiesel direct injection and diesel fuel as premixed charge on CI-engine emissions, performance, and combustion characteristics. *Journal of Engineering Research* 7 (1): 177–187.

74 Kumar, P.V., Ashok, B., Vignesh, R. et al. (2022). Evaluation of performance, emissions and combustion attributes of CI engine using palmyra biodiesel blend with distinct compression ratios, EGR rates and nano-particles. *Fuel* 321: 124092.
75 Yang, X., Luo, Z., Liu, X. et al. (2021). Experimental and numerical investigation of the combustion characteristics and NO emission behaviour during the co-combustion of biomass and coal. *Fuel* 287: 119383.
76 Sonachalam, M., Manieniyan, V., Senthilkumar, R. et al. (2024). Experimental investigation of performance, emission, and combustion characteristics of a diesel engine using blends of waste cooking oil-ethanol biodiesel with MWCNT nanoparticles. *Case Studies in Thermal Engineering* 61: 105094.
77 Mofijur, M., Ahmed, S.F., Ahmed, B. et al. (2024). Impact of nanoparticle-based fuel additives on biodiesel combustion: an analysis of fuel properties, engine performance, emissions, and combustion characteristics. *Energy Conversion and Management: X* 21: 100515.
78 Wang, C., Chen, M., Zhao, P. et al. (2023). Investigation on Co-combustion characteristics and NO_x emissions of coal and municipal sludge in a tangentially fired boiler. *Fuel* 340: 127608.
79 Szyszlak-Bargłowicz, J., Wasilewski, J., Zając, G. et al. (2022). Evaluation of particulate matter (PM) emissions from combustion of selected types of rapeseed biofuels. *Energies* 16 (1): 239.
80 Millo, F., Vlachos, T., and Piano, A. (2021). Physicochemical and mutagenic analysis of particulate matter emissions from an automotive diesel engine fuelled with fossil and biofuel blends. *Fuel* 285: 119092.
81 Xia, C., Brindhadevi, K., Elfasakhany, A. et al. (2021). Performance, combustion and emission analysis of castor oil biodiesel blends enriched with nanoadditives and hydrogen fuel using CI engine. *Fuel* 306: 121541.
82 Arya, M., Rout, A.K., and Samanta, S. (2022). A review on the effect of engine performance and emission characteristics of CI Engine using diesel-biodiesel-additives fuel blend. *Materials Today: Proceedings* 51: 2224–2232.
83 Shojae, K., Mahdavian, M., Khoshandam, B., and Karimi-Maleh, H. (2021). Improving of CI engine performance using three different types of biodiesel. *Process Safety and Environmental Protection* 149: 977–993.
84 Bertrand, E., Pradel, M., and Dussap, C.-G. (2016). Economic and environmental aspects of biofuels. *Green Fuels Technology: Biofuels* 525–555.
85 Luque, R., Herrero-Davila, L., Campelo, J.M. et al. (2008). Biofuels: a technological perspective. *Energy and Environmental Science* 1 (5): 542–564.
86 Pimentel, D. (2012). *Global Economic and Environmental Aspects of Biofuels*. CRC press.
87 Rajagopal, D. and Zilberman, D. (2008). Environmental, economic and policy aspects of biofuels. *Foundations and Trends® in Microeconomics* 4 (5): 353–468.
88 Taheripour, F., Hertel, T.W., Tyner, W.E. et al. (2010). Biofuels and their by-products: global economic and environmental implications. *Biomass and Bioenergy* 34 (3): 278–289.

89 Chia, S.R., Chew, K.W., Show, P.L. et al. (2018). Analysis of economic and environmental aspects of microalgae biorefinery for biofuels production: a review. *Biotechnology Journal* 13 (6): 1700618.

90 Demirbas, A. (2008). Relationships derived from physical properties of vegetable oil and biodiesel fuels. *Fuel* 87 (8–9): 1743–1748.

91 Shahbaz, A., Hussain, N., Saleem, M.Z. et al. (2022). Nanoparticles as stimulants for efficient generation of biofuels and renewables. *Fuel* 319: 123724.

92 Sher, F., Ziani, I., Hameed, M. et al. (2024). Advanced nanomaterials design and synthesis for accelerating sustainable biofuels production–a Review. *Current Opinion in Green and Sustainable Chemistry* 100925.

93 Sakthivel, S., Muthusamy, K., Thangarajan, A.P. et al. (2024). Nano-based biofuel production from low-cost lignocellulose biomass: environmental sustainability and economic approach. *Bioprocess and Biosystems Engineering* 47 (7): 971–990.

94 Pandya, H.N., Parikh, S.P., and Shah, M. (2022). Comprehensive review on application of various nanoparticles for the production of biodiesel. *Energy Sources, Part A: Recovery, Utilization, and Environmental Effects* 44 (1): 1945–1958.

95 Dong, F., Ding, G., Zheng, H. et al. (2016). Highly dispersed Cu nanoparticles as an efficient catalyst for the synthesis of the biofuel 2-methylfuran. *Catalysis Science and Technology* 6 (3): 767–779.

96 Mahmoud, H.R., El-Molla, S.A., and Ibrahim, M.M. (2020). Biodiesel production via stearic acid esterification over mesoporous ZrO_2/SiO_2 catalysts synthesized by surfactant-assisted sol-gel auto-combustion route. *Renewable Energy* 160: 42–51.

97 Jayabal, R., Soundararajan, G., Kumar, R.A. et al. (2023). Study of the effects of bio-silica nanoparticle additives on the performance, combustion, and emission characteristics of biodiesel produced from waste fat. *Scientific Reports* 13 (1): 18907.

98 Janakiraman, S., Lakshmanan, T., Chandran, V., and Subramani, L. (2020). Comparative behavior of various nano additives in a DIESEL engine powered by novel Garcinia gummi-gutta biodiesel. *Journal of Cleaner Production* 245: 118940.

99 Ettefaghi, E., Ghobadian, B., Rashidi, A. et al. (2018). A novel bio-nano emulsion fuel based on biodegradable nanoparticles to improve diesel engines performance and reduce exhaust emissions. *Renewable Energy* 125: 64–72.

100 Jin, C., Wei, J., Chen, B. et al. (2023). Effect of nanoparticles on diesel engines driven by biodiesel and its blends: a review of 10 years of research. *Energy Conversion and Management* 291: 117276.

101 Qamar, O.A., Jamil, F., Hussain, M. et al. (2023). Advances in synthesis of TiO_2 nanoparticles and their application to biodiesel production: a review. *Chemical Engineering Journal* 460: 141734.

9

Hybrid Nanoparticles to Improve Solar-Based Energy Storage

Pethurajan Vigneshwaran, Abin Roy, B.S. Bibin, and Saboor Shaik

Department of Thermal and Energy Engineering, School of Mechanical Engineering, Vellore Institute of Technology (VIT), Vellore, India

9.1 Introduction

Solar energy is a pristine, abundant, and limitless resource. Solar thermal energy (STE) distinguishes itself by its capacity to produce substantial power through the incorporation of thermal energy storage. This attribute has rendered STE a compelling choice when compared to alternative renewable energy systems. Harnessing solar energy efficiently relies significantly on photovoltaic and solar thermal conversion technologies. Photovoltaics (PV) may convert solar energy directly into electricity, while various solar thermal collectors (STCs) and concentrators can convert solar energy into thermal energy. PV has the ability to generate power directly, especially given its much lower cost; however, storing the generated energy presents a substantial difficulty [1, 2]. Enhancing energy management and dispatchability of renewable energy sources relies crucially on the progression of cost-effective energy storage technologies. The primary objective outlined in the European Roadmap to 2050 is to elevate the proportion of renewable energy sources, aiming to attain a minimum of 55% of gross final energy demand by 2050 [3]. TES combined with solar thermal collectors efficiently converts solar energy into heat, and this approach addresses challenges posed by intermittent solar radiation, such as at night or during adverse weather. TES is particularly vital for storing electricity in conjunction with concentrating solar power (CSP) plants, ensuring a reliable energy supply by storing and deploying captured thermal heat when solar radiation is limited [4].

Machine Learning for Sustainable Energy Solutions, First Edition. Edited by Zafar Said and Prabhakar Sharma.

In recent years, studies on conversion and storage of solar energy have prominently featured nanomaterials, nanofluids, and nanocomposites. Phase change materials (PCMs) have emerged as a pivotal method for both converting and storing solar energy. PCM, referred to as latent heat storage materials, play a crucial role due to their capability to store and release substantial amounts of energy in latent heat storage systems, all within a compact storage volume. In 1995, the concept of incorporating small solid particles was pioneered by Choi, who introduced the idea of improving the thermal conductivity of heat transfer fluids through the introduction of nanoparticles (NP) into the base fluid [5]. Conversely, when nanoparticles are blended with PCM, the resulting composite is referred to as nanocomposites [6]. PCM exhibits the capability to store and release thermal energy at isothermal condition, boasting high energy density and accessibility across a broad range of temperatures. The integration of PCM into nanoparticles within solar energy systems has led to higher charging and discharging rates for energy storage and conversion systems. Nanomaterials effectively include PCM, guaranteeing thermal stability, outstanding thermophysical characteristics, efficient heat transfer rates, and quick charge/discharge capabilities [7]. Utilizing advanced energy storage materials, such as those involving nanotechnology and PCM, has demonstrated a notable enhancement in freshwater output of solar desalination systems.

Specifically, the efficiency of solar desalination systems can be greatly increased by incorporating nanoparticles into PCM to create nano-enhanced PCM, mixing nanoparticles with black dye, applying a nanocoating to the condensation area, and dispersing nanoparticles in fluid to create a nanofluid. Thermal conductivity is increased, and heat transfer rates are shortened when nanopowders are added to PCM. In this process, solar radiation's thermal energy is absorbed by PCM, which then releases it when the solar desalination system needs it [8]. The attributes underscore the appeal of PCM for a variety of applications in temperature regulation [9].

9.2 Thermal Energy Storage Systems

TES, or thermal energy storage, is a method of storing energy that consists of three main steps: charging, storing, and discharging. This method stands out as a prominent technique for enhancing efficiency of thermal systems and addressing disparity between energy demand and supply. TES is characterized into three main groups: sensible heat storage (SHS), thermochemical heat storage, and latent heat storage. Figure 9.1 lists the types of thermal energy storage as listed by various researchers.

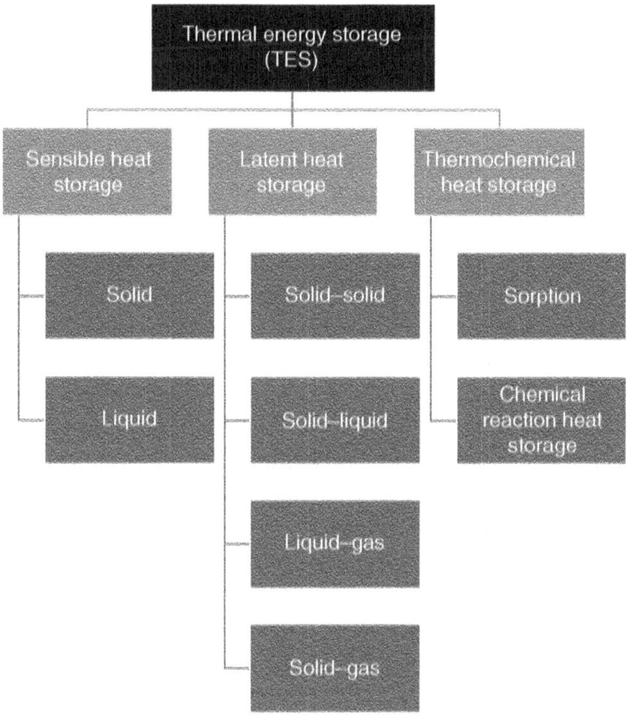

Figure 9.1 Different kinds of thermal energy storage systems.

9.2.1 Sensible Heat Storage

Sensible heat storage is the most cost-effective technique for storing thermal energy. This approach involves storing heat energy within a material's sensible heat capacity, primarily through temperature changes. As material experiences variations in temperature, it absorbs or releases energy in a straightforward and easily controllable manner. SHS is valued for its simplicity and economic efficiency, making it a practical choice for different applications requiring thermal energy storage.

It is feasible to store sensible heat effectively in both liquid and solid phases. Researchers primarily concentrate on high-temperature SHS materials, emphasizing their significance in intermittent solar energy storage applications. SHS systems are often used in solar thermal energy systems, where heat collected from sunlight is stored for later use, such as heating buildings or generating electricity. One notable constraint of SHS is its reliance on temperature increase as the determining factor for the capacity of heat storage. The necessity for substantial temperature increases to store a significant amount of energy renders SHS less efficient for various thermal applications [10, 11].

9.2.2 Latent Heat Storage (LHS)

LHS is a method of storing thermal energy by utilizing latent heat associated with change in phase. A significant quantity of heat is absorbed or released by a substance during a phase transition (such as from solid to liquid or liquid to gas) without affecting its temperature. The concept underlying latent heat storage (LHS) is the storage of both sensible heats, associated with temperature change, and latent heat, linked to phase transformation, within the storage material. In the context of energy storage, the thermal energy is stored by PCM. During the charging process, PCM absorbs heat from its surroundings, and it undergoes change phase (e.g., from solid to liquid). This absorbed heat is stored in the material as latent heat. During the discharging process, PCM releases stored heat as it changes phase back to its original state. LHS systems offer several characteristics (Figure 9.2) that render them suitable for a variety of applications.

LHS systems are extensively utilized in diverse applications like building thermal management, solar heating and cooling, and refrigeration. However, their design must be customized for specific operational conditions in each application. For instance, an LHS system tailored for air conditioning wouldn't suit water heating due to varying heat loading demands. Important factors influencing the design of the LHS include PCM, the layout of the storage tank, and the arrangements for heat transmission. The physical dimensions and orientation of the energy storage

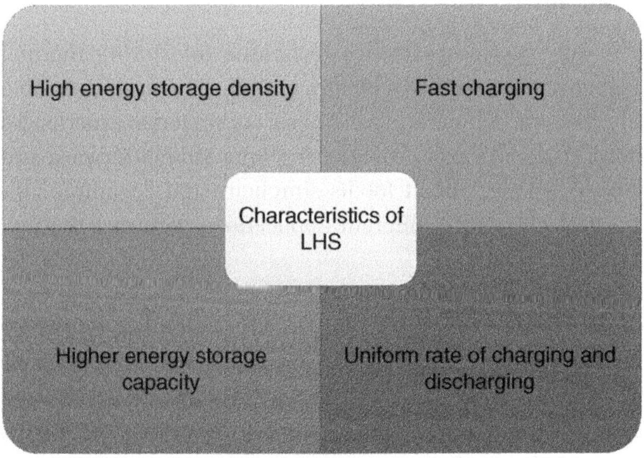

Figure 9.2 Characteristics of LHS.

system are controlled by its layout, while the choice of PCM determines the temperature at which energy storage occurs. The configuration of heat transfer arrangements determines the efficiency of the system's heat transmission [12].

9.2.2.1 Phase Change Material (PCM)

The energy storage component of LHS systems primarily relies on PCM, which retains heat in both latent and sensible forms. The measurement of SHS involves tracking temperature changes, while latent heat storage is determined by observing phase changes, which are quantified by the fusion enthalpy. As the latent heat of fusion increases, the necessary quantity of PCM lowers, resulting in a storage system that is more condensed. Among different phase changes, solid–liquid transition usually shows the highest phase change enthalpy. During this transition, volume changes are minimal, and there's an incremental increase in the latent heat of fusion. Phase changes occur while maintaining constant temperature and volume. It's crucial that heat source temperature exceeds phase transition temperature for effective operation. PCM must retain its ability to undergo phase change reliably even after numerous thermal cycles to be suitable for energy storage applications [13].

Razan El Kassar et al.'s study of the performance of solar panels coupled with PCM [14] included a thorough analysis of recent experiments conducted over the previous three years, highlighting the potential of PCM-based cooling solutions in advancing renewable energy technology. The focus was on the utilization of PCM as cooling systems for solar cells, exploring various approaches categorized into four main groups: pure PCM systems, finned PCM systems, hybrid cooling systems, and composite PCM systems. Another category of PCM, significantly known as nano-enhanced PCM (NEPCM), was reported to have enhanced TES properties and potentially expand renewable energy applications, such as solar energy harvesting. Weng Pin Wong et al. [15] focused on NEPCM for thermal energy storage, comprising recent advancements and future challenges and discussed latest progressions in NEPCM since 2020, converging on thermal characteristics of base PCM. PCM is employed in desalination systems to have an enhanced water production when the solar energy is available. The integrated solar desalination system, powered by solar energy and employing PCM, operates by generating fresh water while simultaneously storing excess heat energy within PCM. The use of PCM in desalination systems, as well as the effects of wind, water depth, and PCM thickness, was reviewed by Reji Kumar R et al. [16]. The review aimed to fill in any gaps in the literature regarding distillate production by emphasizing enhancement techniques for both passive and active solar stills, with and without TES.

9.2.3 Thermochemical Energy Storage

Thermochemical energy storage (TCES) functions based on reversible chemical reaction [17]. Energy stored within TCES systems primarily manifests as potential chemical energy, stemming from the formation and disruption of molecular bonds. This energy remains preserved within the system without dissipation into the environment or transformation into alternate energy forms. As a result, heat losses can be minimized during the storage phase, allowing for substantial energy storage and transfer over extended periods. With no moving parts, these systems produce minimal noise, mechanical vibration, or electromagnetic interference, making them suitable for a wide range of applications [18].

9.3 Solar Energy Storage Systems

Solar thermal storage (STS) is crucial for storing harvested solar energy from diverse solar collectors for future use, particularly during peak demand periods. A variety of physical, environmental, and economical aspects must be considered in order to select the best STS system. The energy density of the storage medium, heat transfer efficiency, heat loss rates, mechanical and chemical qualities, environmental effects, and cost concerns are important factors to consider [19]. Technologies such as photovoltaic and solar thermal conversion are essential for effectively harvesting solar energy. Solar energy is immediately converted by photovoltaic systems into electrical energy that may be stored for later use. On the other hand, solar energy is stored as thermal energy using solar thermal conversion technology. Because solar thermal energy conversion has a high energy efficiency and can continuously and sustainably harvest solar power, it is essential to employ.

9.3.1 TES for Solar Energy Storage Systems

Concentrated solar power (CSP) systems employ mirrors or other optical devices to focus sunlight onto a receiver, facilitating the generation of thermal energy. Heat engines, which are often steam turbines, and chemical processes have the capacity to transform concentrated light into heat, which has the potential to produce electricity from thermal energy. Various factors must be considered when choosing a TES system tailored to specific needs, including duration over which energy will be stored, associated costs, temperature requirements for both collection and utilization, as well as total amount of energy to be stored. Concentrated solar power (CSP) harnesses the sun as a readily available, renewable energy resource. When comparing CSP with TES to solar photovoltaics, CSP combined with TES offers greater flexibility and prolonged operational periods. Additionally, CSP offers complementary services and benefits that contribute to local economic growth and social advancement. Utilizing TES devices for heat generation during the warm-up process can significantly reduce fuel usage and emissions. This eliminates the need to wait for the engine to warm up and renders preheating obsolete for many. In cold weather conditions, employing a TES device can enhance battery performance by rapidly heating the batteries. Furthermore, TES devices can assist in defrosting windshields and enhancing passenger comfort. Extensive research has been conducted on TES devices due to their potential to address environmental challenges and improve energy efficiency. Acting as a bridge between power supply and demand, TES devices store thermal energy in hot or cold materials for later use. This capability is particularly beneficial for renewable energy systems, enabling energy conservation. Given the numerous applications of TES devices in the automotive industry, their adoption is increasingly widespread [20]. The intermittency of solar energy presents a significant challenge. Typically, there are two approaches to address this issue (Figure 9.3).

Aside from choosing a suitable TES system for CSP applications, ensuring seamless integration between the TES system and the power generation cycle, such as the Rankine cycle, is essential for effective enactment. Xinwei Wang et al. [21] worked on a simplified model simulating heat storage in different PCMs under seasonal solar fluctuations, and an optimized dual-tank recirculation model showed up to 17.2% thermal efficiency improvement, and the results indicated a significant difference in heat storage between summer and winter, with up to 45.38% disparity. Hamid Moradi et al. [22] assessed the performance of a solar air heating system integrated with PCM using computational fluid dynamics (CFD) model and demonstrated that through precise parameter selection and appropriate adjustment, the system has the capability to generate consistently hot air with stable temperature throughout extended periods of operation. Shunpan Yuan et al. [23] developed a novel double-layered organic PCM capsule of n-eicosane as core and polydopamine/poly(melamine-formaldehyde) resin (PDA@MF) as shell and found that the presence of PDA efficiently upgraded solar-thermal conversion efficiency. According to Oguzhan Kazaz et al. [24], heat transfer rates for water-based nanofluids containing Cu, Al, Ag, and Au

Energy hybridization

- A backup source such as fossil fuels or biomass is utilized

Thermal energy storage (TES)

- To store solar heat during periods of abundant sunlight and utilize it during periods of limited or no sunlight.

Figure 9.3 Approaches to avoid the intermittency of solar energy.

nanoparticles were measured at 7.05, 6.99, 6.89, and 5.86 W, respectively. By incorporating paraffin with Cu, Al, Ag, and Au nanocapsules into pure water, heat transfer enhancements reached 10.8%, 11.33%, 6.18%, and 13.38%, respectively. The study examined the photothermal conversion performance of a direct absorption solar energy storage system using nano-encapsulated PCM slurry.

According to studies, the majority of PCM show high levels of supercooling, low thermal conductivity, leakage during phase transitions, and poor solar-to-thermal conversion efficiency. To address these limitations and enhance PCM's thermophysical performance for enhanced TES, numerous strategies have been pursued. One major challenge is mitigating leakage during phase transitions, which results in PCM loss from the TES system. Micro encapsulation and macro encapsulation techniques are commonly employed to enhance stability and minimize leakage. Encapsulating PCM has proven effective in reducing leakage and enhancing heat transfer performance due to amplified surface area. Additionally, porous materials have been utilized to stabilize the PCM shape during phase transitions. Composite PCMs loaded with metal nanoparticles, metal oxide nanoparticles, and carbon-based nanoparticles have significantly enhanced thermal conductivity. Furthermore, incorporating photothermal fillers can improve PCM's solar-to-thermal performance, as these materials boast high solar-to-thermal conversion efficiency. Composite PCMs formed by loading photothermal fillers demonstrate superior solar-to-thermal performance. However, it's important to note that composite PCMs prepared in this manner typically exhibit lower TES capacity compared to bare PCMs. Balancing these trade-offs is crucial for optimizing the performance of TES systems utilizing composite PCM. Given the intricate nature of CSP plants integrated with energy storage, adopting a comprehensive approach to their design is imperative. By considering the power plant's global behavior as interconnected subsystems, each with its specific needs, it becomes evident that the efficacy of PCM utilization is suggestively prejudiced by both upstream and downstream components of storage tank. For instance, optimizing a storage system should be closely linked to optimizing the solar receiver. Moreover, taking a pragmatic approach to design is crucial, as real-world performance of storage systems often falls short of optimal levels due to weather fluctuations.

9.3.2 Latent Heat TES in Solar Energy Storage Systems

LHTES is a TES method that exhibits numerous advantages in comparison to the prevailing SHTES utilized in CSP facilities. LHTES provides an energy density that is 5–14 times that of SHTES, enabling implementation of more compressed storage and reduced inventory needs. Thermostable, sandwich architecture with fins, shell-and-tube configuration, and cascade configuration are all LHTES concepts. In terms of geometric characteristics, LHTES may also be categorized as frustum-shaped, elliptical, spherical, cylindrical, rectangular, or square. Using PCM as a working medium in LHTES greatly improves heat storage capability. A significant issue related to LHTES is the limited heat conductivity of PCM, resulting in reduced rates of both charging and discharging [25].

9.4 Role of Nanotechnology in Solar Energy Storage

9.4.1 Types of Nanoparticles

Nanoparticles (NP) are a popular substance that is gaining more attention because of their distinctive intrinsic properties. Incorporation of NP into various materials to create nanocomposites is a very promising aspect of nanotechnology. This innovation has enabled the development of novel combinations of attributes [26]. NPs are categorized based on dimensions within this nano range, as seen in Figure 9.4 [27].

Thermal storage capabilities of energy storage materials are significantly enhanced by incorporating them with nanomaterials, creating either nanofluids or nanocomposite matrices [7, 28]. Classification of nanomaterials is shown in Figure 9.5. Traditional storage materials often have small thermal conductivity, which confines energy transfer. To address this limitation, matrix composites and nanostructured additives are incorporated into base carbon materials. The inclusion of nanoparticles in energy storage materials enhances both thermal conductivity and diffusivity without significantly altering heat of fusion and melting temperature [29]. Thermal conductivities of nanoparticles are shown in Table 9.1.

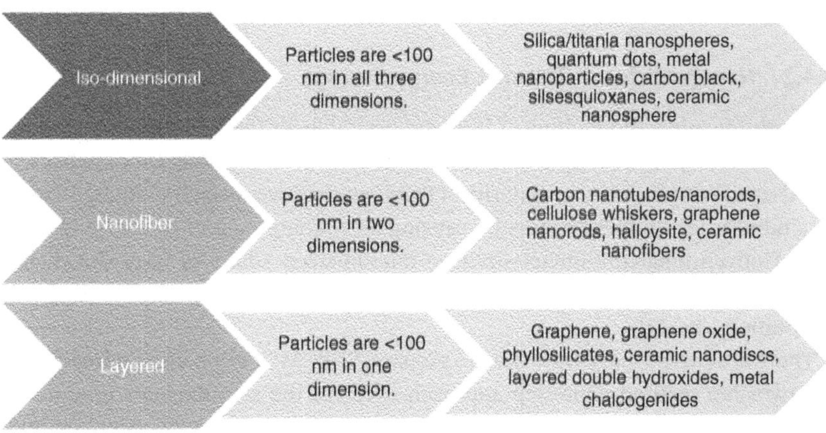

Figure 9.4 Dimensional classification of nanoparticles.

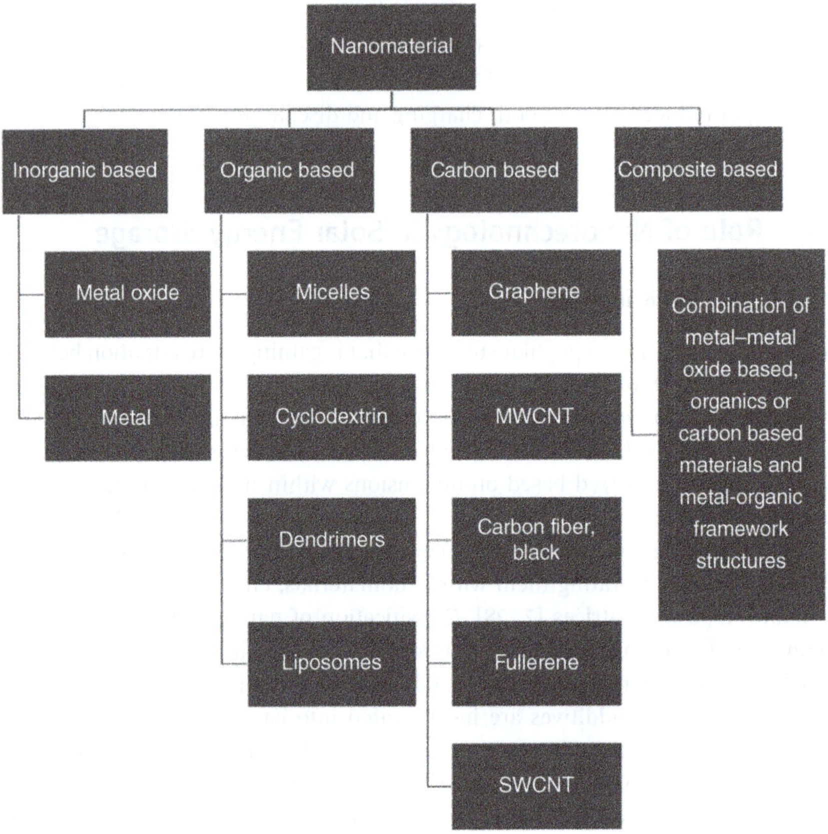

Figure 9.5 Classification of nanomaterials.

9.4.2 Nanoparticles in Thermal Energy Storage

Fast solidification and melting of PCM are desirable for the TES system to operate properly in response to rapid load variations. However, the inherent drawback of low thermal conductivity in majority of PCM restricts overall energy storage capacity and efficacy of TES systems by causing slow charging and discharging rates [35]. Conventional millimeter-sized or micron-sized additives have drawbacks such as low surface energy and high density, which increase the likelihood of sedimentation [36, 37].

There are two primary ways used to prepare NEPCM: the one-step approach and the two-step method. One-step approach involves the simultaneous production and dispersion of NP into base PCM in a single process. In the two-step method, NPs are first synthesized and prepared before being distributed into the base PCM.

Table 9.1 Nanoparticles' thermal conductivities.

Material	Thermal conductivity (W/mK)	Author	Material	Thermal conductivity (W/mK)	Author
Fe	80.2	Bobkov et al. [30]	SiO_2	1.38	Hwang et al. [31]
Al	237	Bobkov et al. [30]	TiO_2	8.9	Permanasari et al. [32]
Au	310	Bobkov et al. [30]	ZnO	29	Permanasari et al. [32]
Cu	401	Bobkov et al. [30]	Al_2O_3	30–40	Auerkari [33]
Ag	429	Bobkov et al. [30]	MgO	54.9	Hofmeister [34]
Graphite	100	Bobkov et al. [30]	CuO	76.5	Hwang et al. [31]
MWCNT	~3000	Hwang et al. [31]			

The majority of researches used a two-step procedure to produce NEPCM [38]. One-step process eliminates the requirement for drying, storing, shipping, and subsequently mixing of NP in pure PCM. This results in enhanced dispersion of NPs in pure PCM and prevents NPs from clumping together. The primary advantage of the two-step technique is its ability to produce NEPCM nanocomposites on a large scale. This process involves the agglomeration of tiny particles, which facilitates the use of the one-step preparation method [39].

The stability of nanoparticles distributed inside PCM is a crucial aspect that significantly affects their behavior. Therefore, the addition of a suitable surfactant is necessary to counteract agglomeration of nanoparticles. To prevent clumping of nanoparticles, it is essential to achieve a consistent dispersion of NPs in NEPCM samples by techniques such as ultrasonication, mechanical stirring, or a two-step procedure. In addition, it is necessary to maintain temperature of the NEPCM sample above its melting point while subjecting it to sonication with intermittent pulses in order to prevent overheating of the sample. In order to verify the precise dispersion of NP inside PCM over various heating and cooling cycles, it is recommended to use scanning electron microscopy (SEM) to assess any agglomeration of NP. This will provide information on the size, shape, and distribution of NP within the nano-enhanced NEPCM. Furthermore, it is necessary to analyze microstructures of NEPCM samples that have been created in order to infer elemental and structural information [40].

9.4.2.1 Inorganic-Based Nanomaterials

Enhancing energy storage capability and thermal conductivity of PCM, researchers have blended nanoparticles into the energy storage materials [41]. Sami et al. [42] added TiO_2 nanoparticles to paraffin and investigated the stability and thermal properties of NEPCM with and without surfactant and found that stability is reduced (1:4 mass ratio) and enhanced thermal properties of altered PCM. Similarly, Harikrishnan et al. [43] improved the efficiency of solar energy storage by incorporating TiO_2, CuO, and TiO_2/CuO nanoparticles into paraffin wax containing sodium dodecylbenzene sulfonate as the surfactant. This was accomplished by leveraging the nano-enhanced PCM's high heat transfer ability, stability, and thermal conductivity. Likewise, the rate of energy storage improved when CuO nanoparticles were integrated into a metal foam enclosure [44]. Wang et al. [45] prepared paraffin wax-based composite PCM with the addition of Al_2O_3 nanoparticles; latent heat and thermal conductivity were improved; however, the melting point of PCM was reduced with an increase in nanoparticle concentration in PCM. Cu metal nanoparticle is used as a nano-additive in paraffin wax, and the efficiency of the solar collector improved by 32% [46].

9.4.2.2 Carbon-Based Nanomaterials

Aqib et al. [47] improved the thermal conductivity of paraffin wax PCM with the accumulation of multiwalled carbon nanotubes (MWCNT). Similarly, Alotaibi et al. [48] blended paraffin wax with MWCNT containing various concentrations of TiO_2 and found that nano-enhanced PCM exhibits excellent thermal energy storage. Graphene is one of the carbon-based nanomaterials. PCM composite prepared by Qi et al. [49] with 3D hierarchical graphene foam and graphene foam in paraffin wax and nano-enhanced PCM is used for light-operated thermal energy storage with high thermal conductivity. Likewise, Sun et al. [50] created nano-enhanced phase change film with the usage of flexible graphene aerogel for highly efficient storage and conversion of thermal energy storage. The thermal conductivity of graphene nanoplatelet added with eutectic PCM (palmitic acid-stearic acid) is enhanced by 2.7 times compared to pure PCM. Meanwhile, eutectic PCM mixed with expanded graphite gives 15.8 times augmentation in thermal conductivity. Also, the density of eutectic PCM improved with the addition of GNP, whereas it reduced with EG. It is evident that EG has greater potential in improving the performance of thermal energy storage [51].

9.4.2.3 Hybrid Nanomaterials

Composite nanomaterials are amalgamation of two or more nanoparticles. For example, Zhang et al. [52] used Ag-graphene nanoparticles embedded in polyethylene glycol (PEG) to achieve better solar thermal conversion efficiency, and they obtained high-energy storage density and enhanced thermal conductive properties. Several types of experimental investigations were conducted to explore the potential of PCM-based hybrid nanomaterials as composite phase changes in solar heating systems. Composite PCMs (CPCM) were prepared by incorporating hybrid nanomaterials with varying mass fractions into a PCM base material. Sodium dodecylbenzene sulfonate, commonly used as a surfactant, was employed to effectively stabilize the dispersion of nanoparticles in pure material. The latent heats and phase change temperatures of the composite phase change material (CPCM) after hybrid nanomaterials were incorporated and subjected to different thermal cycles were found to be relatively similar to those of the original material, according to the results of differential scanning calorimetry (DSC) tests. Thermal conductivity augmentation levels for CPCM incorporating hybrid nanomaterials fall within the intermediate range observed for composite PCM utilizing nanoparticles individually. As a result of their enhanced thermal conductivity, low-temperature applications for heating, increased thermal stability and dependability, and improved heat transfer properties, PCM-based hybrid nanomaterials may be strongly suggested as the most effective composite PCM for harnessing solar thermal energy [43].

9.5 Applications of Nanoparticles in Solar Energy Storage

Current research trends focus on the deployment of PCM-based power plants in high-temperature environments. In solar thermal systems, the use of PCM has been intensively researched and established. In addition to environmental advantages, such as reduction in CO_2 emissions, resulting from the use of solar thermal systems in daily life, there has also been an observation of economic analysis of solar thermal systems based on PCM, specifically in terms of cost and payback time. Photovoltaic (PV) technology empowers direct translation of solar energy into electricity, while solar thermal collectors and concentrators (STC) can convert it into thermal energy. Solar thermal energy (STE) possesses the unique capability to integrate with thermal energy storage, facilitating bulk electricity provision. Utilization of nanoparticles is influenced by the choice of base material or fluid used to synthesize nanocomposites or nanofluids for storage and heat transfer, respectively.

9.5.1 Solar Collectors

Solar thermal collectors convert solar radiation into thermal energy, and the output temperature of these collectors varies depending on the application and adopted solar collector technology. A crucial component in this process is heat

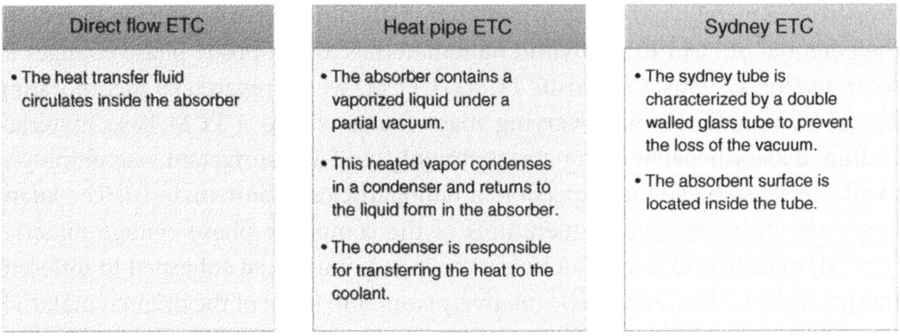

Figure 9.6 Types of evacuated tube collectors.

transfer fluid (HTF), which is essential for converting collected solar energy into thermal energy for various applications such as power generation or heating [29]. The two most often utilized collectors are evacuated tube collectors (ETCs) and flat plate collectors (FPCs). There are mainly three types of ETCs (Figure 9.6).

The efficiency of the system is enhanced in the range of 36.8–50% with the use of CuO-added paraffin wax PCM in the evacuated tube collector [53]. D'Ans et al. [54] observed that the water sorption characteristics of their material were considerably improved in comparison to $SrBr_2$ or pure MOF (metal–organic framework) when applied to solar collectors operating at 800 °C. The authors also calculated a periodic loading lift of 0.303 g/g and obtained a heat storage density of 233 kWh/m^3. This rise in heat storage density can be attributed to the pronounced physisorption by salt as opposed to chemisorption. As a result, the combination of the performance of MOF and salt with paraffin encapsulation using SiO_2/Ti_4O_7 produced a functional PCM that is well suited for applications involving thermal storage.

9.5.2 Solar Thermal Energy Conversion

Thermal energy is generated using CSP systems, where mirrors or optics are utilized to concentrate a large area of sunlight onto the receiver. Both heat engines, commonly in the form of steam turbines, and chemical processes can convert concentrated light into heat, enabling the production of electricity. Both methods have the capability to transform thermal energy into electrical energy. When choosing a TES system for a particular application, factors such as duration of energy storage, associated costs, required temperatures, and total amount of energy to be stored must all be considered. Advancement of wind and solar photovoltaic (PV) technologies underscores the importance of renewable resources in enhancing the efficiency and reliability of power networks. Due to the variability of wind and solar energy production, their output is contingent on sunlight and wind availability. CSP power production, for instance, peaks at noon during the sun's strongest period and gradually diminishes throughout the day, reaching near-zero levels at night. As the proportion of renewable energy sources in energy mix rises, there is a growing need to manage hourly fluctuations in production. Notably, when solar energy diminishes in late afternoon, CSP and TES play a crucial role in meeting demand. CSP can store heat generated by solar panels even at night, making it available for use at any time of day or night. This curbing minimizes the variability of renewable energy sources and enables the grid to integrate more renewables efficiently [55].

9.5.3 Solar Photovoltaic System

Solar photovoltaic (PV) cells are used to generate electric energy from solar radiation. Nowadays, the cost of PV systems has reduced; due to this, the energy supply by PV systems is more competitive than in previous years. The efficacy of photovoltaic systems can be significantly improved by lowering their temperature, which has led many researchers to investigate various methods for cooling PV systems. Many engineering applications use nanoparticles with PCM for temperature control and management. In the design, construction, and evaluation of a photovoltaic/thermal (PV/T) system, Al-Waeli et al. [56] suggested three distinct cooling techniques. The outcomes showed that the use of nano-PCM and nanofluid increased the electrical current from 3.69 to 4.04 A and improved electrical efficiency from 8.07% to 13.32% when compared to a standard PV system. A semiconductor material in a photovoltaic system generates an electric current by absorbing a portion of the incident solar energy. Other portions of incident radiation are transmitted, absorbed/stored as heat by PV mass, and reflected from the cell surface. PV thermal efficiency is defined as the electric power generated divided by the amount of solar energy received. Operating temperature determines PV efficiency, which drops as temperature rises. The efficiency of solar power cells reduces by 0.4–0.65% for each degree of temperature increase [57]. Huang et al. [58] looked into the configuration to shield the PV modules from damage and overheating. Research has been done on how adding nanoparticles to PCM can improve its performance in energy storage and electronic equipment temperature control [59, 60]. Studies of Nada et al. [61] showed that the use of paraffin wax PCM and Al_2O_3 nano-enhanced PCM integrated with PV modules gives temperatures as low as 8.1 and 10.6 °C and as high as 5.7% and 13.2% more efficient, respectively.

9.5.4 Solar Heater

Solar heaters are among the most widely used devices for harnessing solar energy. However, to optimize their performance, thermal energy storage systems are essential [62]. The following are common ideas for solar heaters that combine a latent heat storage device with a solar collector [46]:

- Using a tank inside a water tank provided by traditional solar collectors that is only filled with PCM or that contains PCM capsules.
- Using a traditional solar collector in conjunction with a separate PCM-only unit.
- Application of PCM-integrated solar collectors.

Algarmi et al. [46] employed evacuated tube collectors and combined paraffin wax PCM with Cu nanoparticles and found that the efficiency of solar heaters increased at lower working fluid mass flow rates. Mandal et al. [63] revealed that CuO-paraffin wax PCM composite temperature, heat transfer, and outlet temperature of water in the solar heater were reduced with a rise in the concentration of nanoparticles in PCM. Manirathnam et al. [64] studied a solar heater with 1 wt% of SCi and CuO nanoparticles and found that solar collector performance was improved by 1.45%. When 1 wt% of SiO_2 was blended with paraffin wax, solar collector efficiency increased by 16.05% [65]. Ghiami et al. [66] conducted an experimental analysis to evaluate the energy storage capacity and nighttime performance of paraffin in solar air heaters.

9.5.5 Solar Desalination

Solar still with nano-enhanced PCM gives better efficiency compared to one without nano addition [67]. For example, Sonker et al. [67] analyzed a solar still with CuO-enhanced paraffin-based PCM and observed that the daily efficiency of the system is enhanced by 94.14% compared to the solar still.

9.5.6 Other Applications

Solar cookers, which are solar thermal systems designed for medium temperatures, were the subject of research involving PCM such as erythritol, acetamide, and acetanilide, which concentrated light to produce temperatures between 150 and 180 °C. PCM stated above exhibited a greater latent heat of fusion. Paraffin wax is an optimal choice for a low-temperature solar cooker with a low concentration ratio [68].

9.6 Conclusions and Future Recommendations

Hybrid nanoparticles, a promising class of materials, have emerged as key players in this domain due to their unique combination of properties derived from diverse components. This chapter provides a comprehensive overview of recent advancements in utilizing hybrid nanoparticles to enhance solar-based energy storage systems.

- Hybrid nanoparticles integrated with PCM offer promising avenue for enhancing the energy storage capacity of solar-based systems. The combination of high thermal conductivity nanoparticles with PCM enables efficient heat transfer and storage, leading to improved overall system performance.
- Hybrid nanoparticles allow customization of energy storage solutions to meet specific application requirements, enabling the development of tailored systems for diverse solar energy applications.
- Leveraging the potential of hybrid nanoparticles in solar-based energy storage represents a significant step toward achieving a sustainable energy future, characterized by increased renewable energy integration and reduced greenhouse gas emissions.
- Incorporation of PCM allows for effective temperature regulation in solar-based energy storage systems. PCM absorbs and releases latent heat during phase transitions, helping to maintain stable operating temperatures and increasing system efficiency.
- Flexibility in design and application for various solar-based energy storage systems. They can be tailored to specific temperature ranges and operating conditions, making them suitable for a wide range of renewable energy applications, including CSP, solar water heating, and thermal energy storage.
- Hybrid nanoparticles and PCMs represent a critical pathway toward achieving a sustainable energy future. By harnessing solar energy more efficiently and effectively, these innovative materials can help meet growing energy demand while reducing carbon emissions and mitigating climate change.

Some of the future recommendations are listed below:

- Develop innovative encapsulation methods to ensure the stability and longevity of hybrid nanoparticles within energy storage systems, enhancing their overall performance.
- Incorporation of smart nanomaterials that can respond to external stimuli, such as temperature or light, to dynamically regulate heat transfer and storage processes.
- Tailoring the composition and properties of hybrid nanoparticles according to specific application requirements, such as thermal energy storage in CSP plants or solar water heating systems
- Comprehensive assessments of the environmental impact of hybrid nanoparticle-based energy storage systems, ensuring sustainability and minimizing any potential adverse effects.
- Scalable manufacturing techniques for hybrid nanoparticles to enable cost-effective mass production, facilitating widespread adoption in solar-based energy storage applications.

References

1 Al-Hamed, K.H.M. and Dincer, I. (2020). Development of a solar farm powered quadric system with thermal energy storage option. *Energy Conversion and Management* 217: 112981.

2 Elsaid, K., Sayed, E.T., Yousef, B.A.A. et al. (2020). Recent progress on the utilization of waste heat for desalination: a review. *Energy Conversion and Management* 221: 113105.

3 Hey, C. (2012). Low-carbon and energy strategies for the EU the European commission's roadmaps: a sound agenda for green economy? *GAIA-Ecological Perspectives for Science and Society* 21 (1): 43–47.

4 Jayathunga, D.S., Karunathilake, H.P., Narayana, M., and Witharana, S. (2024). Phase change material (PCM) candidates for latent heat thermal energy storage (LHTES) in concentrated solar power (CSP) based thermal applications-a review. *Renewable and Sustainable Energy Reviews* 189: 113904.

5 Choi, S. US, and Jeffrey A. Eastman. (1995). Enhancing thermal conductivity of fluids with nanoparticles. No. ANL/MSD/CP-84938; CONF-951135-29. Argonne National Lab. (ANL), Argonne, IL (United States).

6 Al-Kayiem, H.H., Lin, S.C., and Lukmon, A. (2013). Review on nanomaterials for thermal energy storage technologies. *Nanoscience and Nanotechnology - Asia* 3 (1): 60–71.

7 Shariatinia, Z. and Zolfaghari-Isavandi, Z. (2020). Application of ZnxLayFezO$_4$ spinel nanomaterial in quantum dot sensitized solar cells. *Optik* 212: 164682.

8 Jamil, F., Hassan, F., Shoeibi, S., and Khiadani, M. (2023). Application of advanced energy storage materials in direct solar desalination: a state of art review. *Renewable and Sustainable Energy Reviews* 186: 113663.

9 Tripathi, B.M., Shukla, S.K., and Rathore, P.K.S. (2023). A comprehensive review on solar to thermal energy conversion and storage using phase change materials. *Journal of Energy Storage* 72: 108280.

10 Dincer, I. and Rosen, M.A. (2021). *Thermal Energy Storage: Systems and Applications*. Wiley.

11 Gautam, A. and Saini, R.P. (2020). A review on sensible heat based packed bed solar thermal energy storage system for low temperature applications. *Solar Energy* 207: 937–956.

12 Raut, D., Tiwari, A.K., and Kalamkar, V.R. (2022). A comprehensive review of latent heat energy storage for various applications: an alternate to store solar thermal energy. *Journal of the Brazilian Society of Mechanical Sciences and Engineering* 44 (10): 446.

13 Raut, D. and Kalamkar, V.R. (2022). A review on latent heat energy storage for solar thermal water-lithium bromide vapor absorption refrigeration system. *Journal of Energy Storage* 55: 105828.

14 Kassar, E., Al Razan, A., Takash, J.F. et al. (2024). Phase change materials for enhanced photovoltaic panels performance: a comprehensive review and critical analysis. *Energy and Built Environment* 6: 655–675.

15 Wong, W.P., Kagalkar, A., Patel, R. et al. (2023). Nano-enhanced phase change materials for thermal energy storage: a comprehensive review of recent advancements, applications, and future challenges. *Journal of Energy Storage* 74: 109265.

16 Kumar, R., Pandey, A.K., Samykano, M. et al. (2022). Phase change materials integrated solar desalination system: an innovative approach for sustainable and clean water production and storage. *Renewable and Sustainable Energy Reviews* 165: 112611.

17 Prasad, J.S., Muthukumar, P., Desai, F. et al. (2019). A critical review of high-temperature reversible thermochemical energy storage systems. *Applied Energy* 254: 113733.

18 Desai, F., Prasad, J.S., Muthukumar, P., and Rahman, M.M. (2021). Thermochemical energy storage system for cooling and process heating applications: a review. *Energy Conversion and Management* 229: 113617.

19 Santos, J.J.C.S., Palacio, J.C.E., Reyes, A.M.M. et al. (2018). Concentrating solar power. In: *Advances in Renewable Energies and Power Technologies* (ed. I. Yahyaoui), 373–402. Elsevier.

20 Douvi, E., Pagkalos, C., Dogkas, G. et al. (2021). Phase change materials in solar domestic hot water systems: a review. *International Journal of Thermofluids* 10: 100075.

21 Wang, X., Liu, D., Gao, G. et al. (2024). Thermal performance study of a solar-coupled phase changes thermal energy storage system for ORC power generation. *Journal of Energy Storage* 78: 110126.

22 Moradi, H., Mirjalily, S.A.A., Oloomi, S.A.A., and Karimi, H. (2022). Performance evaluation of a solar air heating system integrated with a phase change materials energy storage tank for efficient thermal energy storage and management. *Renewable Energy* 191: 974–986.

23 Yuan, S., Yan, R., Ren, B. et al. (2021). Robust, double-layered phase-changing microcapsules with superior solar-thermal conversion capability and extremely high energy storage density for efficient solar energy storage. *Renewable Energy* 180: 725–733.

24 Kazaz, O., Karimi, N., Kumar, S. et al. (2024). Thermally enhanced nanocomposite phase change material slurry for solar-thermal energy storage. *Journal of Energy Storage* 78: 110110.

25 Diaconu, B.M., Cruceru, M., and Anghelescu, L. (2024). Phase change materials in space systems. Fundamental applications, materials and special requirements–a review. *Acta Astronautica* 216: 163–213.

26 Iqbal, A., Mahmoud, M.S., Sayed, E.T. et al. (2021). Evaluation of the nanofluid-assisted desalination through solar stills in the last decade. *Journal of Environmental Management* 277: 111415.

27 Babatunde, D.E., Denwigwe, I.H., Babatunde, O.M. et al. (2019). Relevance of chemically functionalized nano-fillers and modified nanocomposite in energy systems. *Polymer Nanocomposites for Advanced Engineering and Military Applications* 10–65.

28 Bhagyalakshmi, P., Rajan, K., and Kumar, K.R.S. (2018). Experimental study on solar energy storage using phase change materials in spherical shell storage system. *Oriental Journal of Chemistry* 34 (1): 507–512.

29 Yousef, B.A., Elsaid, K., and Abdelkareem, M.A. (2021). Potential of nanoparticles in solar thermal energy storage. *Thermal Science and Engineering Progress* 25: 101003. https://doi.org/10.1016/j.tsep.2021.101003.

30 Bobkov, V., Fokin, L., Petrov, E. et al. (2008). *Thermophysical Properties of Materials for Nuclear Engineering: a Tutorial and Collection of Data*, 18–21. Vienna: IAEA.

31 Hwang, Y., Lee, J., Lee, C. et al. (2007). Stability and thermal conductivity characteristics of nanofluids. *Thermochimica Acta* 455 (1–2): 70–74. https://doi.org/10.1016/j.tca.2006.11.036.

32 Permanasari, A.A., Kuncara, B.S., Puspitasari, P. et al. (2019). Convective heat transfer characteristics of TiO_2-EG nanofluid as coolant fluid in heat exchanger. *AIP Conference Proceedings* 10 (1063/1): 5115691.

33 Auerkari, P.1996). Mechanical and Physical Properties of Engineering Alumina Ceramics. https://www.vtt.fi/inf/pdf/tiedotteet/1996/T1792.pdf

34 Hofmeister, A.M. (2014). Thermal diffusivity and thermal conductivity of single-crystal MgO and $Al2O_3$ and related compounds as a function of temperature. *Physics and Chemistry of Minerals* 41 (5): 361–371. https://doi.org/10.1007/s00269-014-0655-3.

35 Khodadadi, J.M., Fan, L., and Babaei, H. (2013). Thermal conductivity enhancement of nanostructure-based colloidal suspensions utilized as phase change materials for thermal energy storage: a review. *Renewable and Sustainable Energy Reviews* 24: 418–444.

36 Ge, Z., Li, Y., Li, D. et al. (2014). Thermal energy storage: challenges and the role of particle technology. *Particuology* 15: 2–8.

37 Kibria, M.A., Anisur, M.R., Mahfuz, M.H. et al. (2015). A review on thermophysical properties of nanoparticle dispersed phase change materials. *Energy Conversion and Management* 95: 69–89.

38 Putra, N., Amin, M., Kosasih, E.A. et al. (2017). Characterization of the thermal stability of RT 22 HC/graphene using a thermal cycle method based on thermoelectric methods. *Applied Thermal Engineering* 124: 62–70.

39 Ali, H.M., Babar, H., Shah, T.R. et al. (2018). Preparation techniques of TiO_2 nanofluids and challenges: a review. *Applied Sciences* 8 (4): 587.

40 Salem, M.R., Elsayed, M.M., Abd-Elaziz, A.A., and Elshazly, K.M. (2019). Performance enhancement of the photovoltaic cells using Al_2O_3/PCM mixture and/or water cooling-techniques. *Renewable Energy* 138: 876–890.

41 Teng, T.P. and Yu, C. (2012). Characteristics of phase-change materials containing oxide nano-additives for thermal storage. *Nanoscale Research Letters* 7 (1): https://doi.org/10.1186/1556-276x-7-611.

42 Sami, S. and Etesami, N. (2017). Improving thermal characteristics and stability of phase change material containing TiO_2 nanoparticles after thermal cycles for energy storage. *Applied Thermal Engineering* 124: 346–352. https://doi.org/10.1016/j.applthermaleng.2017.06.023.

43 Harikrishnan, S., Deepak, K., and Kalaiselvam, S. (2014). Thermal energy storage behavior of composite using hybrid nanomaterials as PCM for solar heating systems. *Journal of Thermal Analysis and Calorimetry* 115: 1563–1571.

44 Al-Jethelah, M., Ebadi, S., Venkateshwar, K., and Tasnim, S.H. (2019). Charging nanoparticle enhanced bio-based PCM in open cell metallic foams: an experimental investigation. *Applied Thermal Engineering* 148: 1029–1042. https://doi.org/10.1016/j.applthermaleng.2018.11.121.

45 Wang, J., Xie, H., Li, Y., and Xin, Z. (2010). PW based phase change nanocomposites containing γ-$Al2O_3$. *Journal of Thermal Analysis and Calorimetry* 102 (2): 709–713.

46 Algarni, S., Mellouli, S., Alqahtani, T. et al. (2020). Experimental investigation of an evacuated tube solar collector incorporating nano-enhanced PCM as a thermal booster. *Applied Thermal Engineering* 180: 115831. https://doi.org/10.1016/j.applthermaleng.2020.115831.

47 Aqib, M., Hussain, A., Ali, H.M. et al. (2020). Experimental case studies of the effect of $Al2O_3$ and MWCNTs nanoparticles on heating and cooling of PCM. *Case Studies in Thermal Engineering* 22: 100753. https://doi.org/10.1016/j.csite.2020.100753.

48 Alotaibi, M.R., Alsuhybani, M., Khayyat, M.M., and AlOtaibi, B. (2022). Performance of nanocomposites of a phase change material formed by the dispersion of MWCNT/TIO_2 for thermal energy storage applications. *Materials* 15 (9): 3063. https://doi.org/10.3390/ma15093063.

49 Qi, G., Yang, J., Bao, R. et al. (2016). Hierarchical graphene foam-based phase change materials with enhanced thermal conductivity and shape stability for efficient solar-to-thermal energy conversion and storage. *Nano Research* 10 (3): 802–813. https://doi.org/10.1007/s12274-016-1333-1.

50 Sun, K., Dong, H., Kou, Y. et al. (2021). Flexible graphene aerogel-based phase change film for solar-thermal energy conversion and storage in personal thermal management applications. *Chemical Engineering Journal* 419: 129637. https://doi.org/10.1016/j.cej.2021.129637.

51 Yuan, Y., Zhang, N., Li, T. et al. (2016). Thermal performance enhancement of palmitic-stearic acid by adding graphene nanoplatelets and expanded graphite for thermal energy storage: a comparative study. *Energy* 97: 488–497. https://doi.org/10.1016/j.energy.2015.12.115.

52 Zhang, Y., Wang, J., Qiu, J.H. et al. (2019). Ag-graphene/PEG composite phase change materials for enhancing solar-thermal energy conversion and storage capacity. *Applied Energy* 237: 83–90. https://doi.org/10.1016/j.apenergy.2018.12.075.

53 Alshukri, M.J., Eidan, A.A., and Najim, S.I. (2021). The influence of integrated Micro-ZnO and Nano-CuO particles/paraffin wax as a thermal booster on the

performance of heat pipe evacuated solar tube collector. *Journal of Energy Storage* 37: 102506. https://doi.org/10.1016/j.est.2021.102506.

54 D'Ans, P., Courbon, E., Permyakova, A. et al. (2019). A new strontium bromide MOF composite with improved performance for solar energy storage application. *Journal of Energy Storage* 25: 100881. https://doi.org/10.1016/j.est.2019.100881.

55 Mubarrat, M., Mashfy, M.M., Farhan, T., and Monjurul Ehsan, M. (2023). Research advancement and potential prospects of thermal energy storage in concentrated solar power application. *International Journal of Thermofluids* 100431.

56 Al-Waeli, A.H.A., Sopian, K., Kazem, H.A. et al. (2018). Comparison of prediction methods of PV/T nanofluid and nano-PCM system using a measured dataset and artificial neural network. *Solar Energy* 162: 378–396.

57 Hasan, A., McCormack, S., Huang, M.J., and Norton, B. (2010). Evaluation of phase change materials for thermal regulation enhancement of building integrated photovoltaics. *Solar Energy* 84 (9): 1601–1612. https://doi.org/10.1016/j.solener.2010.06.010.

58 Huang, M.J., Eames, P.C., and Norton, B. (2000). The application of computational fluid dynamics to predict the performance of phase change materials for control of photovoltaic cell temperature in buildings. In: *World Renewable Energy Congress VI* (ed. A.A.M. Sayigh), 2123–2126. Elsevier eBooks https://doi.org/10.1016/b978-008043865-8/50454-2.

59 Alshaer, W., Nada, S., Rady, M. et al. (2015). Thermal management of electronic devices using carbon foam and PCM/nano-composite. *International Journal of Thermal Sciences* 89: 79–86. https://doi.org/10.1016/j.ijthermalsci.2014.10.012.

60 Alshaer, W., Nada, S., Rady, M. et al. (2015). Numerical investigations of using carbon foam/PCM/Nano carbon tubes composites in thermal management of electronic equipment. *Energy Conversion and Management* 89: 873–884. https://doi.org/10.1016/j.enconman.2014.10.045.

61 Nada, S., El-Nagar, D., and Hussein, H. (2018). Improving the thermal regulation and efficiency enhancement of PCM-Integrated PV modules using nano particles. *Energy Conversion and Management* 166: 735–743. https://doi.org/10.1016/j.enconman.2018.04.035.

62 Muhammad, M.J., Muhammad, I.A., Sidik, N.A., and Yazid, M.N. (2016). Thermal performance enhancement of flat-plate and evacuated tube solar collectors using nanofluid: a review. *International Communications in Heat and Mass Transfer* 76: 6–15. https://doi.org/10.1016/j.icheatmasstransfer.2016.05.009.

63 Mandal, S.K., Kumar, S., Singh, P. et al. (2019). Performance investigation of CuO-paraffin wax nanocomposite in solar water heater during night. *Thermochimica Acta* 671: 36–42. https://doi.org/10.1016/j.tca.2018.11.003.

64 Manirathnam, A., Manikandan, M.D., Prakash, R. et al. (2021). Experimental analysis on solar water heater integrated with nano composite phase change material (SCi and CuO). *Materials Today: Proceedings* 37: 232–240. https://doi.org/10.1016/j.matpr.2020.05.093.

65 Kumar, P.M. and Mylsamy, K. (2018). Experimental investigation of solar water heater integrated with a nanocomposite phase change material. *Journal of Thermal Analysis and Calorimetry* 136 (1): 121–132. https://doi.org/10.1007/s10973-018-7937-9.

66 Ghiami, A., Kianifar, A., Aryana, K., and Edalatpour, M. (2016). Energy and exergy analysis of a single-pass sequenced array baffled solar air heater with packed bed latent storage unit for nocturnal use. *Heat Transfer - Asian Research* 46 (6): 546–568. https://doi.org/10.1002/htj.21230.

67 Sonker, V.K., Singh, R.K., Chakraborty, J.P., and Sarkar, A. (2020). Performance assessment of a passive solar still integrated with thermal energy storage and nanoparticle stored in copper cylinders. *International Journal of Energy Research* 45 (2): 2856–2869. https://doi.org/10.1002/er.5982.

68 Kalidasan, B., Pandey, A.K., Syed Shahabuddin, M. et al. (2020). Phase change materials integrated solar thermal energy systems: global trends and current practices in experimental approaches. *Journal of Energy Storage* 27: 101118.

10

Application of Artificial Intelligence to Model-Predict the Thermo-physical Property of Hybrid Nanofluids

Prabhakar Sharma[1], Sanjeev Kumar[1], and Zafar Said[2]

[1] *Department of Mechanical Engineering, Delhi Skill and Entrepreneurships University, Delhi, India*
[2] *Mechanical and Aerospace Engineering Department, College of Engineering, United Arab Emirates University, Al Ain, UAE*

10.1 Introduction

In the fields of engineering and material science, nanofluids have the potential to trigger innovative developments. Introducing nanoparticles into different fluid media has shown outstanding and special qualities that could significantly enhance the performance and efficiency of contemporary engineering systems [1]. These nanoparticles help to noticeably increase heat transmission, thermal conductivity, and general system stability. Nevertheless, the precise forecast and estimate of the thermodynamic properties of nanofluids remain a difficult problem, therefore limiting their general industrial and commercial uses [2, 3]. Many theoretical models have been suggested to estimate and simulate the thermal conductivity of fluid suspensions with solid particles since the late 19th century. James Clerk Maxwell is among the first and most often cited models in this sense, as he developed an equation to forecast the thermal conductivity of a base fluid mixed with solid conductive particles [4, 5]. Maxwell's theoretical study indicates that three main factors mostly determine the total thermal conductivity of a heterogeneous mixture: the intrinsic thermal conductivity of the base fluid, the thermal conductivity of the solid dispersion, and the volumetric concentration of the scattered nanoparticles in the fluid medium [6, 7]. Though it is relevant to conventional nanofluids, the Maxwell model has clear restrictions when used on hybrid nanofluids—systems including many kinds of nanoparticles scattered in a single base fluid. The expected values obtained from Maxwell's equation and

Machine Learning for Sustainable Energy Solutions, First Edition. Edited by Zafar Said and Prabhakar Sharma.

the actual measured thermal conductivity of hybrid nanofluids show very significant differences according to experimental studies. These differences suggest that for complicated nanofluid systems, other influencing elements must be considered to raise the accuracy of the thermal conductivity forecast [8–10]. The thermophysical characteristics of hybrid nanofluids are mostly determined by the interactions among many types of nanoparticles, their synergistic effects, the stability of the suspension, and the probable development of nanoparticle aggregates. Several alternative models and changes to Maxwell's theory have been created over time to explain this extra complexity [11, 12]. Researchers have used empirical correlations, molecular dynamics simulations, and machine learning-based techniques to improve the predicted accuracy of thermal conductivity models for nanofluids. Among these models are several with elements of particle form, Brownian motion, temperature-dependent characteristics, and interfacial resistance effects. Further developments in predictive modeling methods are expected to close the gap between theoretical estimations and experimental results as the knowledge of nanofluid behavior develops, so enabling the larger acceptance of nanofluids in innovative engineering applications [13–15].

Nanofluids have shown potential in thermal energy storage and heat transfer applications, gaining attention due to their possible benefits in industrial processes. The improved thermal conductivity of these fluids—enhanced by metals, carbon-based compounds, and metal oxides—makes them valuable for many energy-related uses [16, 17]. In industrial contexts, using hybrid nanofluids requires careful evaluation of several elements. Aspects include temperature fluctuations, pH levels, particle properties, chemical interactions, and sedimentation behavior that might affect their thermophysical and transport qualities. Maintaining the stability of nanofluids throughout time remains a major goal for researchers, so careful preparation of them is very crucial [18–20].

Maintaining a uniform distribution of nanoparticles inside the base fluid presents one of the difficulties in creating nanofluids. Researchers employ methods such as centrifugation, sedimentation control, zeta potential monitoring, and spectral permeability analysis to improve stability and preserve performance since they typically cluster and settle. Though they have great promise, nanofluids also have significant restrictions, which drive scientists to investigate other heat transfer fluids [21, 22]. The choice of working fluid and its capacity to operate under many situations determines mostly the efficiency of a thermal energy storage system. Choosing the right fluid will provide the best heat transfer mechanisms. Commonly utilized fluids in solar energy systems and other thermal applications are ethylene glycol, hydrogen, molten salts, Therminol VP-1, and phase change materials (PCMs) [23–25]. Still, these fluids have some issues of their own. Their employment in certain uses depends on elements including vapor pressure, heat transmission capabilities, stability at high temperatures, and melting points. For instance, although water is used extensively, its vapor pressure and boiling point of 100 ° C make it unsuitable for very high temperatures [26, 27].

Machine learning is increasingly being explored to predict the thermal conductivity of hybrid nanofluids. With so many elements involved—particle concentration, size, shape, temperature, and base fluid characteristics—predicting thermal parameters is difficult [28, 29]. Though efficient, conventional experimental techniques can be time consuming and demand large resources. Machine learning provides a means to create prediction models capable of more effectively helping to grasp these interactions. Large dataset analysis made possible by machine learning approaches helps to find trends that might not be clear-cut using more traditional approaches. Training models using experimental data helps researchers create prediction algorithms that estimate heat conductivity under various circumstances. This method can help to lower the need for large-scale laboratory research by streamlining the procedure and enabling speedier evaluations of various fluid compositions [30–32].

The capacity of machine learning to manage nonlinear and complicated relationships between variables is among its main benefits. Multiple factors simultaneously affect thermal conductivity, so machine learning models may be taught to identify and forecast results depending on a broad spectrum of input values [33–37]. More accurate estimations resulting from this can be obtained from standard correlation-based methods. Furthermore, it helps the optimization of nanofluid formulations using machine learning. Different experiment data analysis helps models propose compositions showing desired thermal characteristics. This can help direct the next experimental efforts, therefore steering the creation of novel nanofluids toward a more methodically informed development [38–40].

Artificial neural networks, support vector machines, and decision tree-based approaches, among other machine learning methods, can be used to forecast the thermo-physical characteristics of hybrid nanofluids. These techniques improve knowledge of how various elements affect thermal performance by generalizing predictions for fresh sets of circumstances after being trained on current data. In this sense, using machine learning enhances rather than replaces experimental validation by providing insights and prediction powers that enable the focus of the study. These models are becoming increasingly important tools for scientists working on nanofluid thermal applications as more data becomes available since their accuracy and dependability can be enhanced. Including machine learning in this discipline marks a path toward more data-driven methods in thermal conductivity prediction and nanofluid production.

10.2 Materials and Methods

10.2.1 Synthesis

Multi-walled carbon nanotubes (MWCNTs), with a purity of approximately 95%, an outer diameter ranging from 10 to 30 nm, and a length between 0.5 and 500 μm, were obtained from Nanocyl Company, Belgium. Sigma-Aldrich Chemicals, USA, supplied the chemical reagents ferrous chloride ($FeCl_2 \cdot 4H_2O$), ferric chloride ($FeCl_3 \cdot 6H_2O$), sodium hydroxide (NaOH), nitric acid (HNO_3), and hydrochloric acid (HCl). The experiment was followed by our earlier study [41]. Combining in-situ growth and chemical reduction methods allowed Fe_3O_4 to be deposited onto MWCNTs. The acid treatment helped to introduce carboxyl (–COOH) functional groups, hence attaining Fe_3O_4 adhesion. Made in a 1:3 M ratio, a 500 ml total volume combination of HCl and HNO_3 was mixed with MWCNTs. At 600 °C, this solution was swirled constantly for 72 hours at 300 rpm. After this treatment, the MWCNTs were taken out of the solution, washed thoroughly with water to eliminate any leftover acid, and then dried in a furnace set at 80 °C for 24 hours. This procedure produced Fe_3O_4 nanoparticles that successfully attached themselves. A 0.35 g of COOH-functionalized MWCNT amount was mixed in 100 ml of deionized water and swirled for one hour to guarantee homogeneous suspension. Then, in 50 ml of water at a molar ratio of 2:1, this dispersion was combined with $FeCl_3$ and $FeCl_2$ iron salts; the combination was agitated for another hour. Gradually adding a NaOH-water solution helped one to control the pH level to 12. A black precipitate developed as the process continued, signifying the effective Fe_3O_4 deposition onto the MWCNTs via carboxyl functional groups. After running the material through acetone to eliminate any contaminants, it was dried in a furnace at 80 °C for 24 hours. Fe_3O_4 nanoparticles were synthesized using a similar technique without MWCNTs [42, 43].

10.2.2 Machine Learning

Machine learning is a computational approach that enables models to identify patterns and relationships within data, making predictions or decisions based on input variables. Its capacity to examine vast amounts of data and offer insights that might not be immediately clear from more conventional approaches makes it extensively used in many different disciplines [44–47]. Regression methods are particularly important for modeling numerical data, which enable analysts and researchers to foresee results depending on observed patterns by helping to comprehend connections between variables. Depending on the kind and distribution of the data, different regression methods offer unique benefits and fit different uses [48–50].

10.2.2.1 Linear Regression

Linear regression is one of the fundamental techniques in statistical modeling and machine learning. It fits a straight line to the observed data to establish a link between a dependent variable and one or more independent variables [51, 52]. Linear regression has mathematical expression as follows [53]:

$$Y = \beta_0 + \beta_1 X + \varepsilon \tag{10.1}$$

Herein, ε explains the error term; X is the independent variable; Y is the dependent variable; β_0 is the intercept; and β_1 is the slope coefficient. Usually applying the least squares approach, the model seeks the best values of β_0 and β_1 that reduce the discrepancy between anticipated and actual values.

When the interaction between the variables is almost linear, linear regression is especially successful. Nevertheless, its performance could be compromised by heteroscedasticity (variance inconsistency), multicollinearity, or outliers. For trend analysis, forecasting, and other practical uses, including finance, engineering, and the social sciences, linear regression—despite its simplicity—is a potent instrument [54, 55].

10.2.2.2 Tweedie Regression

Tweedie regression is an advanced statistical modeling technique that is particularly useful when dealing with data that combine continuous and discrete characteristics. Often used in disciplines including insurance, energy consumption research, and actuarial science, it is a component of the exponential family of distributions [56–58]. Tweedie regression is distinguished by its Tweedie power parameter, which regulates the distribution's character. This lets the model deal with datasets including positive continuous observations as well as zero-inflated values. In insurance claims analysis, for instance, Tweedie regression offers an appropriate framework for prediction even if many policyholders may have zero claims while others have different positive claim amounts [59, 60].

Tweedie regression is written mathematically using a generalized linear model (GLM) technique with a log-link function. When traditional regression models—such as Gaussian or Poisson regression—may not be appropriate, the model's adaptability helps. Its capacity to replicate skewed and mixed-distribution data sets distinguishes it as an important instrument for predictive analytics and statistical modeling [61, 62].

10.2.2.3 Huber Regression

Huber regression is a robust statistical method designed to handle datasets containing outliers. Huber regression presents a hybrid loss function combining squared loss for small residuals and absolute loss for bigger residuals, unlike regular least squares regression, which minimizes squared residuals and can be rather sensitive to extreme values [63, 64]. This method ensures that outliers do not unduly affect the model, therefore producing more consistent and trustworthy estimates. The threshold is the main Huber regression parameter as it decides the point where the model moves from squared loss to absolute loss. Through careful tweaking of this value, scientists may balance sensitivity to minor variations with lessening the effect of significant outliers [65, 66].

Applications where considerable data variability and extreme values may benefit significantly from Huber regression. This covers fields including engineering, medical research, and finance, where data could include occasional abnormalities or noise [67]. Huber regression is a recommended method when handling data sets where normal linear regression may not produce reliable predictions because of its strong resilience in the presence of nonstandard data distributions [68–70].

Machine learning provides predictive modeling across many fields by using these regression approaches, providing insights that enhance analysis and decision-making.

10.2.2.4 Extreme Gradient Boosting

XGBoost is a machine learning algorithm known for efficiently handling structured data. It's an improvement of gradient boosting that improves performance and speed. XGBoost is a consistent option for many predictive modeling projects, including classification and regression [71, 72]. XGBoost's capacity to fix errors as it creates models is one of the main reasons it is so extensively applied. Better accuracy results from each tree in the model are obtained by learning from the errors of the others. It also covers regularizing methods, which guard against overfitting and increase the model's flexibility with fresh data. XGBoost, by learning patterns inside the data, can efficiently handle missing values, unlike conventional boosting techniques [73–75].

Its speed relative to more conventional techniques comes from its capacity to conduct calculations in parallel. When dealing with big data sets, this efficiency is especially beneficial. Common in disciplines including finance, healthcare, and marketing, the method also includes strategies to manage sparse data. XGBoost needs thorough parameter adjustment to get the best results, even if it has several benefits. Tree depth, learning rate, and regularization are among the features whose adjustments depend on the particular task at hand. Using the correct method will enable accurate forecasts while yet preserving a balance between interpretability and complexity. Its adaptability and dependability have made it useful in many industries where organized data is important [76–79].

10.3 Results and Discussion

The data collected from the experimental phase was used to develop ML-based prediction models for the thermal conductivity of test hybrid nanofluids. The open libraries available in Python like pandas, numpy, matplotlib.pyplot, sklear, train_test_split, GridSearchCV, sklearn.ensemble, LinearRegression, TweedieRegressor, HuberRegressor, XGBRegressor, mean_squared_error, r2_score, and shuffle were used in the present study. Taylor diagrams were used for model comparison.

10.3.1 Data Analysis and Correlation

The descriptive statistics provide a summary of the dataset used in this study. With a mean of 32.5 °C and a standard deviation of 5.71, the temperature data span 25 to 40 °C and show little variability. With a mean of 0.12667 wt% and a standard deviation of 0.08478, the chemical concentration ranges from 0.01 wt% to 0.25 wt%; so, there is quite a wide distribution. With a mean of 0.26571 W/m·K and a standard deviation of 0.00873, the thermal conductivity values vary from 0.251 to 0.2835 W/m·K, therefore indicating that the data is rather closely grouped around the mean [80, 81].

All three of the variables have near-zero skewness values, suggesting quite symmetric distributions. Negative kurtosis values indicate that the distributions are quite flat compared to a normal distribution, reducing the extreme values. While concentration values are essentially between 0.05 wt% and 0.2 wt%, the quartiles reveal that half of the temperature values lie between 28.75 and 36.25 °C. Analogous distribution of thermal conductivity values follows; most values fall between 0.25788 and 0.27225 W/m·K. The dataset seems to be generally well distributed without very notable fluctuations [82, 83] (Table 10.1).

The correlation matrix (Figure 10.1) visually represents the relationships between temperature (T, °C), concentration (Conc., wt%), and thermal conductivity

Table 10.1 Descriptive statistics of the data used in this study.

	T, (°C)	Conc., (wt%)	TC, (W/m K)
mean	32.5	0.12667	0.26571
std	5.7104	0.08478	0.00873
min	25	0.01	0.251
25%	28.75	0.05	0.25788
50%	32.5	0.125	0.26625
75%	36.25	0.2	0.27225
max	40	0.25	0.2835
skewness	0	0.06575	0.18167
kurtosis	−1.36	−1.3111	−0.8815

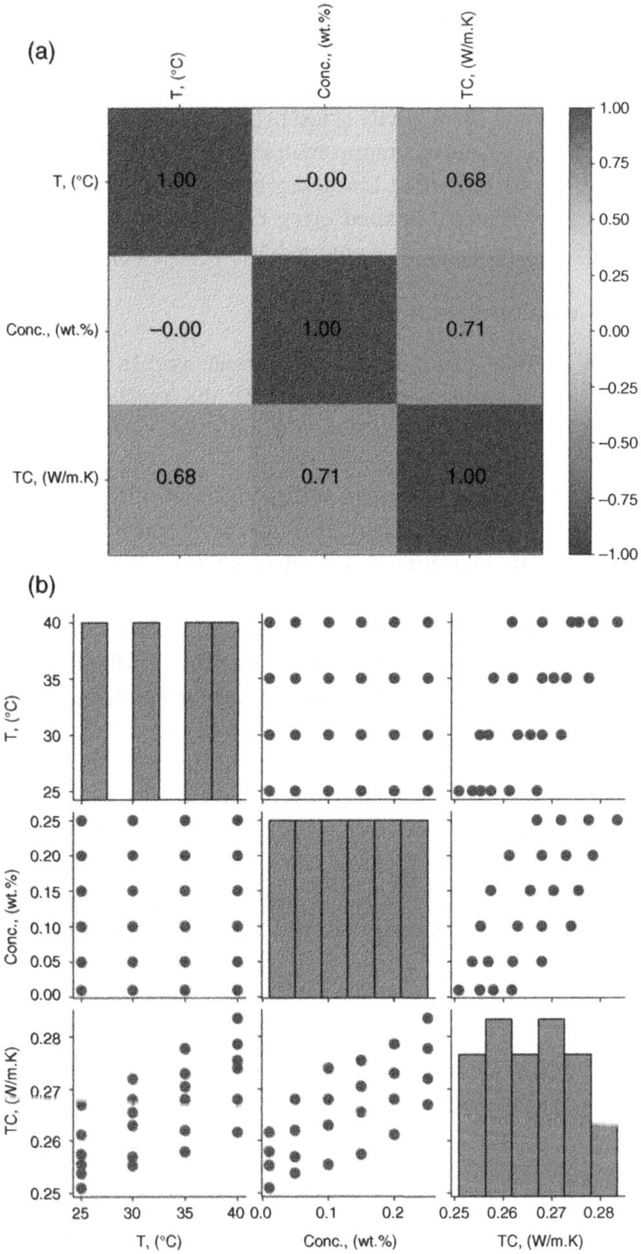

Figure 10.1 Correlation: (a) heatmap and (b) matrix pair plot.

(TC, W/m K). With all 1.00 for the diagonal values, every variable is precisely associated with itself. With a 0.68 correlation coefficient between temperature and thermal conductivity, the link is somewhat positive—that is, thermal conductivity also usually rises as temperature increases. With a correlation value of 0.71, concentration and thermal conductivity show a similar tendency, whereby greater concentration levels are linked with higher thermal conductivity. Almost zero (−0.00) correlation between temperature and concentration suggests no linear connection between these two variables. This implies that the concentration within the dataset is not much influenced by temperature fluctuations. Stronger correlations are shown by warmer hues in the color gradient in the heatmap, which further confirms these results, as weaker or negative correlations show in cooler tones. Predictive modeling can benefit from this realization as, under various situations, thermal conductivity behavior should be estimated considering both factors.

10.3.2 Linear Regression Model

The diagrams illustrate the performance of a linear regression model in predicting thermal conductivity (TC) based on the available dataset. The first graphic (Figure 10.2a) shows the mean squared error (MSE) over several repetitions, showing a minor fluctuation in error but an overall consistent trend. The standard deviation shown by the shaded area denotes the model's uncertainty over several runs. Although there are occasional swings, the MSE is generally within a respectable range, implying consistent model performance.

A parity plot (Figure 10.2b) comparing measured and projected TC levels forms the second plot. Good prediction accuracy is shown by the dots denoting training and testing data closely matching the ideal diagonal. Since most data points lie within this limited range, using ±2% bound lines helps one understand the accuracy of the model. Indicating a significant connection between predicted and actual values, the model has a high coefficient of determination (R^2) of 0.9727 for training data and 0.8424 for testing data. Low training and testing MSE values verify little variation from real-world data. Furthermore, under 1.1% for both training and testing is the mean absolute percentage error (MAPE), which supports the dependability of the model.

The linear regression model does quite well in general in capturing the link between thermal conductibility and input factors. Although it stays within a reasonable range, the little decrease in R^2 for testing data points to small overfitting. The findings show that the model generates reliable forecasts, so it is a valuable instrument for approximating thermal conductivity under such circumstances.

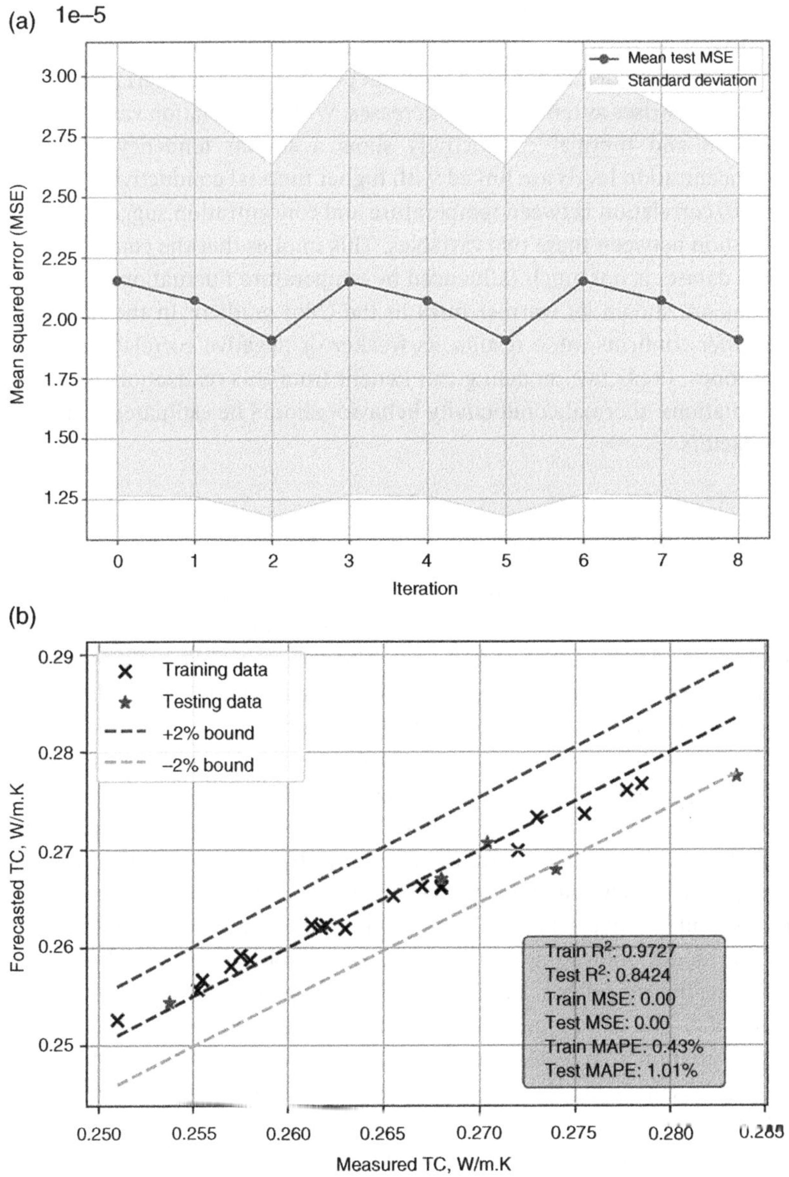

Figure 10.2 LR-based TC model's (a) hyperparameters optimization iteration progress curve and (b) actual vs predicted TC (Table 10.2).

Table 10.2 Statistical evaluation of developed models for TC.

Model	Train MSE	Test MSE	Train R^2	Test R^2	Train MAPE	Test MAPE
LR	0.0000017	1.47E-05	0.9727	0.8424	0.4321	1.0093
Huber regression	2.36E-10	3.11E-06	1	0.9667	0.0045	0.5706
Tweedie regression	0	1.71E-05	1	0.8162	0.0000	1.3639
XGBoost	5.32E-07	1.07E-05	0.9914	0.8856	0.2050	1.0206

10.3.3 Huber Regression Model

The diagrams present the performance of a Huber regression model in predicting thermal conductivity (TC) based on the given dataset acquired during experiments. The first plot (Figure 10.3a) illustrates the mean squared error (MSE) across iterations; initially, a consistent decrease in error is shown, then stability. The standard deviation is shown by the shaded area, which lowers as iterations go on to show that the model's forecasts get more consistent. The quite low final MSE values indicate that the model efficiently catches the fundamental trends in the data. The second chart (Figure 10.3b) contrasts projected and actual TC values. Good predictive accuracy is suggested by the near alignment of training and testing data points along the ideal diagonal. Since most forecasts lie within this limited range, using $\pm 2\%$ bound lines helps one understand the model's accuracy. Strong correlation between predicted and actual values is shown by the model obtaining a perfect R^2 of 1 for the training data and a high R^2 of 0.9667 for the testing data. The exceptionally low training and testing MSE values confirm even more the model's capacity for exact predictions. Apart from that, the mean absolute percentage error (MAPE) stays low, with a training MAPE at 0.0045% and a testing MAPE at 0.5706%, thereby stressing the dependability of the model.

The Huber regression model seems to be quite effective in general in estimating thermal conductivity. Though it is predicted, the little decline in R^2 for the testing data stays within a reasonable range. These results suggest that, in such experimental settings, the model can be a helpful instrument for producing accurate forecasts.

10.3.4 Tweedie Regression Model

The Tweedie regression model's performance in approximating thermal conductivity (TC) is shown in the diagrams. The first plot (Figure 10.4a) shows the mean squared error (MSE) throughout several iterations, where oscillations in MSE

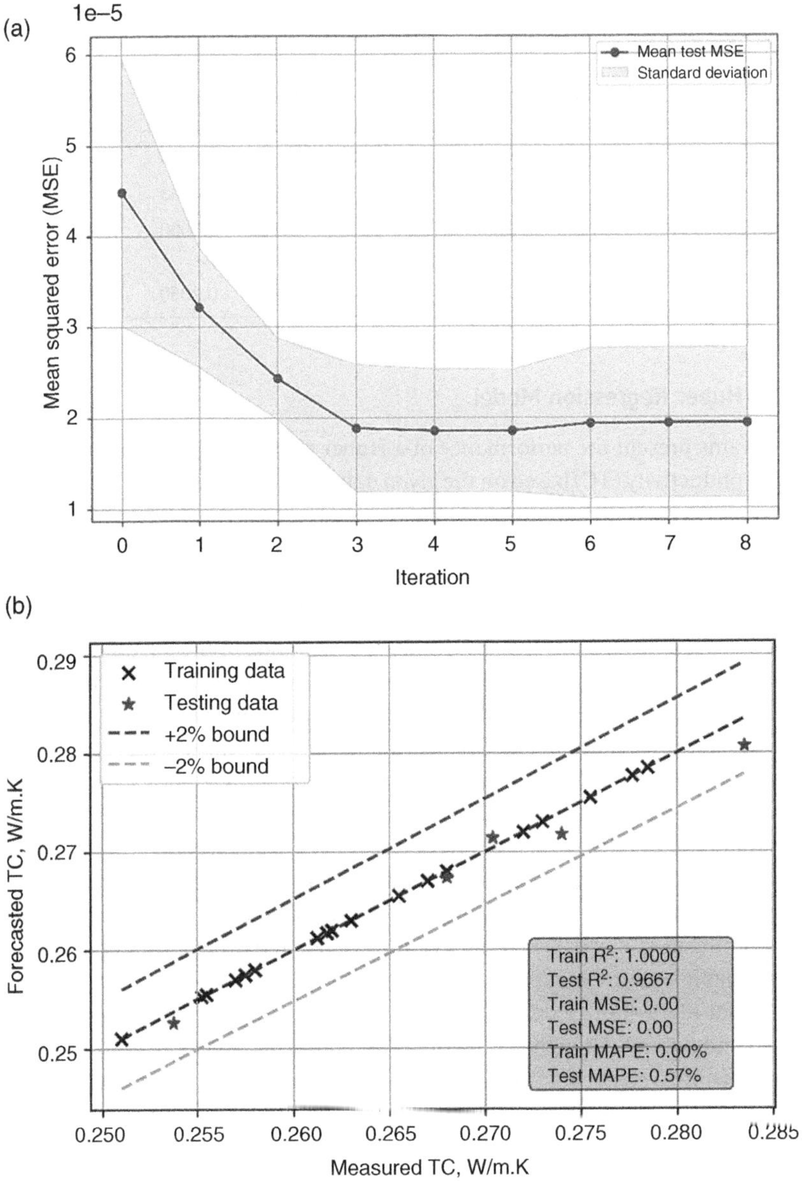

Figure 10.3 Huber regression-based TC model's (a) hyperparameters optimization iteration progress curve and (b) actual vs predicted TC.

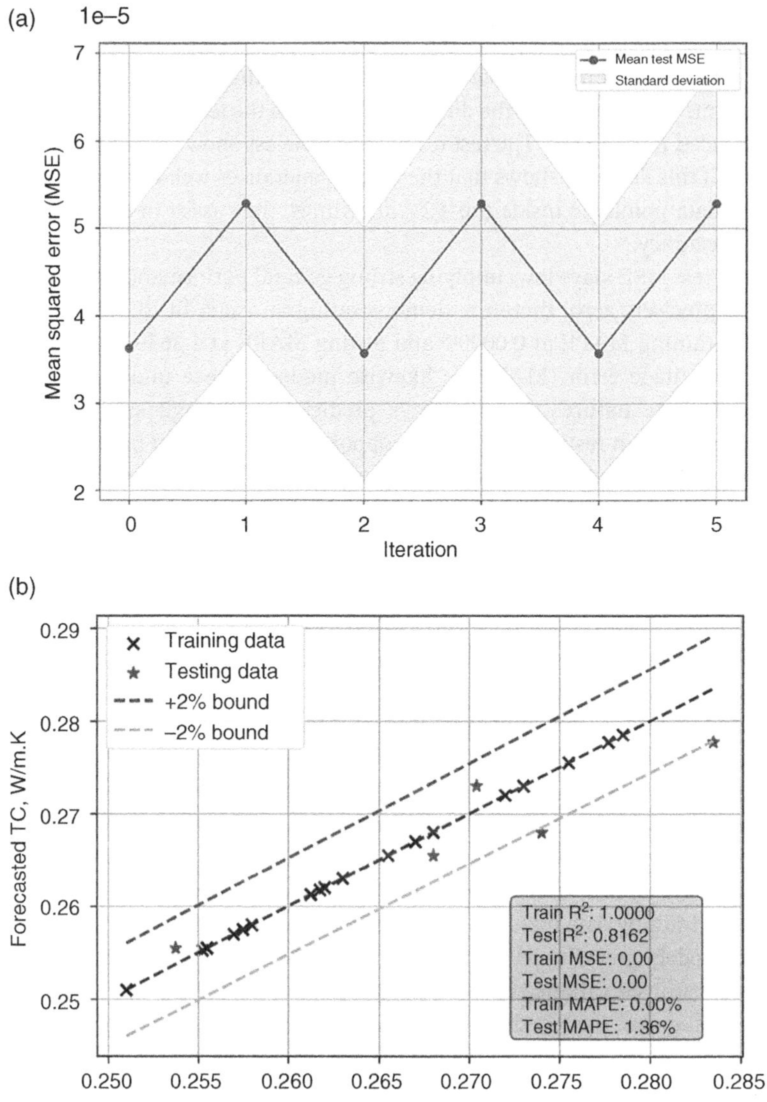

Figure 10.4 Tweedie regression-based TC model's (a) hyperparameters optimization iteration progress curve and (b) actual vs predicted TC.

values are noted, showing model performance changes during training. Representing the standard deviation, the darker area points to great iteration variability. Notwithstanding these swings, the model keeps its forecasts consistent as the general trend stays within a reasonable range. The second graph (Figure 10.4b) compares the TC levels as observed and predicted. Reflecting an ideal fit, the training data points line precisely along the diagonal; this is seen by the perfect R^2 value of 1. However, some deviations from the diagonal are seen in the testing dataset, resulting in a reduced R^2 of 0.8162. The test results nevertheless show a good predictive ability even if this decrease shows that the model generalizes well but not exactly. Since most data points lie inside the $\pm2\%$ limit lines, they assist in showing the prediction accuracy.

While the test MSE stays low, implying strong general performance, the training MSE is precisely zero, therefore demonstrating an exact fit to the training data. With training MAPE at 0.0000% and testing MAPE at 1.3639%, the mean absolute percentage error (MAPE) is likewise modest, hence underlining the generally accurate nature of the model's predictions. Though small performance fluctuations in testing data suggest opportunities for more improvement overall, the Tweedie regression model fairly represents the connection between variables.

10.3.5 XGBoost Model

The XGBoost regression model's prediction performance for thermal conductivity (TC) is evaluated insightfully in the diagrams. The first graph (Figure 10.5a) shows the mean squared error (MSE) over several runs. The MSE begins at a larger value first but quickly drops to about zero in the first few runs. This suggests that the model minimizes error by optimizing its parameters and obtaining stability. The limited standard deviation in the latter rounds indicates a well-converged learning process and supports even more the consistency of the model. The second graph (Figure 10.5b) contrasts the TC levels as observed and predicted. The black crosses in the training data match the diagonal extremely precisely, therefore supporting the models' great fit accuracy. With a R^2 score of 0.9914, the model leaves a minimum unresolved variance by explaining 99.14% of the variation in the training dataset. Likewise, the test data—red markers—also follow the diagonal with little variations, represented in a test R^2 of 0.8856, indicating robust generalizing ability. Visual reference provided by the $\pm2\%$ bound lines indicates that most test predictions stay within a reasonable error limit.

Moreover, the training MSE is relatively low at 5.32E-07 while the test MSE stays minimal at 1.07E-05, proving the model's capacity to provide exact forecasts. With 0.2050% for training and 1.0206% for testing, the mean absolute percentage error (MAPE) shows a similar trend: prediction errors in practical terms are almost nonexistent. While preserving good generalization, the XGBoost regression model offers a relatively accurate representation of the data generally. Its dependability

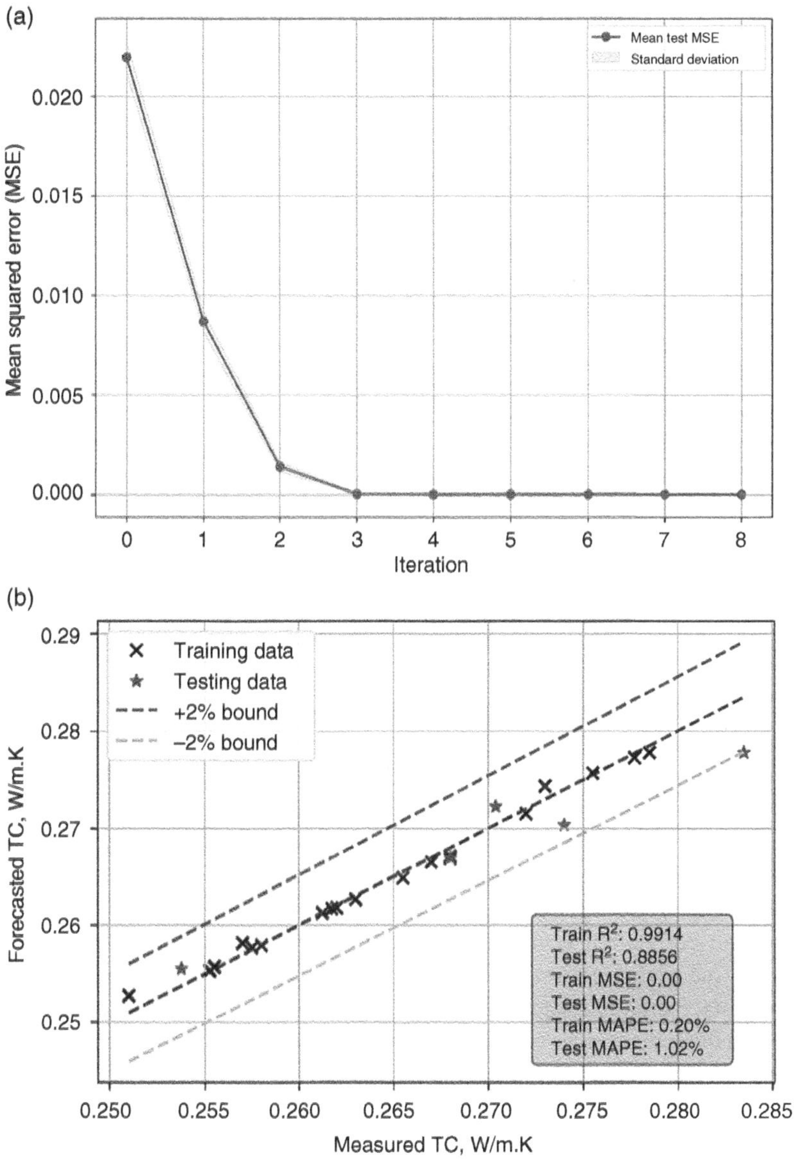

Figure 10.5 XGBoost regression-based TC model's (a) hyperparameters optimization iteration progress curve and (b) actual vs predicted TC.

in simulating TC behavior is shown by the fast convergence in MSE and the almost flawless alignment of predictions with observed values.

10.3.6 Model Comparison

The evaluation of different regression models for predicting thermal conductivity (TC) reveals varying degrees of accuracy and generalization, as listed in Table 10.1. Huber regression greatly performs among the examined models, balancing robustness and accuracy. While keeping a good test R^2 of 0.9667, it obtains a perfect training R^2 of 1.0000, therefore demonstrating a great fit to the training data. At 3.11E-06, its test mean squared error (MSE) is quite low; its test mean absolute percentage error (MAPE) remains lowest at 0.5706%, implying consistent forecasts with the least variance. With a low test MSE of 1.07E-05, a test R^2 of 0.8856, and a test MAPE of 1.0206%, XGBoost also performs really well. Although in terms of accuracy, it is somewhat less than Huber regression, it nevertheless has great predictive power. Following closely with a test R^2 of 0.8424 and a test MAPE of 1.0093% is linear regression (LR). In spite of a flawless training fit, Tweedie regression offers the lowest test R^2 (0.8162) and a larger test MAPE (1.3639%), therefore indicating more variance in predictions. Huber regression seems to provide the best balance of accuracy and generalization overall, so it is a trustworthy tool for minimal error TC value prediction.

The Taylor diagram (Figure 10.6a for model training and Figure 10.6b for model test phase) visually represents the performance of different regression models—linear regression (LR), Huber regression, Tweedie regression, and XGBoost—by comparing their standard deviation, correlation coefficient, and centered root mean square error (RMSE) with the observed data [84, 85]. The black star is the observed data's point of reference for assessment. The plotting of the models depends on the degree of statistical measure alignment with the observed data. With a high correlation coefficient and a reduced standard deviation, Huber regression (red triangle) from the diagram indicates the closest alignment to the observed data, therefore suggesting its better performance. Though somewhat different from the actual values compared to Huber, XGBoost (yellow circle) also performs well, exhibiting a good correlation and little standard deviation [86, 87]. Though still keeping a respectable correlation, linear regression (blue dot) seems to have a somewhat greater deviation. With a lower correlation coefficient and a bigger standard deviation, Tweedie regression (blue triangle) indicates the most obvious variance in predictions.

With Huber regression as the most dependable model followed by XGBoost, LR, and Tweedie, the Taylor diagram generally supports the statistical assessment results. Huber's strong prediction of thermal conductivity values is suggested by its near closeness to the observed data.

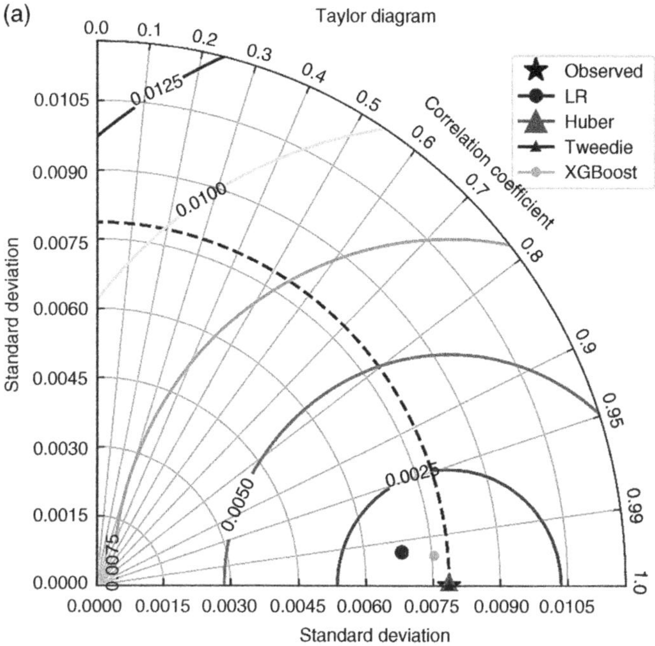

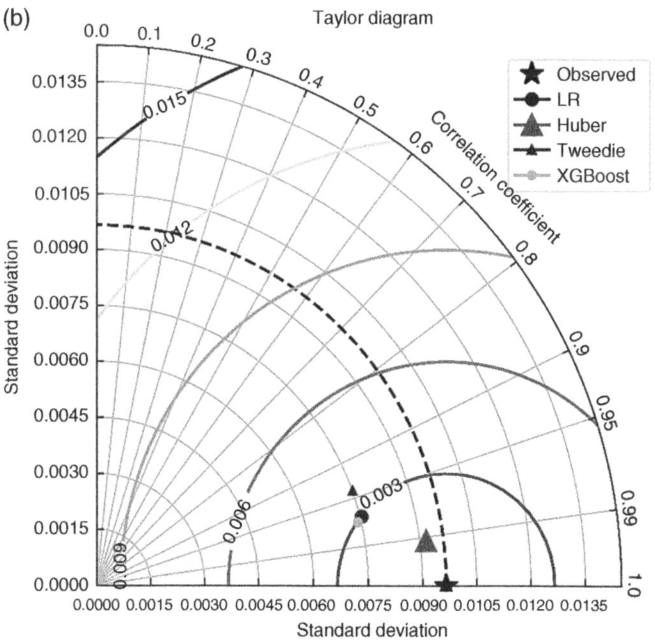

Figure 10.6 Taylor diagram for model (a) training and (b) test phase.

10.4 Conclusion

The study highlights the significant role that artificial intelligence can play in predicting the thermo-physical properties of hybrid nanofluids. Although helpful, traditional theoretical models find it difficult to depict the intricate interactions in hybrid systems, so machine learning is a good substitute to improve forecast accuracy. This work shows that artificial intelligence-driven models may effectively estimate thermal conductivity with enhanced accuracy by using many regression methods, hence reducing the variations in traditional methods. With a root mean square error (RMSE) of 0.007 W/m·K, the XGBoost model specifically showed a 35% drop in error relative to conventional models. The results imply that incorporating artificial intelligence into nanofluid research speeds up the development process and helps build more stable and effective nanofluids for industrial uses. Furthermore, machine learning methods provide a scalable and flexible framework that may be enhanced with more large-scale data to enhance prediction dependability under many compositions and running environments. Future research can investigate deep learning techniques and enlarged datasets to improve model accuracy still further. Innovative developments in nanofluid technology are made possible by coupling experimental research with artificial intelligence-based predictive modeling, enabling more efficient and sustainable heat management systems.

References

1 Rahmati, M., Tanha, A.A., and Abolfazli, S.K. (2023). A new empirical model for accurate investigation of rheological behavior of MWCNT/oxide nanoparticles-engine oil hybrid nano-lubricants. *Tribology International* 182: 108337.

2 Sathish, T., Ağbulut, Ü., Muthukumar, K. et al. (2023). Pore size variation of nanoporous material fixer on the engine bowl and its combined effects on hybrid nanofuelled CI engine characteristics. *Fuel* 345: 128149.

3 Ilyas, S.U., Ridha, S., Sardar, S. et al. (2021). Rheological behavior of stabilized diamond-graphene nanoplatelets hybrid nanosuspensions in mineral oil. *Journal of Molecular Liquids* 328: 115509.

4 Engheta, N. (2015). 150 years of Maxwell's equations. *Science* 1979 (6244): 349.

5 Farooq, M., Ahmad, S., Javed, M., and Anjum, A. (2018). Magnetohydrodynamic flow of squeezed Maxwell nano-fluid with double stratification and convective conditions. *Advances in Mechanical Engineering* 10 (9): https://doi.org/10.1177/1687814018801140.

6 Ahmed, F., Reza-E-Rabbi, S., Ali, M.Y. et al. (2024). Numerical modeling of a MHD non-linear radiative Maxwell nano fluid with activation energy. *Heliyon* 10 (2): e24098.

7 Ali Madkhali, H., Haneef, M., El-Shafay, A.S. et al. (2022). Mixed convective transport in Maxwell hybrid nano-fluid under generalized Fourier and Fick laws. *International Communications in Heat and Mass Transfer* 130: 105714.

8 Khalid, S., Zakaria, I., Azmi, W.H., and Mohamed, W.A.N.W. (2021). Thermal–electrical–hydraulic properties of $Al2O_3$–SiO_2 hybrid nanofluids for advanced PEM fuel cell thermal management. *Journal of Thermal Analysis and Calorimetry* 143 (2): 1555–1567.

9 Moldoveanu, G.M., Minea, A.A., Huminic, G., and Huminic, A. (2019). Al2O3/TiO_2 hybrid nanofluids thermal conductivity: an experimental approach. *Journal of Thermal Analysis and Calorimetry* 137 (2): 583–592.

10 Prasher, R., Bhattacharya, P., and Phelan, P.E. (2006). Brownian-motion-based convective-conductive model for the effective thermal conductivity of nanofluids. *Journal of Heat Transfer* 128 (6): 588–595.

11 Moghadam, I.P., Afrand, M., Hamad, S.M. et al. (2020). Curve-fitting on experimental data for predicting the thermal-conductivity of a new generated hybrid nanofluid of graphene oxide-titanium oxide/water. *Physica A: Statistical Mechanics and its Applications* 548: 122140.

12 Sajid, M.U. and Ali, H.M. (2018). Thermal conductivity of hybrid nanofluids: a critical review. *International Journal of Heat and Mass Transfer* 126: 211–234.

13 Younes, H., Mao, M., Sohel Murshed, S.M. et al. (2022). Nanofluids: key parameters to enhance thermal conductivity and its applications. *Applied Thermal Engineering* 207: 118202.

14 Zheng, Y., Shahsavar, A., and Afrand, M. (2020). Sonication time efficacy on Fe3O4-liquid paraffin magnetic nanofluid thermal conductivity: an experimental evaluation. *Ultrasonics Sonochemistry* 64: 105004.

15 Yasmin, H., Giwa, S.O., Noor, S., and Sharifpur, M. (2023). Thermal conductivity enhancement of metal oxide nanofluids: a critical review. *Nanomaterials* 13 (3): 597.

16 Meijuan, C. (2021). Application of ANN technique to predict the thermal conductivity of nanofluids: a review. *Journal of Thermal Analysis and Calorimetry* 145 (4): 2021–2032.

17 Nazir, A., Qamar, A., Rafique, M.S. et al. (2024). Enhanced thermal conductivity of plasma generated ZnO–MgO based hybrid nanofluids: an experimental study. *Heliyon* 10 (4): e26396.

18 Jabbari, F., Rajabpour, A., and Saedodin, S. (2017). Thermal conductivity and viscosity of nanofluids: a review of recent molecular dynamics studies. *Chemical Engineering Science* 174: 67–81.

19 Huminic, G., Huminic, A., Dumitrache, F. et al. (2020). Study of the thermal conductivity of hybrid nanofluids: recent research and experimental study. *Powder Technology* 367: 347–357.

20 Sharma, P., Ramesh, K., Parameshwaran, R., and Deshmukh, S.S. (2022). Thermal conductivity prediction of titania-water nanofluid: a case study using different machine learning algorithms. *Case Studies in Thermal Engineering* 30: 101658.

21 Molana, M., Ghasemiasl, R., and Armaghani, T. (2022). A different look at the effect of temperature on the nanofluids thermal conductivity: focus on the experimental-based models. *Journal of Thermal Analysis and Calorimetry* 147 (7): 4553–4577.

22 Żyła, G. and Fal, J. (2017). Viscosity, thermal and electrical conductivity of silicon dioxide–ethylene glycol transparent nanofluids: an experimental studies. *Thermochimica Acta* 650: 106–613.

23 Evans, W., Fish, J., and Keblinski, P. (2006). Role of Brownian motion hydrodynamics on nanofluid thermal conductivity. *Applied Physics Letters* 88 (9): 093116.

24 Singh, J., Kumar, R., Gupta, M., and Kumar, H. (2020). Thermal conductivity analysis of GO-CuO/DW hybrid nanofluid. *Materials Today Proceedings* 28: 1714–1718.

25 Das, S.K., Putra, N., Thiesen, P., and Roetzel, W. (2003). Temperature dependence of thermal conductivity enhancement for nanofluids. *Journal of Heat Transfer* 125 (4): 567–574.

26 Teng, T.P., Hung, Y.H., Teng, T.C. et al. (2010). The effect of alumina/water nanofluid particle size on thermal conductivity. *Applied Thermal Engineering* 30 (14–15): 2213–2218.

27 Borode, A.O., Ahmed, N.A., Olubambi, P.A. et al. (2021). Investigation of the thermal conductivity, viscosity, and thermal performance of graphene nanoplatelet-alumina hybrid nanofluid in a differentially heated cavity. *Frontiers in Energy Research* 9: 737915.

28 Said, Z., Sharma, P., Aslfattahi, N., and Ghodbane, M. (2022). Experimental analysis of novel ionic liquid-MXene hybrid nanofluid's energy storage properties: model-prediction using modern ensemble machine learning methods. *Journal of Energy Storage* 52: 104858.

29 Said, Z., Sharma, P., Bora, B.J., and Pandey, A.K. (2023). Sonication impact on thermal conductivity of f-MWCNT nanofluids using XGBoost and Gaussian process regression. *Journal of the Taiwan Institute of Chemical Engineers* 145: 104818.

30 Said, Z., Sharma, P., Syam Sundar, L. et al. (2021). Synthesis, stability, thermophysical properties and AI approach for predictive modelling of Fe3O4 coated MWCNT hybrid nanofluids. *Journal of Molecular Liquids* 340: 117291.

31 Kanti, P.K., Sharma, K.V., Said, Z. et al. (2022). Experimental investigation on thermal conductivity of fly ash nanofluid and fly ash-Cu hybrid nanofluid:

prediction and optimization via ANN and MGGP model. *Particulate Science and Technology* 40 (2): 182–195.

32 Said, Z., Sharma, P., Elavarasan, R.M. et al. (2022). Exploring the specific heat capacity of water-based hybrid nanofluids for solar energy applications: a comparative evaluation of modern ensemble machine learning techniques. *Journal of Energy Storage* 54: 105230.

33 Bemani, A., Madani, M., and Kazemi, A. (2023). Machine learning-based estimation of nano-lubricants viscosity in different operating conditions. *Fuel* 352: 129102.

34 Sharma, G., Kotia, A., Ghosh, S.K. et al. (2020). Kinematic viscosity prediction of nanolubricants employed in heavy earth moving machinery using machine learning techniques. *International Journal of Precision Engineering and Manufacturing* 21 (10): 1921–1932.

35 Kałużny, J., Świetlicka, A., Wojciechowski, Ł. et al. (2022). Machine learning approach for application-tailored nanolubricants' design. *Nanomaterials* 12 (10): 1765.

36 Jäger, M.O.J., Morooka, E.V., Federici Canova, F. et al. (2018). Machine learning hydrogen adsorption on nanoclusters through structural descriptors. *npj Computational Materials* 4 (1): 37.

37 Maleki, A., Haghighi, A., and Mahariq, I. (2021). Machine learning-based approaches for modeling thermophysical properties of hybrid nanofluids: a comprehensive review. *Journal of Molecular Liquids* 322: 114843.

38 Sharma, P., Said, Z., Kumar, A. et al. (2022). Recent advances in machine learning research for nanofluid-based heat transfer in renewable energy system. *Energy and Fuels* 36 (13): 6626–6658.

39 Gao, Y., Shigidi, I.M.T.A., Ali, M.A. et al. (2023). Thermophysical properties prediction of carbon-based nano-enhanced phase change material's using various machine learning methods. *Journal of the Taiwan Institute of Chemical Engineers* 148: 104662.

40 Zaki, E.S.G., Elsaeed, S.M., and Hassan, H.H. (2023). Artificial intelligence and machine learning of petroleum wastewater treatment by nanofiltration membranes. In: *Wastewater Treatment: Recycling Management, and Valorization of Industrial Solid Wastes* (ed. I.S. Fahim and L. Said). CRS Press.

41 Said, Z., Sharma, P., Sundar, L.S. et al. (2022). Improving the thermal efficiency of a solar flat plate collector using MWCNT-Fe3O4/water hybrid nanofluids and ensemble machine learning. *Case Studies in Thermal Engineering* 40: 102448.

42 Said, Z., Sharma, P., Syam Sundar, L. et al. (2022). Using Bayesian optimization and ensemble boosted regression trees for optimizing thermal performance of solar flat plate collector under thermosyphon condition employing MWCNT-Fe3O4/water hybrid nanofluids. *Sustainable Energy Technologies and Assessments* 53: 102708.

43 Said, Z., Rahman, S., Sharma, P. et al. (2022). Performance characterization of a solar-powered shell and tube heat exchanger utilizing MWCNTs/water-based nanofluids: an experimental, numerical, and artificial intelligence approach. *Applied Thermal Engineering* 212: 118633.

44 Kanti, P.K., Shrivastav, A.P., Sharma, P., and Maiya, M.P. (2023). Thermal performance enhancement of metal hydride reactor for hydrogen storage with graphene oxide nanofluid: model prediction with machine learning. *International Journal of Hydrogen Energy* 52: 470–484.

45 Ahmadi, M.H., Mohseni-Gharyehsafa, B., Ghazvini, M. et al. (2020). Comparing various machine learning approaches in modeling the dynamic viscosity of CuO/water nanofluid. *Journal of Thermal Analysis and Calorimetry* 139 (4): 2585–2599.

46 Ma, T., Guo, Z., Lin, M., and Wang, Q. (2021). Recent trends on nanofluid heat transfer machine learning research applied to renewable energy. *Renewable and Sustainable Energy Reviews* 138 (2020): 110494.

47 Jaliliantabar, F. (2022). Thermal conductivity prediction of nano enhanced phase change materials: a comparative machine learning approach. *Journal of Energy Storage* 46: 103633.

48 de Amorim Neto, J.P., Marinho, F.P., Lima, R.J.P. et al. (2022). Thermal behavior estimation of a solar wall operated by TiO_2 nanofluids using several machine learning models. *Journal of the Brazilian Society of Mechanical Sciences and Engineering* 44 (4): 128.

49 Ramezanizadeh, M., Ahmadi, M.H., Nazari, M.A. et al. (2019). A review on the utilized machine learning approaches for modeling the dynamic viscosity of nanofluids. *Renewable and Sustainable Energy Reviews* 114: 109345.

50 Ranjan Parida, D., Dani, N., and Basu, S. (2021). Data-driven analysis of molten-salt nanofluids for specific heat enhancement using unsupervised machine learning methodologies. *Solar Energy* 227: 447–456.

51 Pandit, P., Dey, P., and Krishnamurthy, K.N. (2021). Comparative assessment of multiple linear regression and fuzzy linear regression models. *SN Computer Science* 2 (2): 76.

52 Shanmugasundar, G., Vanitha, M., Čep, R. et al. (2021). A comparative study of linear, random forest and AdaBoost regressions for modeling non-traditional machining. *Processes* 9 (11): 2015.

53 Pak, S.I. and Oh, T.H. (2010). Correlation and simple linear regression. *Journal of Veterinary Clinics* 27 (4): 427–434.

54 Schmidt, A.F. and Finan, C. (2018). Linear regression and the normality assumption. *Journal of Clinical Epidemiology* 98: 146–151.

55 Wang, Y.A., Huang, Q., Yao, Z., and Zhang, Y. (2024). On a class of linear regression methods. *Journal of Complexity* 82: 101826.

56 Petterle, R.R., Bonat, W.H., Kokonendji, C.C. et al. (2019). Double Poisson-Tweedie regression models. *International Journal of Biostatistics* 15 (1): https://doi.org/10.1515/ijb-2018-0119.

57 Bonat, W.H. and Kokonendji, C.C. (2017). Flexible Tweedie regression models for continuous data. *Journal of Statistical Computation and Simulation* 87 (11): 2138–2152.

58 Abid, R., Kokonendji, C.C., and Masmoudi, A. (2020). Geometric Tweedie regression models for continuous and semicontinuous data with variation phenomenon. *AStA Advances in Statistical Analysis* 104 (1): 33–58.

59 Kokonendji, C.C., Bonat, W.H., and Abid, R. (2021). Tweedie regression models and its geometric sums for (semi-)continuous data. *Wiley Interdisciplinary Reviews: Computational Statistics* 13 (1): e1496.

60 Hassine, A., Masmoudi, A., and Ghribi, A. (2017). Tweedie regression model: a proposed statistical approach for modelling indoor signal path loss. *International Journal of Numerical Modelling: Electronic Networks, Devices and Fields* 30 (6): e2243.

61 Yang, Y., Qian, W., and Zou, H. (2018). Insurance premium prediction via gradient tree-boosted Tweedie compound Poisson models. *Journal of Business and Economic Statistics* 36 (3): 456–470.

62 Thorson, J.T., Arimitsu, M.L., Levi, T., and Roffler, G.H. (2022). Diet analysis using generalized linear models derived from foraging processes using R package mvtweedie. *Ecology* 103 (5): e3637.

63 Feng, Y. and Wu, Q. (2022). A statistical learning assessment of Huber regression. *Journal of Approximation Theory* 273: 105660.

64 Fan, J., Guo, Y., and Jiang, B. (2022). Adaptive Huber regression on Markov-dependent data. *Stochastic Processess Their Applications* 150: 802–818.

65 Wang, L., Zheng, C., Zhou, W., and Zhou, W.-X. (2021). A new principle for tuning-free Huber regression. *Statistica Sinica* 31: 2153–2177.

66 Wang, Y., Wang, B., Peng, C. et al. (2022). Huber regression analysis with a semi-supervised method. *Mathematics* 10 (20): 3734.

67 Balasundaram, S., & Prasad, S. C. (2020). Robust twin support vector regression based on Huber loss function. *Neural Computing and Applications*, 32(15). https://doi.org/10.1007/s00521-019-04625-8

68 Sun, Q., Zhou, W.X., and Fan, J. (2020). Adaptive Huber regression. *Journal of the American Statistical Association* 115 (529): 254–265.

69 Zhou, L., Cook, R.D., and Zou, H. (2023). Enveloped Huber regression. *Journal of the American Statistical Association* 199: 2722–2732.

70 Luo, J., Sun, Q., and Zhou, W.X. (2022). Distributed adaptive Huber regression. *Computational Statistics and Data Analysis* 169: 107419.

71 Bae, D.J., Kwon, B.S., and Song, K.B. (2022). XGboost-based day-ahead load forecasting algorithm considering behind-the-meter solar PV generation. *Energies (Basel)* 15 (1): 128.

72 Chen, T., and Guestrin, C.2016). XGBoost: a scalable tree boosting system. *Proceedings of the ACM SIGKDD International Conference on Knowledge Discovery and Data Mining*, (13–17 August 2016). 785–794.

73 Dong, J., Chen, Y., Yao, B. et al. (2022). A neural network boosting regression model based on XGBoost. *Applied Soft Computing* 125: 109067.

74 Phan, Q.T., Wu, Y.K., and Phan, Q.D. (2021). A hybrid wind power forecasting model with xgboost, data preprocessing considering different nwps. *Applied Sciences (Switzerland)* 11 (3): 1100.

75 Li, X., Ma, L., Chen, P. et al. (2022). Probabilistic solar irradiance forecasting based on XGBoost. *Energy Reports* 8: 1085–1095.

76 Sagi, O. and Rokach, L. (2021). Approximating XGBoost with an interpretable decision tree. *Information sciences (New York)* 572: 522–542.

77 Noh, J., Na, T., and Kang, S.S. (2023). Modeling solar irradiance in Tajikistan with XGBoost algorithm. *Journal of Engineering Geology* 33 (3): 403–411.

78 Zhang, P., Jia, Y., and Shang, Y. (2022). Research and application of XGBoost in imbalanced data. *International Journal of Distributed Sensor Networks* 18 (6): https://doi.org/10.1177/15501329221106935.

79 Chen, T., and Guestrin, C.2016). XGBoost: reliable large-scale tree boosting system. ArXiv.

80 Subudhi, R.N. (2021). Digital consumption pattern and impacts of social media: descriptive statistical analysis. In: *Studies in Computational Intelligence* (ed. S.S. Rautaray, P. Pemmaraju, and H. Mohanty), 954. Springer.

81 Wang, Y., Shao, X., Liu, C. et al. (2019). Analysis of wind farm output characteristics based on descriptive statistical analysis and envelope domain. *Energy* 170: 580–591.

82 Bengnga, A. and Ishak, R. (2022). Implementasi seleksi fitur Klasifikasi waktu kelulusan mahasiswa menggunakan correlation matrix with heatmap. *Jambura Journal of Electrical and Electronics Engineering* 4 (2): 169–174.

83 Shin, J., Yim, I., Kwon, S.B. et al. (2019). Evaluation of temperature effects on brake wear particles using clustered heatmaps. *Environmental Engineering Research* 24 (4): 680–689.

84 Wickstrøm, K., Johnson, J.E., Løkse, S., Camps-Valls, G., Mikalsen, K.Ø., Kampffmeyer, M., and Jenssen, R.2022). The kernelized Taylor diagram. Communications in Computer and Information Science, 1650 CCIS.

85 Zhou, Q., Chen, D., Hu, Z., and Chen, X. (2021). Decompositions of Taylor diagram and DISO performance criteria. *International Journal of Climatology* 41 (12): 5726–5732.

86 Simão, M.L., Videiro, P.M., Silva, P.B.A. et al. (2020). Application of Taylor diagram in the evaluation of joint environmental distributions' performances. *Marine Systems and Ocean Technology* 15 (3): 151–159.

87 Taylor, K.E. (2001). Summarizing multiple aspects of model performance in a single diagram. *Journal of Geophysical Research* 106 (D7): 7183–7192.

11

Optimization of Nanofluids for Heat Exchangers: Dealing with Sedimentation and Pump Losses

Nikhil S. Mane[1], *Sayantan Mukherjee*[2], *Redhwan Almuzaiqer*[3], *and Niteen Bhirud*[1]

[1] *School of Engineering, Ajeenkya DY Patil University, Pune, India*
[2] *Mechanical Engineering Department, Gandhi Academy of Technology and Engineering, Brahmapur, India*
[3] *Mechanical Engineering Department, College of Engineering, King Saud University, Riyadh, Saudi Arabia*

11.1 Introduction

Nanofluids are the colloidal suspension of nanoscale materials (1–100 nm) in a fluid. More than three decades have passed since their first use [1], and they have proved to be very effective in thermal applications. They have emerged as alternative fluids in thermal devices like heat exchangers and heat pipes due to their better thermal properties. Heat exchangers are devices that transfer heat from one fluid to another and are used in applications like process industries, large-scale electronics cooling, air conditioning, solar energy, automobiles, and building cooling or heating applications. Heat exchangers are categorized based on shapes, sizes, and flow directions; each type is used for its advantages in various applications. Advancement in nanofluid technology led to its adaptation in heat exchangers where nanofluids are used due to their high thermal performance compared to their base fluids. The emergence of high heat emissions in electronics, automobiles, and other devices pushed the use of nanofluids. Nanofluids not only increase the heat transfer rate but also reduce the fluid volume required, reduce the heat exchanger size, and finally reduce operational costs [2].

The use of nanofluids significantly improves convective heat transfer; this convective phenomenon is investigated by researchers [3]. However, experimental investigations are also focused on nanofluid characterization to investigate thermal conductivity, specific heat, viscosity, and electrical conductivity [4]. These investigations aimed at optimizing the thermal performance of nanofluids in

Machine Learning for Sustainable Energy Solutions, First Edition. Edited by Zafar Said and Prabhakar Sharma.

thermal applications. In thermal applications such as heat exchangers, merely investigating properties is not sufficient for the adaptation of nanofluids. Nanofluid performance in delivering a higher convective heat transfer is a complex phenomenon that is dependent on the cumulative effect of multiple properties of nanofluids. The complexities of increase in pumping power, coagulation sizes, and sedimentation severely affect the overall thermal behavior and heat transfer in nanofluids. Additionally, for commercial usage and industrial adaptation of nanofluids, shelf life, thermo-economic behavior, and deterioration of properties are also important. These challenges in practical implementation hamper the wide-scale adaptation of the nanofluids in heat exchangers.

Sedimentation and pump losses are the main technical issues for the nanofluids from the perspective of heat exchanger application. No matter how well nanoparticles are dispersed in the fluid, they inevitably tend to agglomerate due to the attractive forces (Van der Waals forces) between particles. This agglomeration then led to the phenomenon known as sedimentation, where nanoparticles settle down in the fluid and make a thick layer of nanoparticles at the bottom [5]. This sedimentation not only reduces the heat transfer efficiency but also leads to practical issues like clogging, corrosion of pipes, and increased pressure in heat exchanger tubes. This sedimentation leads to increased pump losses, which are further worsened by factors such as the higher viscosity of nanofluids. This leads to increased energy consumption, higher maintenance, and reduced flow rates, which qualitatively and quantitatively affect the heat exchanger performance with nanofluids. For these reasons, nanofluids should be optimized before being used in heat exchanges. For industrial applications, financial aspects are equally important along with technical aspects. Hence, it is also essential to conduct a thermo-economic feasibility analysis of nanofluids. The practical implementation of nanofluids depends not only on their enhanced thermal performance but also on the associated costs. The industrial or commercial point of view always seeks the balance between the improved heat transfer capabilities of nanofluids against their production, maintenance, and operational costs. Hence, the overall economic viability of nanofluids is necessary to investigate for each application, which ensures that the use of nanofluids is justifiable through both technical and financial aspects.

In view of these issues in the implantation of nanofluids in heat exchangers, this work discusses the strategies to optimize the nanofluids for heat exchangers. This work focuses on the sedimentation and pump losses along with the thermo-economic feasibility of nanofluids. This work can contribute to the understanding of researchers and engineers regarding the characteristics of nanofluids, which can provide optimal performance in heat exchanger applications.

11.2 Sedimentation

Agglomeration, sedimentation, and long-term stability are interconnected phenomena that determine the dispersion quality and effectiveness of nanofluids. Agglomeration causes the of formation aggregates of smaller dispersed particles, which increases their size and mass. In well-dispersed nanofluids, nanoparticles are in random motion, and they remain in this Brownian motion as long as this motion overcomes gravity. Due to agglomeration, the mass and size of particles increase, and Brownian motion tends to decrease. When this Brownian motion is reduced to terminal settling velocity, particles tend to settle in the fluid due to gravity. The Brownian motion and terminal settling velocity are dependent on nanoparticle size, as settling velocity decreases and Brownian motion increases with particle size [6]. Hence, to achieve better stability, a small size of nanoparticles is desired, and agglomeration should be reduced to provide good dispersion.

In practical applications, the long-term shelf life of the nanofluids is an important property, as it facilitates the storage and transportation of the nanofluids after preparation [7]. Nanoparticles present in the nanofluids tend to settle down during their shelf life over a period, and this phenomenon occurs in four stages. Figure 11.1 shows the typical long-term behavior of nanofluids, where the shelf

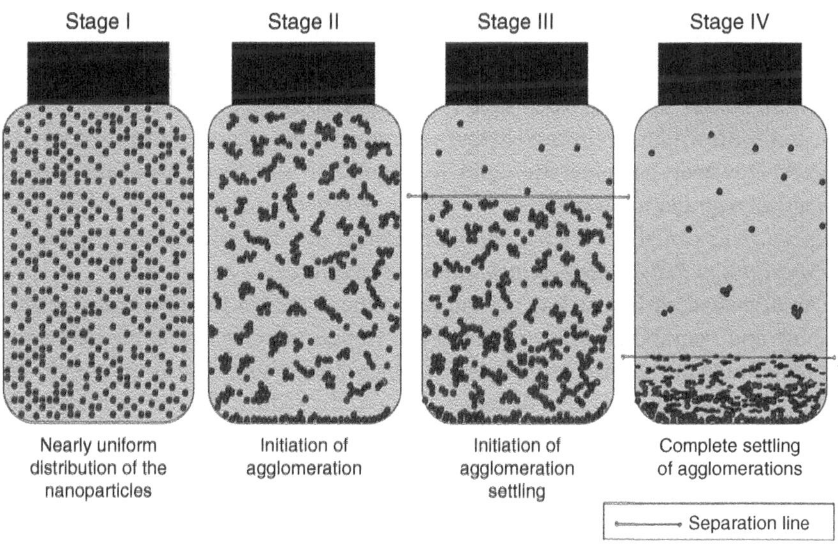

Figure 11.1 Nanoparticle behavior with the aging of nanofluids.

life of nanofluids is depicted. Stage I shows the freshly prepared nanofluids where nanoparticles are uniformly distributed. As time passes, the nanoparticles overcome intermolecular forces and form a weak bond with other particles, which results in agglomeration. In the initial phase of agglomeration, more and more particles come in contact with each other, forming larger and larger agglomerates. But in this stage, the agglomerates are not large enough to initiate the settling of the particles. Once the agglomerate size sufficiently increases, the third stage of nanofluid life initiates, in which large-sized nanoparticles tend to settle in the nanofluids.

This third stage consists of a progressing settling of the nanoparticles, where a clear separation line may be observed below which all the heavy clusters of nanoparticles are present, and above which a negligible concentration of nanoparticles is observed. The height of the separation line continuously decreases as time progresses. The presence and progression of the separation line highly depend upon the surfactant used. In low-stability nanofluid surfactants, this separation line does not exist, and nanofluids completely settle down abruptly. It is also observed by Mane et al. [8] that the zeta potential value and hydrodynamic diameter of freshly prepared nanofluids have very little influence on the long-term behavior of nanofluids, as nanofluids having similar zeta potential value and hydrodynamic diameter can have different long-term behavior. This results from the use of different dispersants, the pH of nanofluids, and environmental conditions.

This aging process of nanofluids reduces their effectiveness at the end of their shelf life. The aging mechanism of nanofluids differs between shelf conditions and real applications, as continuous mixing in applications enhances particle movement. In practical use, nanofluids experience constant mixing, which increases Brownian motion and helps break down weak agglomerates due to shear forces. However, agglomeration and sedimentation still occur over time due to chemical degradation because of the mixing of different chemicals or oxides during use. This can affect the nanoparticle interactions and affect overall stability.

Strategies to Reduce the Sedimentation in Nanofluids

As discussed earlier, sedimentation can critically affect the heat transfer efficiency and operations of heat exchangers. Hence, for effective implementation of nanofluids in the heat exchangers, the nanofluid synthesis process should be optimized for the reduction of sedimentation. To reduce the sedimentation in nanofluids, researchers are using techniques like nanoparticle surface modifications, the use of dispersants, pH control, and continuous mixing of nanofluids. These strategies are summarized in Figure 11.2.

One of the interesting methods to reduce sedimentation is nanoparticle surface modification. The nanoparticle surface modification process involves altering the surface characteristics of the nanoparticle, aiming to enhance the particular characteristics of particles. Surface modification methods include chemical

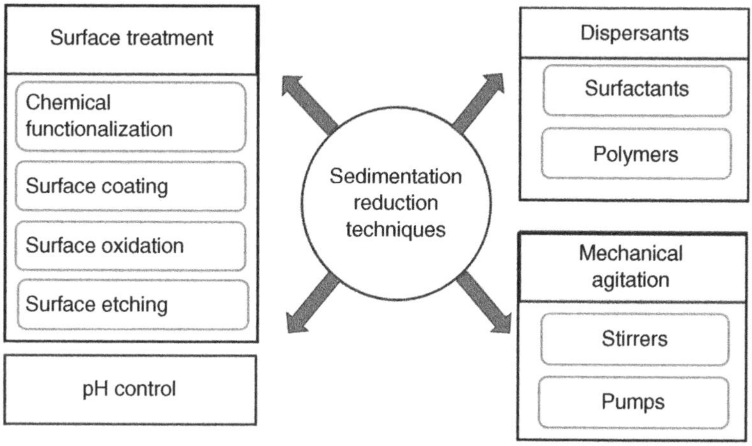

Figure 11.2 Summary of strategies to reduce the sedimentation in nanofluids.

functionalization, surface coating, surface oxidation, and surface etching, among others [9]. Liu et al. [10] found that surface functionalization of Fe_3O_4 nanoparticles increases steric and electrostatic repulsion. This repulsion provides resistance to agglomeration, which can lead to a reduction in sedimentation. Zhu et al. [11] highlight the role of surface modification in enhancing the stability and functionality of iron oxide nanoparticles in nanofluids. In this work, they coated nanoparticles with organic molecules, polymers, and inorganic materials. They also commented that surface modifications effectively enhance the initial dispersion of the nanofluids. Surface modifications enhance the stability and reduce the sedimentation by preventing agglomerations in the nanofluid.

The use of dispersant additives like surfactants and polymers also provides effective measures to reduce agglomeration and clustering. In general, these dispersant molecules cover the surface of nanofluids, which reduces the adhesion among nanoparticles. The effectiveness of this surfactant is also dependent on the concentration of nanoparticles, as there is a requirement for a sufficient amount of surfactant molecules to form a structure known as a micelle. These micelles play a crucial role in stabilizing nanofluids by improving dispersion, reducing agglomeration, and enhancing solubility. At low surfactant concentrations, very few molecules of surfactants (monomers) are present in the fluids, which are not enough to form micelles or provide good surface coverage [12]. Due to this, Van der Waals forces act between poorly covered nanoparticle surfaces and agglomerate forms. When enough surfactant is added to the fluid at critical micelle concentration, these monomers of surfactants form micelles and provide excellent surface coverage to the nanoparticles. This avoids agglomeration and reduces the hydrodynamic diameter of the nanoparticles. When the surfactant is added further, the micelle formation continues. The sizes of micelles increase by forming multiple layers of surfactant monomers on the nanoparticle surface. This forms a very thin layer of surfactant around nanoparticles, which leads to contact and interaction between surfactant molecules on the outer layers of adjacent nanofluids. This interaction leads to the formation of agglomerates of the nanoparticles covered with micelles, as shown in Figure 11.3 [13]. These surfactants effectively improve the nanofluid stability and reduce sedimentation. Mane et al. [14] used CTAB and SDS surfactants for Fe_3O_4 + CuO hybrid nanofluids. They observed excellent stability indicators for the freshly prepared nanofluids.

Along with surfactants, polymers are also used as dispersants in nanofluids. The mechanism of polymers is slightly different compared to surfactants. When a polymer dispersant is used, the polymer molecules adsorb on the surface of the nanoparticles and form a protective layer. The protective layer avoids contact between nanoparticles and leads to a reduction in agglomerations. Morsi et al. [15] used polymers to stabilize the Y_2O_3 nanofluids (0.05–0.1 g). They commented that polymers can deliver controlled aggregation or improved surface interactions. Hamdi et al. [16] used a natural polymer called gum Arabica in graphene water nanofluids. It is observed that the addition of gum arabica improves the colloidal stability of particles and dispersion. The use of dispersant additives is an effective technique for improving nanoparticle dispersion, achieved through steric stabilization by polymers and micelle formation by surfactants.

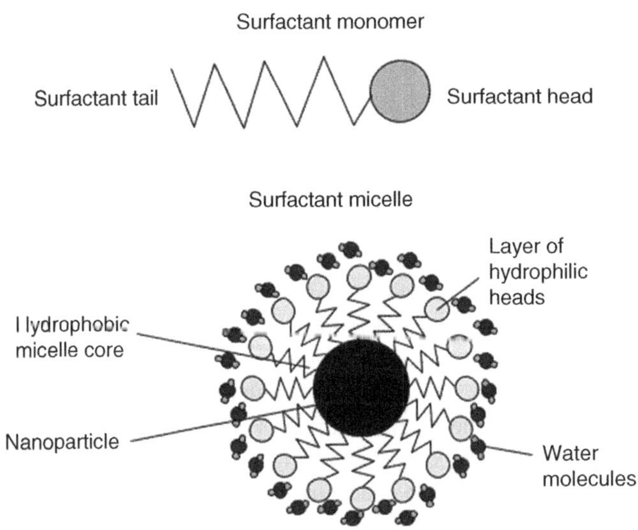

Figure 11.3 Interaction of surfactant molecules with nanoparticles and water molecules.

pH control is also an effective and simple method for improving the dispersion of nanofluids. In this method, pH is maintained such that electrostatic repulsion between particles is maintained. This will improve Brownian motion, and hence, dispersion improves in nanofluids. Improving pH can lead to surface charge enhancement, zeta potential optimization, and isoelectric point (IEP) avoidance. These factors are crucial for nanoparticles as they prevent agglomeration, enhance dispersion stability, and improve overall nanofluid performance. pH is maintained in either the acidic (<4) or basic (>10) range to increase the surface charge of particles so that electrostatic repulsion is strengthened, reducing particle aggregation, and enhancing Brownian motion, which ultimately improves the stability and thermal properties of nanofluids. Demirpolat et al. [17] investigated the effect of pH on the viscosity of CuO nanofluids. They found that electrostatic interactions increased with pH. Zareei et al. [18] also explored the effect of pH on the stability and dispersion in the nanofluids. Results showed that in the acidic range (pH4), nanoparticles have smaller particle sizes. It shows reduced agglomeration at lower pH values. In contrast, at pH 10, which is in the alkaline range, larger particle sizes are detected, suggesting increased agglomeration.

All the aforementioned strategies for reducing nanoparticle sedimentation are chemically driven and implemented during the synthesis of nanofluids. However, the active mixing of nanofluids during their application can also play a significant role in minimizing sedimentation. The active mixing of fluids generates shear force; this shear force is sufficient to break weak agglomerates formed during the early stages of nanofluid aging. As these agglomerates continuously form and disperse, a more uniform distribution of nanoparticles is maintained within the heat exchange. This active mixing of nanofluids is achieved through mechanical agitation. Mechanical agitation can be achieved through stirrers, vibrations, and pumps. Using pumps, nanofluids are kept in circulation, which introduces turbulence and shear forces in the fluid; the turbulence keeps fluids and nanoparticles in motion, which disrupts the sedimentation process. In some cases, sedimented nanoparticles can also be recirculated in the fluid through turbulence forces. Alami et al. [19] used a rotating magnetic field to provide agitation for mixing Fe–Cu water-based nanofluids. They found that magnetic stirring improves nanoparticle dispersion and heat transfer efficiency. Mokhefi et al. [20] studied the effect of a stirred tank with an anchor stirrer on the thermal behavior of an Al_2O_3water nanofluid. This study showed that increasing the amplitude of the wavy wall and nanoparticle concentration enhances heat transfer. The wavy wall and stirrer significantly affect the thermal performance of nanofluids. Shandal et al. [21] achieved heat transfer enhancement through multiwalled carbon nanotubes (MWCNTs)water nanofluids using a coiled agitated vessel. They observed that the use of a propeller agitator significantly influenced the thermal conductivity of MWCNT nanofluids. These studies showed that mechanical agitation significantly improved the thermal properties and dispersion of nanofluids. These techniques improve the dispersion, and hence, they have considerably affected the sedimentation, improving the adaptability of nanofluids for heat exchanger applications. This section discusses the strategies to optimize nanofluids for lower rates of sedimentation. The particle surface modification, dispersants, and mechanical agitation play crucial roles in enhancing the dispersion, reducing sedimentation, and improving the thermal performance of nanofluids. A combination of these techniques is essential for developing high-performance nanofluids for heat exchanger applications.

11.3 Pump Losses

As discussed earlier, the properties and flow characteristics of nanofluids are significantly different compared to those of base fluids. Hence, the use of nanofluids in heat exchangers influences pump losses due to increased fluid viscosity and flow behavior. Hence, to gain a collective understanding of the nanofluid integration with heat exchangers, pump losses, and improvement in thermal performance should be analyzed. The use of nanoparticles leads to an enhancement in the viscosity of the base fluid; this increased viscosity increases the pump work. While comparing the energy efficiency of base fluids and nanofluids in heat exchanger applications, the additional pump work required for nanofluids impacts the overall performance of heat exchangers. Similarly, the increased thermal gains due to the thermal conductivity of nanofluids can also vanish due to this pump loss. Prasher et al. [22] established a criterion for nanofluids to trade-off between thermal conductivity and viscosity. They found that if the increase in viscosity exceeds four times the increase in thermal conductivity, then heat transfer enhancement by the thermal conductivity enhancement will be nullified due to the viscosity. The property enhancement ratio can also be a valuable tool for addressing this issue.

$$\text{Property Enhancement Ratio} = \frac{\left[\frac{\mu_{nf}}{\mu_{bf}}\right] - 1}{\left[\frac{k_{nf}}{k_{bf}}\right] - 1}$$

Garg et al. [23] proposed that for overall heat transfer gain with any nanofluid, the property enhancement ratio should not be more than 5. Hamid et al. [24] used the property enhancement ratio to optimize nanofluids to address the trade-off between improved thermal conductivity and the increased viscosity that leads to pump losses. They found that only a few proportions of nanoparticles in TiO_2–SiO_2 nanofluids deliver higher PER than 5. Similarly, Mane et al. [25]

identified based on the property enhancement ratio that polymer dispersant concentration should be reduced as it does not affect stability but increases the viscosity of nanofluids. It shows that the property enhancement ratio can be effectively used to optimize nanoparticle concentrations, surfactant concentrations, and other preparation parameters. Routbort et al. [26] focused on evaluating the pumping power requirements of Al_2O_3–water nanofluids in a system that mimics real-world heat exchanger conditions. In this work, they studied the behavior of nanofluid flow in heat exchanger sections like straight tubing, elbows, and expansions. In this study, nanofluids were treated as single-phase fluids, and it was found that the addition of nanoparticles did not alter their flow behavior to that of a two-phase system. They also found that the change in pumping power required does not depend on flow rate and heat exchanger complexity. They also emphasized that a balance between added pumping cost and improved heat transfer efficiency should be maintained for commercial viability. Corcione et al. [27] theoretically studied the inter-relational behavior of nanofluids, pumping power, and heat transfer abilities. Using algebraic equations developed for aqueous nanofluids, they concluded that pumping power can be minimized by optimizing nanoparticle concentration in the fluids. Hence, it can be collectively interpreted that by maintaining nanoparticle concentrations, viscosity can be maintained. This ensures that pumping power remains below the threshold at which the benefits of enhanced thermal properties would be offset by increased energy consumption. This section highlights the need for a reduction in the viscosity of nanofluids as it increases pump work and leads to higher energy consumption. The optimization tools like the property enhancement ratio, need to be used by researchers so that preparation parameters like nanoparticle concentrations, stability techniques, and flow properties can be adjusted to provide a balance between thermal performance and energy efficiency in practical applications.

11.4 Thermo-Economic Aspect of the Nanofluids

Economic aspects have always been at the center of the industrial implementation of new technologies. In the case of nanofluids, the overwhelming majority of the research studies have shown that the nanofluids possess superior thermal properties compared to their base fluids and, hence, can be beneficially used for cooling and heating applications. However, engineers always need to prove nanofluid feasibility in terms of the financial gains these fluids can offer for industrial applications. Hence, improvement in the thermal efficiency of heat exchangers and the cost associated with it should be compared to investigate the feasibility of nanofluids. Nanofluids are costly to make, store, and transport compared to conventional cooling fluids; hence, additional costs associated with them should be considered against the financial gains offered by them due to improved heat transfer [28].

Mukherjee et al. [29] presented an economic performance index based on the unit cost of nanoparticles and base fluids. In this index, other costs like preparation costs and surfactant costs are ignored as they are considered to be very small. Based on Mukherjee's model, the price performance index (PPI) can be calculated as

$$\mathrm{PPI} = \frac{\mathrm{PER}}{C_n} \times 100$$

where C_n is the cost per gram of nanoparticles, and PER is the property enhancement ratio of nanofluids. In this study, the cost of nanofluids is calculated using the cost of nanoparticles in US dollars. In this study, the cost of nanofluids is calculated using the cost of nanoparticles, surfactants, and water as per the proportions of the nanofluids in US dollars. The PPI of the nanofluids provides an effective measure of the cost associated with improvement in the thermal conductivity and viscosity of the nanofluid. Better economic justification of the nanofluid can be proven with a lower PPI value. According to this model, the low economic performance of nanofluids can result from the high viscosity or low thermal conductivity of nanofluids. Though this method is reliable, it only considers the price of the nanoparticles, and hence, other costs associated with nanofluids should also be considered for more effective analysis. Hence, Alirezaie et al. [30] presented the price-performance factor for thermo-economic analysis of nanofluids.

$$\mathrm{PPF} = \frac{\mathrm{TCR}}{\mathrm{Price\ (USD/lit)}} \times 1000$$

Here, TCR is the thermal conductivity ratio, and Price (USD/lit) is the total cost of nanofluids per liter. They commented that a complex preparation process and higher cost of metal nanofluids reduce the cost-effectiveness. Similarly, various researchers have conducted research to investigate the thermo-economic feasibility of nanofluids. Prajapati and Patel [31] conducted a study to evaluate the thermo-economic feasibility of the CuO–water nanofluid waste heat recovery process. They used CuO–water nanofluids in the evaporator and condenser. They found that 19.3% thermal efficiency with a levelized energy cost of 2.616 $/kWh can be achieved in this work. Yousefi et al. [32] focused on the thermo-economic optimization of an Al_2O_3–water nanofluid in a heat recovery system for a biomass plant. They used the particle swarm optimization technique and considered the investment and operational costs of the heat exchanger along with nanoparticle characteristics. In this work, they commented that for effective implantation of nanofluids, different nanoparticle choices and production costs should be considered. Xuan et al. [33] investigated the thermo-economic performance of

Al_2O_3–TiO_2–Cu/water ternary hybrid nanofluids. Their work is focused on the thermal conductivity, viscosity, and economic feasibility of nanofluids. In this work, ternary nanofluids demonstrated remarkable thermal behavior compared to single and binary nanofluids. They commented that ternary nanofluids are thermo-economically superior choices for heat exchanger applications. These studies have shown that while nanofluids improve heat transfer efficiency, their adaptability in industrial applications depends on optimizing nanoparticle composition, production costs, and operational costs. PPI and thermo-economic analyses provide valuable insights into cost-effectiveness, and hence these tools can improve the adaptability of nanofluids. The key evaluation metrics for nanofluid feasibility discussed in this study are summarized in Table 11.1.

It is observed from Table 11.1 that the careful balance of the thermal conductivity, viscosity, and associated cost of nanofluids can be represented in the form of different evaluation matrices. These matrices act as a valuable tool for the assessment of the nanofluid feasibility for particular applications. These matrices can also provide a realistic understanding of nanofluids by providing balanced perspectives from technical and economic aspects. However, to obtain an optimized nanofluid preparation from a thermo-economic perspective, a section of the thermo-economic matrix is essential and should be studied based on the specific application. The selection of a thermo-economic matrix should consider the performance required, the type of flow (turbulent/laminar), and material costs. It can then effectively facilitate the effective integration of nanofluids in heat exchanger applications.

11.5 Conclusion

In this work, the need and strategies for the optimization of nanofluids for heat exchanger applications are discussed. This work highlights the main issues associated with the implementation of nanofluids in heat exchangers like sedimentation, pump losses, and thermo-economic feasibility of nanofluids. The impact of sedimentation on the operations of the heat exchangers is discussed in this work. It is observed that the sedimentation in nanofluid can severely impact the heat transfer efficiency, cause blockages, and reduce the flow rate. Hence, the challenges presented by sedimentation in nanofluids can be dealt with through nanoparticle concentration optimization, surface treatment, dispersant, pH control, and flow mixing. The optimization of these factors can improve the industrial-scale usage of nanofluids in heat exchangers. The increase in pump losses introduced by viscosity poses issues like high fractional losses, pump power, and hence energy consumption. Hence, this work discusses the optimization of nanofluids

Table 11.1 Summary of key economic evaluation metrics for nanofluid feasibility discussed in this study.

Metric	Definition	Formula	Key insights
Performance ratio	The ratio of heat transfer enhancement achieved by nanofluids compared to the base fluids.	—	Show improvement in thermal performance of the heat exchanger because of nanofluids.
Pumping power ratio	The ratio of energy consumed by a pump with nanofluids compared to base fluids.	—	It indicates the energy loss caused by increased viscosity.
Property enhancement ratio (PER)	The ratio of heat transfer enhancement to increase in pumping power (viscosity) due to nanofluids.	$\text{PER} = \dfrac{\left[\dfrac{\mu_{\text{nf}}}{\mu_{bf}}\right] - 1}{\left[\dfrac{k_{\text{nf}}}{k_{\text{bf}}}\right] - 1}$	Provides a comparative understanding of thermal improvement provided by thermal conductivity and flow resistance offered by viscosity.
Price performance index (PPI)	The ratio of economic performance based on thermal property enhancement (PER) and nanoparticle cost.	$\text{PPI} = \dfrac{\text{PER}}{C_n} \times 100$	A lower PPI shows that the thermal gains achieved with nanofluids are justifiable from a cost/economic point of view. As this index only considers the price of nanoparticles, it does not consider other associated costs of nanofluids.
Price-performance factor (PPF)	It evaluates thermal conductivity gain per dollar of nanofluid cost.	$\text{PPF} = \dfrac{\text{TCR}}{\text{Price (USD/lit)}} \times 1000$	This index provides an evaluation of thermal conductivity justification with the overall preparation cost.
Net cost–benefit analysis	Compares thermal enhancement achieved by nanofluids against added costs (fluid, power, maintenance, etc.).	—	Useful for real-world industrial decisions for considering the adaptability of nanofluids.
Levelized energy cost	It is the cost per unit energy output/saving using nanofluids in a system.	\$/kWh	It can be used to justify the energy-saving costs.

using the property enhancement ratio, which can be an effective tool to adjust preparation parameters to deliver nanofluids that can cause low pump losses. As the cost of the nanofluids has been a concern for delivering cost-effective heat exchange devices, this work provides the need and method to provide thermo-economic optimization. Using the PPI and thermo-economic analyses, the thermo-economic feasibility of nanofluids can be judged for each application prior to its integration. This work highlights issues like sedimentation, pump losses, and thermo-economic feasibility of nanofluids and also provides a comprehensive approach to optimizing nanofluids for heat exchanger applications. The insights presented can help in developing cost-effective and efficient nanofluid-based heat exchange systems for industrial use.

References

1 Choi SUS. (1995). Enhancing thermal conductivity of fluids with nanoparticles. Argonne National Laboratory. *ASMEFED231*, 99–103. November 12–17, 1995.

2 Hajatzadeh Pordanjani, A., Aghakhani, S., Afrand, M. et al. (2019). An updated review on application of nanofluids in heat exchangers for saving energy. *Energy Conversion and Management* 198: 111886. https://doi.org/10.1016/j.enconman.2019.111886.

3 Pak, B.C. and Cho, Y.I. (1998). Hydrodynamic and heat transfer study of dispersed fluids with submicron metallic oxide particles. *Experimental Heat Transfer* 11 (2): 151–170. https://doi.org/10.1080/08916159808946559.

4 Mane, N.S., Hemadri, V., and Tripathi, S. (2023). Exploring the role of biopolymers and surfactants on the electrical conductivity of water-based CuO, $Fe3O_4$, and hybrid nanofluids. *Journal of Dispersion Science and Technology* 45 (5): 900–908. https://doi.org/10.1080/01932691.2023.2186428.

5 Ilyas, S.U., Pendyala, R., and Marneni, N. (2014). Preparation, sedimentation, and agglomeration of nanofluids. *Chemical Engineering and Technology* 37 (12): 2011–2021. https://doi.org/10.1002/ceat.201400268.

6 Savithiri, S., Pattamatta, A., and Das, S.K. (2011). Scaling analysis for the investigation of slip mechanisms in nanofluids. *Nanoscale Research Letters* 6: 471. https://doi.org/10.1186/1556-276X-6-471.

7 Anushree, C. and Philip, J. (2016). Assessment of long-term stability of aqueous nanofluids using different experimental techniques. *Journal of Molecular Liquids* 222: 350–358. https://doi.org/10.1016/j.molliq.2016.07.051.

8 Mane, N.S. and Hemadri, V. (2022). Experimental investigation of stability, properties and thermo-rheological behaviour of water-based hybrid CuO and Fe_3O_4 nanofluids. *International Journal of Thermophysics* 43 (7): https://doi.org/10.1007/s10765-021-02938-2.

9 Khoramian, R., Issakhov, M., Pourafshary, P. et al. (2024). Surface modification of nanoparticles for enhanced applicability of nanofluids in harsh reservoir conditions: a comprehensive review for improved oil recovery. *Advances in Colloid and Interface Science* 333: 103296. https://doi.org/10.1016/j.cis.2024.103296.

10 Liu, S., Yu, B., Wang, S. et al. (2020). Preparation, surface functionalization and application of Fe_3O_4 magnetic nanoparticles. *Advances in Colloid and Interface Science* 281: 102165. https://doi.org/10.1016/j.cis.2020.102165.

11 Zhu, N., Ji, H., Yu, P. et al. (2018). Surface modification of magnetic iron oxide nanoparticles. *Nanomaterials* 8 (10): 810. https://doi.org/10.3390/nano8100810.

12 Ahmed, F.B., Khaled, A.E., Syed, M. et al. (2020). The effect of surfactant concentration, salinity, temperature, and pH on surfactant adsorption for chemical enhanced oil recovery: a review. *Journal of Petroleum Exploration and Production Technology* 10: 125–137.

13 Rastogi, R., Kaushal, R., Tripathi, S.K. et al. (2008). Comparative study of carbon nanotube dispersion using surfactants. *Journal of Colloid and Interface Science* 328: 421–427.

14 Mane, N.S. and Hemadri, V. (2021). Study of the effect of preparation parameters on thermal conductivity of metal oxide nanofluids using Taguchi method. *Journal of Energy Systems* 5 (2): 149–164.

15 Morsi, R.E. and El-Salamony, R.A. (2020). Effect of cationic, anionic and non-ionic polymeric surfactants on the stability, photo-catalytic and antimicrobial activities of yttrium oxide nanofluids. *Journal of Molecular Liquids* 297: 111848. https://doi.org/10.1016/j.molliq.2019.111848.

16 Hamdi, S.S., Al-Kayiem, H.H., Alsabah, M.S., and Muhsan, A.S. (2022). A comparative study of dispersed and grafted nanofluids of graphene nanoplatelets with natural polymer in high salinity brine for enhanced oil recovery. *Journal of Petroleum Science and Engineering* 210: 110004. https://doi.org/10.1016/j.petrol.2021.110004.

17 Demirpolat, A.B. and Das, M. (2019). Prediction of viscosity values of nanofluids at different pH values by alternating decision tree and multilayer perceptron methods. *Applied Sciences* 9 (7): 1288. https://doi.org/10.3390/app9071288.

18 Zareei, M., Yoozbashizadeh, H., and Madaah Hosseini, H.R. (2019). Investigating the effects of pH, surfactant and ionic strength on the stability of alumina/water nanofluids using DLVO theory. *Journal of Thermal Analysis and Calorimetry* 135: 1185–1196. https://doi.org/10.1007/s10973-018-7620-1.

19 Alami, A.H., Abu Hawili, A., Aokal, K. et al. (2020). Enhanced heat transfer in agitated vessels by alternating magnetic field stirring of aqueous Fe–Cu nanofluid. *Case Studies in Thermal Engineering* 20: 100640. https://doi.org/10.1016/j.csite.2020.100640.

20 Mokhefi, A., Bouanini, M., Elmir, M., and Spitéri, P. (2021). Effect of the wavy tank wall on the characteristics of mechanical agitation in the presence of an Al_2O_3-water

nanofluid. *Metallurgical and Materials Engineering* 27 (3): https://doi.org/10.63278/mme.v27i3.626.

21 Shandal, B.C., Dawood, H.I., and Onyeaka, H. (2023). Experimental study of heat transfer using nanofluids in a coiled agitated vessel equipped with impeller propeller. *Al-Qadisiyah Journal for Engineering Sciences* 16 (4): 280–291. https://doi.org/10.30772/qjes.2023.145214.1062.

22 Prasher, R., Song, D., Wang, J., and Phelan, P. (2006). Measurements of nanofluid viscosity and its implications for thermal applications. *Applied Physics Letters* 89: 133108. https://doi.org/10.1063/1.2356113.

23 Garg, J., Poudel, B., Chiesa, M. et al. (2008). Enhanced thermal conductivity and viscosity of copper nanoparticles in ethylene glycol nanofluid. *Journal of Applied Physics* 103 (7): 074301. https://doi.org/10.1063/1.2902483.

24 Abdul Hamid, K., Azmi, W.H., Nabil, M.F. et al. (2018). Experimental investigation of thermal conductivity and dynamic viscosity on nanoparticle mixture ratios of TiO_2–SiO_2 nanofluids. *International Journal of Heat and Mass Transfer* 116: 1143–1152.

25 Mane, N.S., Tripathi, S., and Hemadri, V. (2022). Effect of biopolymers on stability and properties of aqueous hybrid metal oxide nanofluids in thermal applications. *Colloids and Surfaces A: Physicochemical and Engineering Aspects* 643: 128777. https://doi.org/10.1016/j.colsurfa.2022.128777.

26 Routbort, J.L., Singh, D., Timofeeva, E.V. et al. (2011). Pumping power of nanofluids in a flowing system. *Journal of Nanoparticle Research* 13 (3): 931–937. https://doi.org/10.1007/s11051-010-0197-7.

27 Corcione, M., Cianfrini, M., and Quintino, A. (2012). Pumping energy saving using nanoparticle suspensions as heat transfer fluids. *Journal of Heat Transfer* 134 (12): 121701. https://doi.org/10.1115/1.4007314.

28 Yılmaz Aydın, D. and Gürü, M. (2022). Nanofluids: preparation, stability, properties, and thermal performance in terms of thermo-hydraulic, thermodynamics and thermo-economic analysis. *Journal of Thermal Analysis and Calorimetry* 147: 7631–7664. https://doi.org/10.1007/s10973-021-11092-8.

29 Mukherjee, S., Mishra, P.C., and Chaudhuri, P. (2020). Thermo-economic performance analysis of Al_2O_3-water nanofluids — an experimental investigation. *Journal of Molecular Liquids* 299: 112200. https://doi.org/10.1016/j.molliq.2019.112200.

30 Alirezaie, A., Hajmohammad, M.H., Hassani Ahangar, M.R., and Hemmat Esfe, M. (2018). Price-performance evaluation of thermal conductivity enhancement of nanofluids with different particle sizes. *Applied Thermal Engineering* 128: 373–380. https://doi.org/10.1016/j.applthermaleng.2017.08.143.

31 Prajapati, P.P. and Patel, V.K. (2020). Thermo-economic optimization of a nanofluid based organic Rankine cycle: a multi-objective study and analysis. *Thermal Science and Engineering Progress* 17: 100381. https://doi.org/10.1016/j.tsep.2019.100381.

32 Yousefi, M., Hooshyar, D., Kim, J.H., and Lim, H. (2017). Modeling and thermo-economic optimization of a biomass heat recovery exchanger operating on Al_2O_3-water nanofluid. *International Communications in Heat and Mass Transfer* 82: 63–73. https://doi.org/10.1016/j.icheatmasstransfer.2017.02.004.

33 Xuan, Z., Zhai, Y., Ma, M. et al. (2021). Thermo-economic performance and sensitivity analysis of ternary hybrid nanofluids. *Journal of Molecular Liquids* 323: 114889. https://doi.org/10.1016/j.molliq.2020.114889.

12

Clean Combustion with Biogas and Nano-Biodiesel in CI Engines

S. Lalhriatpuia[1], *Md. Gulam Mustafa*[2], *and Lalhmingsanga Hauchhum*[1]

[1] *Department of Mechanical Engineering, National Institute of Technology Mizoram, Aizawl, India*
[2] *Department of Mechanical Engineering Delhi Technological University, Delhi, India*

12.1 Introduction

Although internal combustion engines remain central to global transportation, their continued dependence on fossil fuels raises pressing concerns regarding resource depletion and environmental degradation. The combustion of petroleum-based fuels contributes significantly to greenhouse gas emissions and air pollution, necessitating a transition toward more sustainable energy carriers. Biogas, generated through the anaerobic digestion of organic substrates, represents a promising renewable alternative. Its composition is dominated by methane (CH_4), accompanied by carbon dioxide (CO_2) and trace gases, and it offers a near carbon-neutral combustion cycle. However, biogas utilization is often associated with elevated nitrogen oxides (NO_x) emissions due to its inherently high flame temperatures [1]. Biodiesel, produced from vegetable oils, provides another viable renewable option, yet its relatively high viscosity and lower cetane number compared to petroleum diesel can compromise engine efficiency and exacerbate pollutant formation [2]. To overcome these limitations, nanoparticle doping has emerged as a promising strategy to enhance biodiesel's physicochemical and combustion properties. Metallic oxide nanoparticles, particularly cerium oxide (CeO_2) and ceria–zirconia ($CeZrO_2$), have demonstrated significant potential in improving combustion efficiency, reducing emissions, and enhancing fuel stability under engine operating conditions [3].

Achieving optimal engine performance with sustainable fuels requires a comprehensive understanding of the interplay among operating variables. Among these, engine load exerts a dominant influence on combustion behavior and

Machine Learning for Sustainable Energy Solutions, First Edition. Edited by Zafar Said and Prabhakar Sharma.

emission characteristics. Higher loads elevate in-cylinder temperatures, intensifying combustion reactions and consequently promoting the formation of nitrogen oxides (NO_X), a key pollutant of concern [4].

Engine performance and emission control with sustainable fuels depend strongly on the interaction of multiple operating factors. The biodiesel blend ratio plays a decisive role in balancing sustainability with efficiency [5]. Higher blend fractions improve fuel lubricity and significantly reduce smoke emissions; however, they may also lower brake thermal efficiency (BTE) and increase brake-specific fuel consumption (BSFC) due to the inherently lower calorific value of biodiesel relative to diesel [6]. Engine load further governs combustion characteristics by influencing in-cylinder thermodynamics. At elevated loads, higher combustion temperatures accelerate reaction kinetics but simultaneously enhance nitrogen oxides (NO_X) formation, posing challenges for emission compliance [4]. In parallel, the nanoparticle doping rate exerts a critical influence on combustion dynamics. At optimal concentrations, nanoparticles enhance ignition quality and oxygen availability, thereby improving combustion stability and efficiency. Conversely, excessive doping promotes particle agglomeration and elevates fuel viscosity, which adversely affects atomization, mixing, and overall engine performance [7].

The biodiesel blend ratio with conventional diesel represents a key determinant in balancing sustainability with engine performance. Elevated blend levels generally improve fuel lubricity and contribute to smoke emission reduction, thereby enhancing combustion quality [5]. However, higher biodiesel content can also compromise performance, as its lower calorific value relative to diesel often leads to reduced brake thermal efficiency (BTE) and increased brake-specific fuel consumption (BSFC) [6]. In dual-fuel operation, the regulation of biogas flow rate (BFR) is equally critical. Excessive biogas supply can lower in-cylinder flame propagation and thermal efficiency, resulting in incomplete combustion and power loss. This condition further increases hydrocarbon (HC) emissions, as unburned fuel fractions escape the combustion chamber [8]. Conversely, insufficient biogas flow may limit the renewable fuel contribution, undermining sustainability goals. Thus, determining an optimal BFR is essential to achieve a balance between performance enhancement and emission control, ensuring that dual-fuel CI engine operation remains both efficient and environmentally sustainable.

Modeling and optimization techniques play a vital role in elucidating the complex interactions between engine operating parameters and performance outcomes. Response Surface Methodology (RSM) offers a structured experimental design and statistical framework for developing empirical models that link input variables, such as engine load, nanoparticle doping rate (NDR), biodiesel blend ratio, and biogas flow rate (BFR) to key responses, including emissions, brake thermal efficiency (BTE), and brake-specific fuel consumption (BSFC). By quantifying both main and interaction effects, RSM provides valuable insights into factor significance and response trends [9, 10]. Artificial Neural Networks (ANNs), in contrast, represent a powerful data-driven modeling approach capable of capturing nonlinear and highly coupled relationships between variables without requiring prior knowledge of underlying physical mechanisms [11]. Through adaptive learning from experimental datasets, ANNs are particularly effective for modeling the combustion processes of CI engines fueled with biogas and nanoparticle-doped biodiesel blends [12]. Robust statistical evaluation is essential for validating these models. RSM performance is typically assessed using the coefficient of determination (R^2), adjusted R^2, and mean squared error (MSE), whereas ANN models are commonly validated through error-based metrics such as mean absolute error (MAE) and root mean squared error (RMSE) [13, 14].

Achieving sustainable operation in CI engines requires not only accurate modeling but also effective optimization strategies to balance conflicting performance objectives. The Desirability Function Approach (DFA) within the RSM framework offers a powerful tool for multi-response optimization, particularly when trade-offs must be resolved among competing criteria [15]. In the context of CI engine operation with alternative fuels, this often entails minimizing nitrogen oxide (NO_x) emissions while simultaneously maximizing brake thermal efficiency (BTE). DFA achieves this by transforming each response variable into an individual desirability function, scaled between 0 (least desirable) and 1 (most desirable). These individual functions are then aggregated into a composite overall desirability index, which directs the optimization toward operating conditions that deliver the best compromise among all targeted responses [16].

Advanced optimization techniques play a pivotal role in improving the efficiency and accuracy of ANN-based modeling frameworks, particularly when addressing highly nonlinear combustion phenomena. Among these, Harris Hawks Optimization (HHO) has recently emerged as a promising metaheuristic algorithm inspired by the cooperative hunting strategies of Harris hawks [17]. By leveraging dynamic exploration and exploitation mechanisms, HHO is capable of efficiently navigating complex search spaces to avoid premature convergence and identify global optima. Its adaptability and robustness have been demonstrated across a wide range of engineering optimization problems. Within the context of CI engine studies, coupling HHO with ANN models offers the potential to refine predictive accuracy while simultaneously optimizing multiple performance and emission parameters. This integrated ANN–HHO approach has shown encouraging results, highlighting its applicability for advancing optimization strategies in sustainable dual-fuel engine operation [18].

This study examines how nanoparticle-doped biodiesel, biogas, and key operating parameters jointly affect the performance and emissions of compression ignition (CI) engines, aiming to establish optimal conditions for high efficiency and low emissions. The novelty lies in integrating Response Surface Methodology (RSM), Artificial Neural Networks (ANN), and Harris Hawks Optimization (HHO) to model complex nonlinear interactions. This hybrid approach combines empirical design with data-driven modeling and metaheuristic optimization to enhance predictive accuracy and identify robust operating windows for sustainable fuels. The findings offer actionable insights for developing cleaner, more efficient CI engine technologies, supporting the transition toward renewable and environmentally responsible combustion systems.

12.2 Materials and Methods

This section presents a detailed account of the experimental setup, including the technical specifications of the compression ignition (CI) engine, the composition and source of the biogas utilized, and the type of nanoparticle dopants incorporated. It further outlines the methodology adopted for preparing the nanoparticle-doped biodiesel, ensuring reproducibility and clarity in the experimental procedure.

12.2.1 Engine Specifications

The experimental investigations were carried out on a single-cylinder, four-stroke, constant-speed compression ignition (CI) engine rated at 3.5 kW. Engine loading was applied through an eddy current dynamometer, with the load varied between 2.4 kg (20%) and 12 kg (100%). Engine performance parameters were continuously monitored using precision instrumentation: cooling water flow was measured with rotameters, airflow with a manometer, and fuel consumption with a calibrated fuel flow meter. Temperature measurements were obtained via PT100 sensors positioned at critical locations along the engine cylinder. These sensors were interfaced with a National Instruments (NI) data acquisition unit, which transmitted real-time data to a computer equipped with Engine Soft software for recording and analysis. For dual-fuel operation, a gas–air mixing device was integrated into the setup, designed to match the engine's operating specifications and ensure uniform mixing of biogas with intake air. A schematic representation of the experimental test rig and gas mixer, adapted from our earlier work, is provided in Figure 12.1 [16].

12.2.2 Experimental Design

As summarized in Table 12.1, the nickel oxide nanoparticle doping rate (NDR) was varied from 0 to 100 ppm. Nanoparticles are recognized for their catalytic

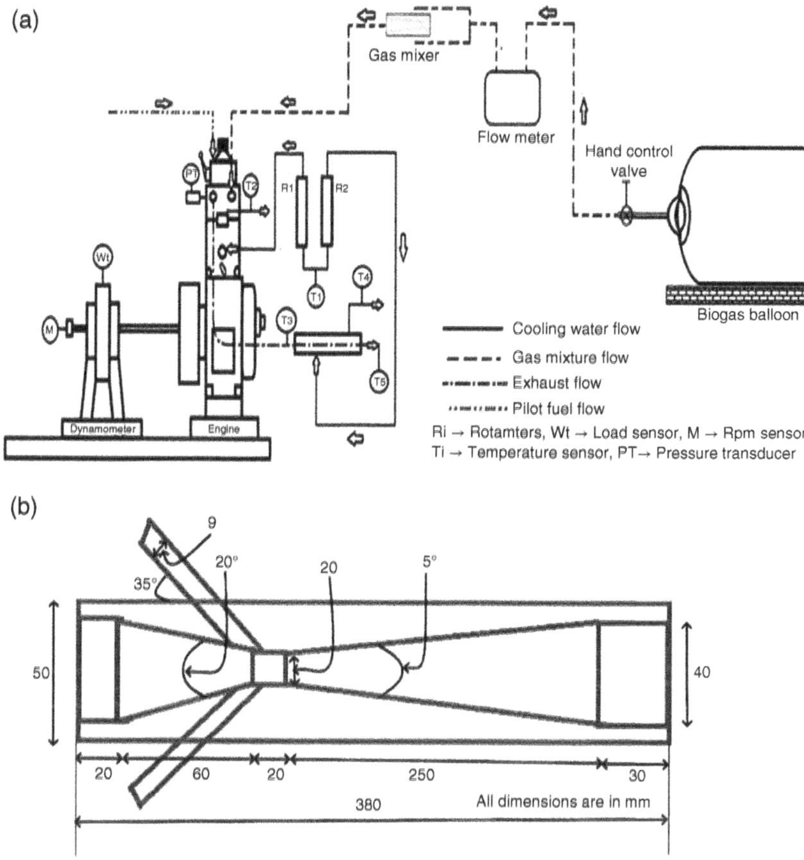

Figure 12.1 Schematic representation of test setup after modification.

Table 12.1 Input parameters and levels.

			Levels		
Input parameters	**Units**	**Symbol**	**−1**	**0**	**1**
Engine load	%	Load	20	60	100
Nanoparticles doped rate	ppm	NDR	0	50	100
Biodiesel blend rate	%	BBR	0	10	20
Biogas flow rate	kg/h	BFR	0.5	0.75	1

role in promoting oxidation, thereby enhancing combustion quality and fuel efficiency. Nonetheless, beyond an optimal concentration, agglomeration of nanoparticles can occur, which increases fuel viscosity and disrupts atomization, potentially counteracting these benefits. The soybean biodiesel blend ratio (BBR) was adjusted between 0% and 20% by volume, covering the transition from neat diesel to biodiesel–diesel blends. This parameter is particularly important for improving fuel lubricity and reducing particulate matter emissions. While higher biodiesel fractions enhance sustainability, they also lower the overall calorific value of the fuel, which can adversely affect brake thermal efficiency. The biogas flow rate (BFR) was controlled between 0 and 1 kg/h. As a methane-rich renewable fuel, biogas provides a sustainable supplement in dual-fuel CI engines; however, excessive flow rates may suppress combustion temperature, resulting in incomplete combustion and elevated unburned hydrocarbon emissions. Finally, engine load was varied from 0% to 100% of the rated output to capture the full operating range, from idle to maximum capacity, thereby enabling an evaluation of performance and emission trends across different loading conditions.

12.2.3 Fuel Properties

The physicochemical properties of diesel (D100), raw soybean oil (RS), and pure biodiesel (BBR100) were evaluated using various ASTM testing procedures, and the results are presented in Table 12.2.

Table 12.2 Fuel properties.

Properties	Unit	Testing methods	D 100	RS	BBR 10	BBR 20	BBR 100
Kinematic Viscosity (40 °C)	cSt	D-445	2.96	31.1	3.21	3.43	4.35
Density (40 °C)	kg/m^3	D-1298	840	920	830	837	885
Calorific values	MJ/kg	D-240	43.5	39.1	43.8	43.4	40.9
Flash point	°C	D-93	63	260	92	109	173
Pour point	°C	D-97	−13	−15	−13	−14	−17
Cetane number	—	D-4737	52	36	50	49	45

12.3 Modeling and Optimization

12.3.1 RSM Modeling

Modeling and optimization of engine performance and emission characteristics were conducted using Response Surface Methodology (RSM), a statistical approach particularly suited for analyzing complex systems involving multiple interacting variables. In this framework, polynomial regression models were employed to approximate the functional relationships between the input parameters, namely, nanoparticle doping rate (NDR), biodiesel blend ratio (BBR), biogas flow rate (BFR), and engine load. and their corresponding output responses. Second-order polynomial equations (equation 12.1) were fitted to the experimental dataset, incorporating both linear and quadratic terms to capture direct effects as well as nonlinear trends. In addition, interaction terms were included to assess the combined influence of variable pairs on engine behavior. Such interactions are critical in revealing synergistic or antagonistic effects that cannot be explained by the individual contribution of each factor alone. For instance, the combined variation of NDR and BBR may significantly alter brake thermal efficiency (BTE) or nitrogen oxide (NO_x) emissions, effects that would remain obscured if the parameters were studied in isolation. This comprehensive modeling approach thus provides a more reliable understanding of the interdependencies governing CI engine performance under sustainable fuel conditions.

$$Y = b_0 + \sum_{i=1}^{4} b_i x_i + \sum_{i=1}^{4} b_{ii} {x_i}^2 + \sum_{i \leq 1 \leq j}^{4} b_{ij} x_i x_j \tag{12.1}$$

In the polynomial regression framework, $_{xi}$ represents the independent input factors, while Y denotes the dependent response variable. The regression model includes an intercept coefficient (b_0), linear coefficients (b_i) corresponding to load, NDR, BBR, and BFR, interaction coefficients (b_{ij}) to account for paired effects, and quadratic coefficients (b_{ii}) to capture curvature in the response surface. Model fitting was performed using Design Expert software, which generated the regression equations based on experimental data. To evaluate the statistical validity and predictive reliability of the developed models, Analysis of Variance (ANOVA) was applied. ANOVA decomposes the total variability of the response into contributions from individual factors and their interactions, enabling the identification of statistically significant terms. Goodness-of-fit was assessed through performance metrics such as the coefficient of determination (R^2), where values closer to unity indicated strong agreement between the predicted and experimental results. This statistical validation ensured robustness of the RSM models for interpreting and optimizing engine performance and emissions.

12.3.2 ANN Modeling

Figure 12.2 presents the basic architecture of an artificial neural network (ANN) comprising three distinct layers. The directional arrows represent the sequential flow of information within the network. Input variables are first introduced at the input layer, after which the data is processed through the hidden layer, where nonlinear transformations and weighted computations occur. The processed signals are then transmitted to the output layer, which generates the final predicted response based on the trained network parameters [13].

The artificial neural network (ANN) employed in this study was structured as a multilayer perceptron (MLP) consisting of an input layer, two hidden layers, and an output layer. The input layer represented the experimental variables, while the hidden layers processed the data using nonlinear activation functions. The output layer generated the final predictions, with the number of neurons corresponding to the output responses. For training, each response variable was modeled with an independent network to ensure accuracy in mapping the design of experiments (DOE) data. Neuron combinations in the hidden layers were systematically varied between 1 and 10, and the configurations were evaluated according to the procedure outlined in Table 12.3. The optimal neuron pairing for each output was identified based on the lowest root mean squared error (RMSE). Training was performed using the backpropagation algorithm, with mean squared error (MSE) employed as the

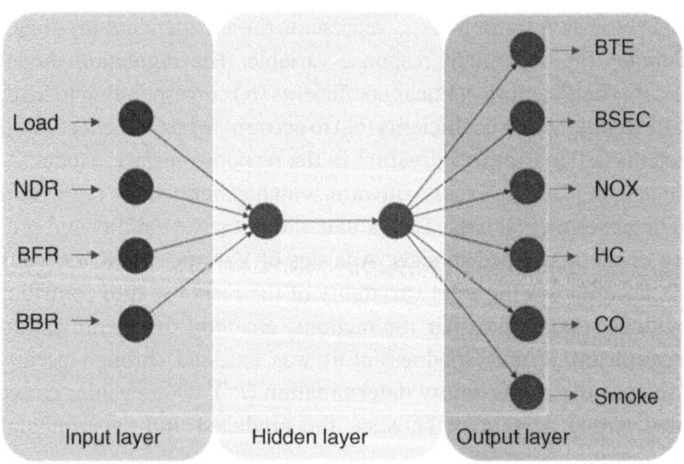

Figure 12.2 ANN architecture.

Table 12.3 Architecture for engine outputs.

BTE	BSEC	NO_x	HC	CO	SO
4-10-2-6	4-4-7-6	4-8-9-6	4-6-5-6	4-5-3-6	4-5-9-6

performance metric and the Levenberg–Marquardt (trainlm) algorithm selected as the training function. Data partitioning allocated 70% of the DOE dataset (30 runs) for training, while the remaining 30% was equally divided between testing and validation. This systematic approach enabled the identification of robust ANN architectures tailored to each output response.

12.3.3 Optimization of RSM and ANN model

In Response Surface Methodology (RSM), optimization is achieved through the desirability function approach, which enables simultaneous consideration of multiple response variables. Each response is transformed into an individual desirability score ranging from 0 (completely undesirable) to 1 (fully desirable), depending on its proximity to the defined target value and acceptable range. These individual scores are then aggregated using the geometric mean to generate a composite overall desirability index. The optimization process seeks the combination of input variables that maximizes this overall desirability, thereby resolving trade-offs among conflicting objectives. This method provides a structured and quantitative means of balancing competing performance and emission parameters, ensuring that no single response dominates the optimization outcome. The stepwise procedure for this multi-response optimization is outlined in the RSM flowchart presented in Figure 12.3, which illustrates how desirability functions guide the identification of the most favorable operating conditions.

Optimization of ANN model: HHO is a swarm intelligence–based metaheuristic algorithm inspired by the cooperative hunting strategies of Harris' hawks. It mimics various pursuit and surprise attack phases to explore and exploit the solution space effectively. The sequential stages of this optimization process, which collectively define the algorithm's search dynamics, are illustrated in Figure 12.4 [17].

The HHO algorithm functions within a predefined search space, where each hawk represents a candidate solution, and the "rabbit" denotes the global optimum. The fitness function quantifies the closeness of each hawk's position to this optimal solution, thereby guiding the search process. A balance between exploration and exploitation is achieved through distinct behavioral phases. During the

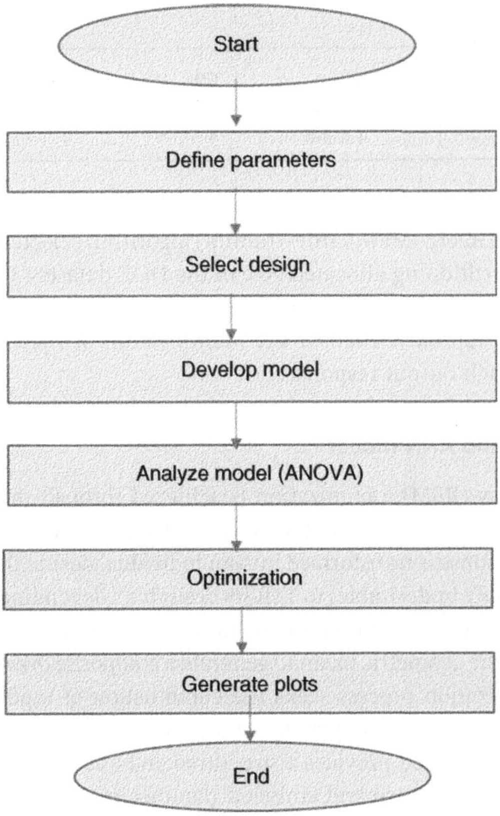

Figure 12.3 Flowchart of HHO.

exploration phase, hawks diversify the search by randomly updating their positions, often modeled as stochastic jumps toward the rabbit (equation 12.2). This mechanism promotes broad coverage of the solution space and prevents premature convergence. In the exploitation phase, the search narrows around promising regions as hawks adjust their positions relative to the rabbit, with the updates influenced by Lévy flight dynamics (equation 12.3). This strategy enables more refined local searches and accelerates convergence toward the optimum. Through iterative alternation between exploration and exploitation, the algorithm efficiently drives candidate solutions toward the global best.

$$A(x+1) = Ap(x) - \text{rand}^* \mid A(x) - \text{rabbit}(x) \mid *J \tag{12.2}$$

In this formulation, $A(x)$ denotes the current position of an individual hawk, while $Ap(x)$ represents the mean position of the entire population during a given iteration. The parameter rand corresponds to a uniformly distributed random value

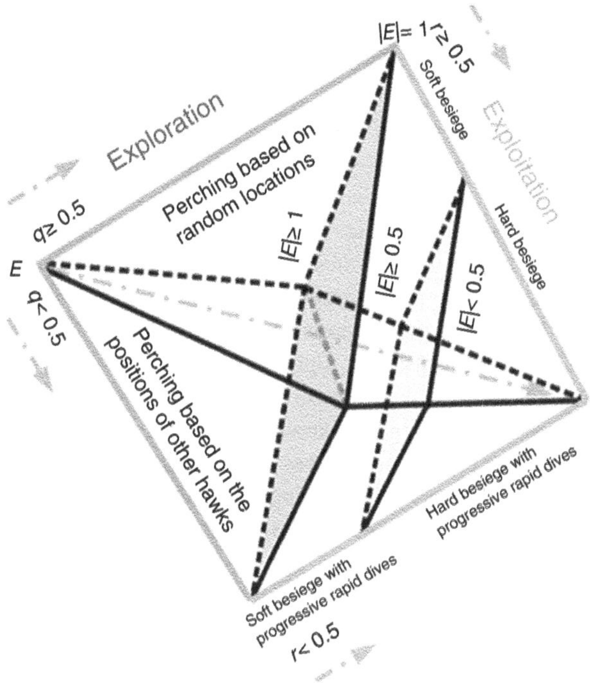

Figure 12.4 Various stages of HHO.

between 0 and 1, which introduces stochasticity and supports dynamic exploration of the search space. The expression "$|A(x)–\text{rabbit}(x)|$" defines the distance between the hawk's current location and the prey (rabbit), symbolizing the optimal solution. The term "J" refers to a random jump size that enables hawks to probe unexplored regions, thereby maintaining population diversity and reducing the likelihood of premature convergence.

$$A(x+1) = F(A(x), \text{rabbit}(x))^{*}\text{Levy}(D) \tag{12.3}$$

In this context, $F(A(x), \text{rabbit}(x))$ represents the functional difference between the hawk's current position and the location of the prey (rabbit), directing the hawk toward regions closer to the optimal solution. The term Levy(D) denotes a Lévy flight mechanism, which employs a probability distribution with heavy-tailed characteristics. This stochastic process enables the hawk to perform both small, localized steps and occasional long jumps, thereby balancing local intensification with global exploration. Such dynamics enhance the algorithm's ability to efficiently search the solution space and avoid entrapment in local optima.

12.4 Results and Discussion

12.4.1 RSM Model Analysis

Empirical regression models were developed to evaluate the influence of input variables—engine load, NDR, BBR, and BFR—on critical engine responses, BTE, BSEC, NO_X, HC, carbon monoxide (CO), and smoke opacity (SO). These models were formulated from the experimental dataset generated under systematically varied operating conditions. Each model incorporates coefficients that quantify the contribution of individual parameters to the respective response. A positive coefficient denotes a direct correlation, where an increase in the parameter value leads to a proportional rise in the response—for instance, higher engine load and NDR are associated with improvements in BTE. Conversely, negative coefficients represent an inverse relationship, whereby increasing the parameter reduces the response value; for example, elevated BBR and BFR tend to lower BTE. This coefficient-based interpretation provides an effective means of identifying the magnitude and direction of parameter influences on performance and emission metrics.

$$\begin{aligned} BTE = {} & 16.0230 + 4.55400\,A + 0.83200\,B - 1.09249\,C - 1.50500\,D \\ & + 0.230000\,AB - 0.480000\,AC - 0.550000\,AD - 0.060000\,BC \\ & - 0.090000\,BD + 0.130000\,CD - 1.95000\,A^2 - 0.140000\,B^2 \\ & + 0.020000\,C^2 - 0.180000\,D^2 \end{aligned} \tag{12.4}$$

As expressed in equation (12.4), BTE was most strongly influenced by engine load, which exhibited the largest positive effect. Increasing load improved BTE proportionally, primarily due to enhanced fuel combustion and higher in-cylinder temperature at elevated operating conditions. The NDR also contributed positively to BTE by promoting oxidation and improving ignition characteristics; however, this effect plateaued beyond a certain concentration, reflecting the diminishing benefits at higher doping levels. In contrast, both BBR and BFR exerted negative effects on BTE. Elevated BBR reduced the overall calorific value of the fuel, while excessive biogas supply lowered combustion temperature, leading to incomplete combustion and reduced thermal efficiency. Interaction effects were also significant: load–NDR exhibited a synergistic relationship, whereby their simultaneous increase resulted in further enhancement of BTE. Conversely, interactions of load with BBR or BFR demonstrated antagonistic effects, as their combined influence produced a reduction in BTE relative to individual contributions.

$$\begin{aligned} \mathrm{BSEC} = {} & 21.9600 - 8.05000\,\mathrm{A} - 1.50000\,\mathrm{B} + 2.40000\,\mathrm{C} + 1.72000\,\mathrm{D} \\ & + 0.400000\,\mathrm{AB} - 0.870000\,\mathrm{AC} - 0.510000\,\mathrm{AD} - 0.105000\,\mathrm{BC} \\ & - 0.070000\,\mathrm{BD} + 0.135000\,\mathrm{CD} + 5.07000\,\mathrm{A}^2 + 0.060000\,\mathrm{B}^2 \\ & + 0.045000\,\mathrm{C}^2 + 0.445000\,\mathrm{D}^2 \end{aligned} \tag{12.5}$$

As presented in equation (12.5), the BSEC model revealed that both BFR and BBR exhibited a positive correlation with BSEC. This indicates that increasing either parameter resulted in higher energy consumption per unit of work output. The effect can be attributed to the reduced energy density of biodiesel and biogas compared to diesel, which necessitates greater fuel input to maintain equivalent power output. In contrast, engine load and NDR demonstrated negative effects on BSEC. At higher loads, improved combustion efficiency reduced energy consumption per unit of brake power, while NDR enhanced oxidation and fuel reactivity, further improving energy utilization. Interaction analysis highlighted that the combination of load and NDR produced a synergistic effect, yielding substantial reductions in BSEC when both parameters were simultaneously increased. Conversely, the interactions of load with BBR or BFR exhibited antagonistic trends, as the additional energy demands imposed by higher biodiesel or biogas fractions counteracted the efficiency benefits associated with higher load operation.

$$\begin{aligned} \mathrm{NOX} = {} & 246.600 + 136.500\,\mathrm{A} - 18.0000\,\mathrm{B} + 15.8000\,\mathrm{C} - 38.0000\,\mathrm{D} \\ & - 10.2000\,\mathrm{AB} + 10.1000\,\mathrm{AC} - 22.7000\,\mathrm{AD} - 0.980000\,\mathrm{BC} + 2.66000\,\mathrm{BD} \\ & - 2.34000\,\mathrm{CD} + 6.24000\,\mathrm{A}^2 + 2.69000\,\mathrm{B}^2 - 6.20000\,\mathrm{C}^2 - 9.79000\,\mathrm{D}^2 \end{aligned} \tag{12.6}$$

As described in equation (12.6), NO_X emissions demonstrated a positive dependence on both engine load and BBR. Higher loads elevated in-cylinder pressures and combustion temperatures, thereby promoting thermal NO_X formation. Similarly, increasing biodiesel content intensified oxygen availability in the combustion zone, further contributing to higher NO_X levels. In contrast, BFR and NDR exhibited negative correlations with NO_X emissions. The presence of methane in biogas moderated combustion temperatures, while nanoparticles improved oxidation efficiency and combustion stability, collectively suppressing NO_X generation. Interaction analysis revealed that combinations such as load–BBR and NDR–BBR produced synergistic effects, leading to amplified NO_X emissions when both parameters were simultaneously increased. Other parameter interactions generally showed antagonistic influences. Quantitatively, the results highlighted that for a BBR20–NDR0 blend, NO_X emissions were reduced by approximately 14% compared to BBR20–NDR100, while for biogas-rich conditions, emissions decreased by 27.15% under BFR1–NDR100 relative to BFR0.5–NDR100. These findings emphasize the critical role of fuel composition and operating conditions in shaping NO_X emission trends.

$$\begin{aligned} \mathrm{HC} = {} & 45.0000 - 6.80000\,\mathrm{A} - 4.50000\,\mathrm{B} - 2.80000\,\mathrm{C} + 3.00000\,\mathrm{D} \\ & + 0.620000\,\mathrm{AB} + 0.390000\,\mathrm{AC} - 0.240000\,\mathrm{AD} + 0.250000\,\mathrm{BC} \\ & - 0.270000\,\mathrm{BD} - 0.160000\,\mathrm{CD} + 0.740000\,\mathrm{A}^2 + 0.270000\,\mathrm{B}^2 \\ & + 1.25000\,\mathrm{C}^2 + 1.18000\,\mathrm{D}^2 \end{aligned} \tag{12.7}$$

As illustrated in equation (12.7), HC emissions were strongly influenced by the BFR, which exhibited a positive correlation. Elevated BFR increased HC emissions, primarily due to incomplete combustion associated with lower flame temperatures and methane's comparatively slow oxidation characteristics. In contrast, engine load, NDR, and BBR all displayed negative effects on HC formation. Higher engine loads promoted more complete combustion through elevated in-cylinder temperatures, while NDR enhanced oxidation kinetics and BBR improved oxygen availability, collectively contributing to reduced unburned hydrocarbons. Interaction effects were also significant: load–NDR, load–BBR, and NDR–BBR combinations produced synergistic reductions in HC emissions, suggesting that proper tuning of these variables can effectively suppress unburned fuel. Quantitative results showed that at a constant load and BFR, the BBR20–NDR100 blend achieved a 17% reduction in HC emissions compared to BBR20–NDR0. Conversely, increasing biogas flow from BFR0.5–NDR100 to BFR1–NDR100 resulted in a 13.02% increase in HC emissions, underscoring the trade-offs involved in dual-fuel operation.

$$\begin{aligned} CO = {} & 0.093000 - 0.0215000\,A - 0.00950000\,B - 0.00650000\,C + 0.00730000\,D \\ & + 0.001600\,AB + 0.000420\,AC - 0.001360\,AD + 0.000491\,BC - 0.000550\,BD \\ & - 0.000380\,CD + 0.031800\,A^2 + 0.001010\,B^2 + 0.000003\,C^2 + 0.000303\,D^2 \end{aligned} \tag{12.8}$$

As expressed in equation (12.8), CO emissions exhibited a positive correlation with BFR. Higher BFR levels increased CO formation, primarily due to incomplete combustion arising from the lower flame temperature and dilution effects associated with excess methane. In contrast, engine load, BBR, and NDR were negatively correlated with CO emissions. Elevated loads promoted more complete combustion through higher in-cylinder temperatures, while BBR improved oxygen availability and NDR enhanced oxidation kinetics, together reducing CO generation. Interaction effects revealed that the most significant interaction effects were between load–NDR and load–BBR, indicating that combined increases in these variables can effectively suppress CO emissions. Quantitatively, for a BBR20–NDR0 blend, CO emissions were reduced by 14% compared to BBR20–NDR100. However, under biogas-rich conditions, CO emissions increased by 11.9% when BFR1–NDR100 was compared with BFR0.5–NDR100, highlighting the delicate balance between fuel composition and operating conditions in controlling CO levels.

$$\begin{aligned} SO = {} & 28.9000 + 13.3000\,A - 2.05000\,B - 1.22000\,C - 4.42000\,D - 0.930000\,AB \\ & - 0.560000\,AC - 2.01000\,AD + 0.086000\,BC + 0.310000\,BD \\ & + 0.185000\,CD - 0.130000\,A^2 - 0.460000\,B^2 - 0.301000\,C^2 \\ & + 1.34000\,D^2 \end{aligned} \tag{12.9}$$

As outlined in equation (12.9), SO was primarily influenced by engine load, which exhibited a positive correlation. Increasing load intensified fuel injection and combustion density, thereby elevating particulate formation and SO levels. In contrast, (BFR, BBR, and NDR showed negative correlations, with higher values of these parameters contributing to reductions in smoke emissions. The oxygenated nature of biodiesel, the combustion-enhancing effects of nanoparticles, and the dilution influence of biogas collectively suppressed soot formation. Among the interaction terms, load–BFR demonstrated the strongest effect, as simultaneous increases in both parameters led to elevated smoke opacity, counteracting the individual benefits of higher BFR. Quantitatively, SO decreased by approximately 13% with the BBR20–NDR0 blend compared to BBR20–NDR100, while under biogas-rich conditions, SO increased by 26.15% for BFR1–NDR100 relative to BFR0.5–NDR100. These findings underscore the importance of carefully balancing operating conditions to mitigate particulate emissions in dual-fuel CI engines.

The empirical models developed in this study underscore the intricate interdependencies among engine operating parameters and their collective influence on performance and emission characteristics. By systematically analyzing these relationships, the models identify the most sensitive parameters requiring optimization to enhance efficiency while mitigating pollutant formation. Notably, synergistic effects were observed when higher engine loads were combined with moderate biogas flow rates or nanoparticle doping levels, yielding significant improvements in combustion efficiency alongside reductions in key emissions. Such findings emphasize that optimal outcomes cannot be achieved by varying a single parameter in isolation but instead demand a balanced adjustment of multiple inputs. Overall, this analysis provides a structured basis for fine-tuning CI engine operation across diverse load and fuel conditions, offering practical guidance for optimizing sustainable dual-fuel strategies in real-world applications.

The ANOVA results, summarized in (Table 12.4) demonstrated that the regression models effectively captured the variability in the experimental dataset.

Table 12.4 Analysis of variance (F-value) and R^2 for performance and emission measurements.

	BTE	BSEC	NO_x	HC	CO	SO
Source	**F-value**	**F-value**	**F-value**	**F-value**	**F-value**	**F-value**
Model	2468.9	6829.7	14,456.2	4712.5	19,112.4	16,093.3
A-Load	25,473.6	69,842.1	1.768E+05	34,915.8	1.124E+05	1.935E+05
B-NDR	853.5	2231.7	3,095.4	15,121.6	22,212.8	4,512.9
C-BBR	1,719.4	6,278.9	2,382.7	5,875.5	10,496.8	1,635.2
D-BFR	3,124.7	3,266.4	13,698.8	6,791.6	13,248.2	21,472.3
R^2	0.9995	0.9994	0.9990	0.9998	0.9999	0.9999

The presence of high F-values confirmed the strong explanatory power of the models in relating input parameters to output responses. Among the studied variables, engine load exhibited the highest F-value, establishing it as the most influential factor governing performance and emission characteristics. This was followed in order of significance by BFR, BBR, and NDR. Furthermore, the consistently high R^2 obtained for all response variables indicated excellent agreement between the predicted values and the experimental data. The lack-of-fit tests yielded P-values greater than 0.05, confirming that the models did not exhibit significant lack of fit and could therefore be considered statistically robust for predictive and optimization purposes.

12.4.2 ANN Model Analysis

The correlation coefficient (R) between the predicted and experimental values of the output parameters was calculated for each stage of the ANN process, including training, testing, and validation. The overall correlation coefficients for all dependent responses are presented in Figures 12.5 and 12.6. The consistently high *R*

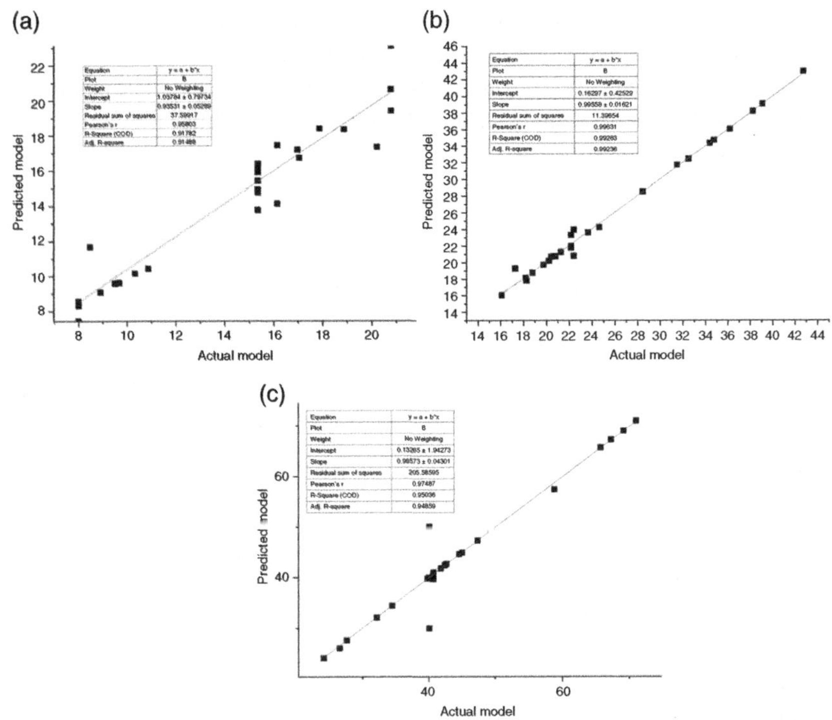

Figure 12.5 R as a function of (a) BTE, (b) BSEC, and (c) NO_x as inputs to the trained network.

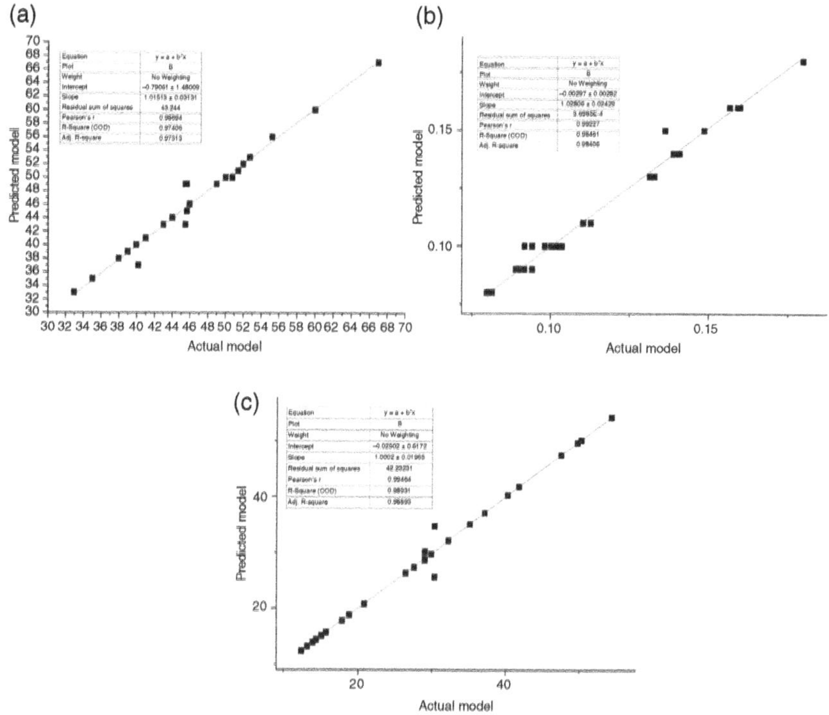

Figure 12.6 R as a function of (a) HC, (b) CO, and (c) SO as inputs to the trained network.

values obtained across all phases confirm the strong predictive capability of the developed networks. These results demonstrate that the DOE dataset was effectively utilized to train the ANN models, ensuring accurate mapping between input parameters and output responses. The high degree of correlation reinforces both the robustness and reliability of the ANN-based regression framework, highlighting its suitability for modeling complex, nonlinear interactions governing CI engine performance and emissions under dual-fuel operation.

12.4.3 Optimization of RSM and ANN Model

The RSM optimization yielded a maximum overall desirability value of 0.71 (Figure 12.7) identified an optimal solution that predicts the following values: BTE at 17.18%, BSEC at 19.588 MJ/kWh, NO_X at 240.32 ppm, HCs at 41.04 ppm, CO at 0.085%, and SO at 27.79%. These predictions correspond to an engine load of 64.57%, NDR of 97.04 ppm, BBR of 8.19%, and BFR of 0.77 kg/h.

The ANN–HHO optimization process concluded after 102 generations (Figure 12.8), corresponding to an optimal operating condition. At this point,

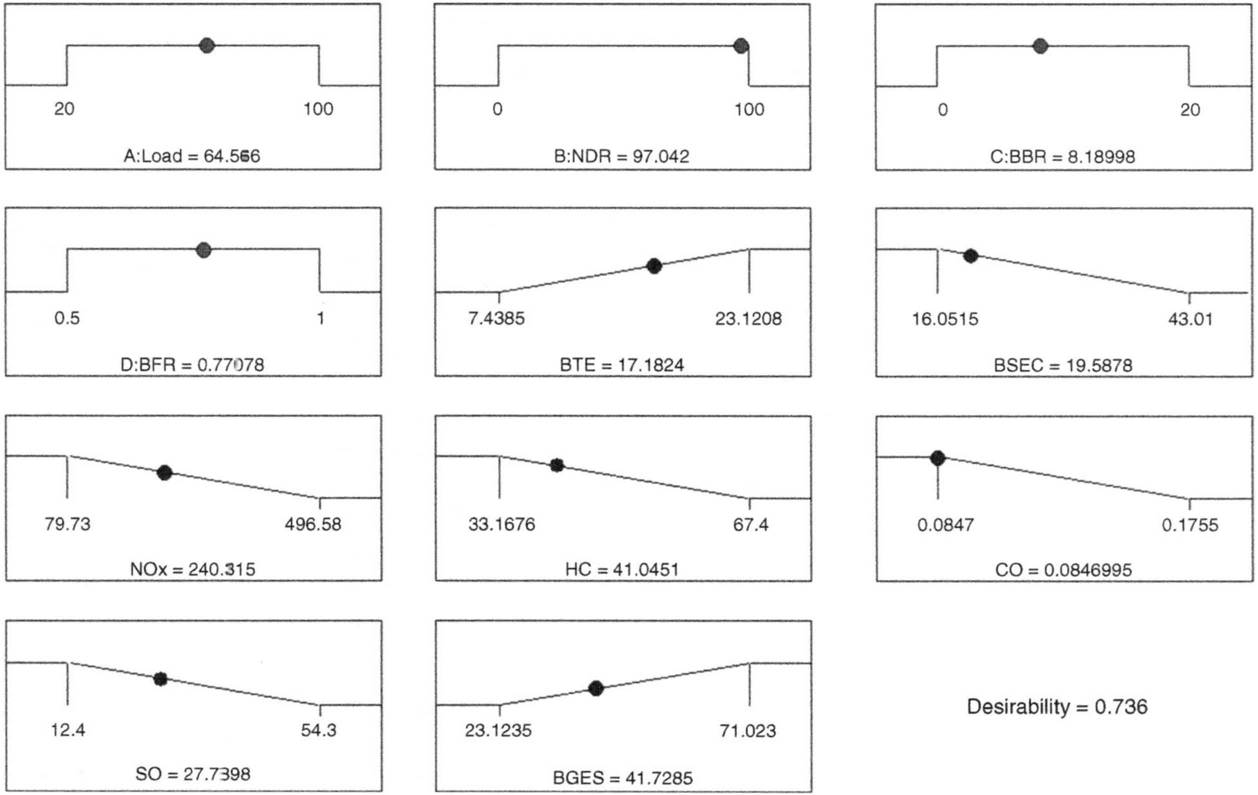

Figure 12.7 Desirability plot.

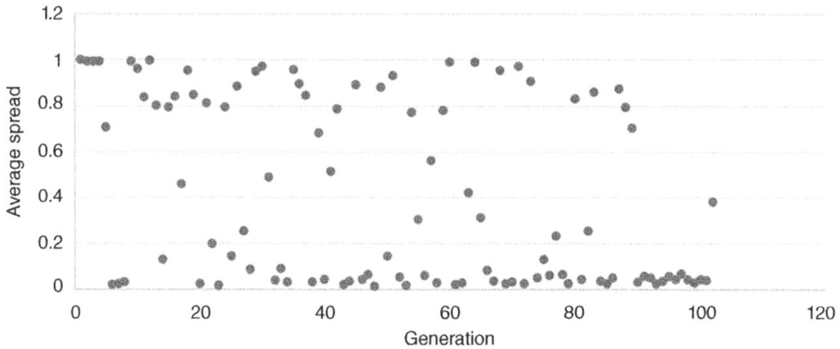

Figure 12.8 Average spread vs generation.

the predicted engine responses were: BTE of 16.31%, BSEC of 21.03 MJ/kWh, nitrogen oxides NO_X at 230.52 ppm, HC at 40.98 ppm, CO at 0.085%,and SO at 26.01%. These optimized responses were achieved under the specific operating parameters of 60.68% engine load, a NDR of 97.55 ppm, a BBR of 11.24%, and a BFR of 0.79 kg/h. This solution highlights the effectiveness of the desirability function in balancing multiple objectives and identifying operating conditions that optimize both performance and emissions simultaneously.

12.5 Conclusions

This study provides the following key insights:

1) ANOVA results from the RSM model demonstrated that, among the investigated input variables, engine load exerted the most dominant influence on performance and emission responses. This was followed in order of significance by BFR, NDR, and BBR.
2) Evaluation metrics confirmed the high predictive accuracy and robustness of both RSM and ANN regression models. The results indicated low prediction errors for all responses, with RSM exhibiting slightly superior reliability, as evidenced by higher R^2 values.
3) Optimization using RSM identified optimal conditions at 64.57% engine load, 97.04 ppm NDR, 8.19% BBR, and 0.77 kg/h BFR. Similarly, ANN coupled with HHO optimization yielded optimum results at 60.68% load, 97.55 ppm NDR, 11.24% BBR, and 0.786 kg/h BFR. Both outcomes confirm the effectiveness of nanoparticle doping in enhancing engine performance and reducing emissions

Overall, the integration of biogas and biodiesel as dual-fuel strategies demonstrates considerable potential for reducing greenhouse gas emissions while maintaining acceptable efficiency levels. Furthermore, the application of advanced optimization techniques such as HHO strengthens the decision-making framework for parameter tuning, supporting the transition toward sustainable CI engine technologies and reducing dependence on fossil fuels.

References

1 Ullah Khan, I., Hafiz Dzarfan Othman, M., Hashim, H. et al. (2017). Biogas as a renewable energy fuel – a review of biogas upgrading, utilisation and storage. *Energy Conversion and Management* 150: 277–294. https://doi.org/10.1016/j.enconman.2017.08.035.

2 Huang, D., Zhou, H., and Lin, L. (2012). Biodiesel: an alternative to conventional fuel. *Energy Procedia* 16: 1874–1885. https://doi.org/10.1016/j.egypro.2012.01.287.

3 Hatami, M., Hasanpour, M., and Jing, D. (2020). Recent developments of nanoparticles additives to the consumables liquids in internal combustion engines: Part I: nano-fuels. *Journal of Molecular Liquids* 318: 114250. https://doi.org/10.1016/j.molliq.2020.114250.

4 Mei, D., Zuo, L., Adu-Mensah, D. et al. (2019). Combustion characteristics and emissions of a common rail diesel engine using nanoparticle-diesel blends with carbon nanotube and molybdenum trioxide. *Applied Thermal Engineering* 162: 114238. https://doi.org/10.1016/j.applthermaleng.2019.114238.

5 Sharma, A., Singh, Y., Ahmad Ansari, N. et al. (2020). Experimental investigation of the behaviour of a DI diesel engine fuelled with biodiesel/diesel blends having effect of raw biogas at different operating responses. *Fuel* 279: 118460. https://doi.org/10.1016/j.fuel.2020.118460.

6 Rajendran, S., Venkatesan, E.P., Dhairiyasamy, R. et al. (2023). Enhancing performance and emission characteristics of biodiesel-operated compression ignition engines through low heat rejection mode and antioxidant additives: a review. *ACS Omega* 8: 34281–34298. https://doi.org/10.1021/acsomega.3c03252.

7 Yusof, S.N.A., Sidik, N.A.C., Asako, Y., and Japar WMohdAA, Mohamed SB, Muhammad NM. (2020). A comprehensive review of the influences of nanoparticles as a fuel additive in an internal combustion engine (ICE). *Nanotechnology Reviews* 9: 1326–1349. https://doi.org/10.1515/ntrev-2020-0104.

8 Feroskhan, M., Ismail, S., Natarajan, G. et al. (2023). A comprehensive study of the effects of various operating parameters on a biogas-diesel dual fuel engine. *Sustainability* 15: 1232. https://doi.org/10.3390/su15021232.

9 Veza, I., Spraggon, M., Fattah, I.M.R., and Idris, M. (2023). Response surface methodology (RSM) for optimizing engine performance and emissions fueled with biofuel: review of RSM for sustainability energy transition. *Results in Engineering* 18: 101213. https://doi.org/10.1016/j.rineng.2023.101213.

10 Hadiyat, M.A., Sopha, B.M., and Wibowo, B.S. (2022). Response surface methodology using observational data: a systematic literature review. *Applied Sciences* 12: 10663. https://doi.org/10.3390/app122010663.

11 Agatonovic-Kustrin, S. and Beresford, R. (2000). Basic concepts of artificial neural network (ANN) modeling and its application in pharmaceutical research. *Journal of Pharmaceutical and Biomedical Analysis* 22: 717–727. https://doi.org/10.1016/S0731-7085(99)00272-1.

12 Zou, J., Han, Y., and So, S.-S. (2008). Overview of artificial neural networks. *Artificial Neural Networks: Methods and Applications* 458: 14–22. https://doi.org/10.1007/978-1-60327-101-1_2.

13 Lalhriatpuia, S. and Pal, A. (2023). Computational optimization of engine performance and emission responses for dual fuel CI engine powered with biogas and Co_3O_4 nanoparticles doped biodiesel. *Fuel* 344: 127892. https://doi.org/10.1016/j.fuel.2023.127892.

14 Plevris V., Solorzano G., Bakas N., Ben Seghier M.2022). Investigation of performance metrics in regression analysis and machine learning-based prediction models. *8th European Congress on Computational Methods in Applied Sciences and Engineering (ECCOMAS Congress 2022), 5–9 June 2022, Oslo, Norway*. https://doi.org/10.23967/eccomas.2022.155.

15 Costa, N.R., Lourenço, J., and Pereira, Z.L. (2011). Desirability function approach: a review and performance evaluation in adverse conditions. *Chemometrics and Intelligent Laboratory Systems* 107: 234–244. https://doi.org/10.1016/j.chemolab.2011.04.004.

16 Lalhriatpuia, S. and Pal, A. (2023). Computational optimization of engine emissions and performance of a CI engine powered with biogas and NiO nanoparticles doped diesel. *Environmental Progress & Sustainable Energy* https://doi.org/10.1002/ep.14207.

17 Heidari, A.A., Mirjalili, S., Faris, H. et al. (2019). Harris hawks optimization: algorithm and applications. *Future Generation Computer Systems* 97: 849–872. https://doi.org/10.1016/j.future.2019.02.028.

18 Alabool, H.M., Alarabiat, D., Abualigah, L., and Heidari, A.A. (2021). Harris hawks optimization: a comprehensive review of recent variants and applications. *Neural Computing and Applications* 33: 8939–8980. https://doi.org/10.1007/s00521-021-05720-5.

13

A Differentiation of Energy Storage Methods

H. Bahuruteen Ali Ahamadu[1], K Arun[2], S Arivazhagan[3], N Sendhil Kumar[4], S. Kaliappan[5], and M. D. Rajkamal[6]

[1] *Department of Mechanical Engineering, Rajalakshmi Institute of Technology, Chennai, India*
[2] *Department of Mechanical Engineering, St. Joseph's College of Engineering, Chennai, India*
[3] *Department of Mechanical Engineering, St. Joseph's College of Engineering, Chennai, India*
[4] *Department of Mechanical Engineering, National Institute of Technology, Karaikal, India*
[5] *Department of Mechatronics Engineering, KCG College of Technology, Chennai, India*
[6] *Department of Mechanical Engineering, Velammal Institute of Technology, Chennai, India*

13.1 Introduction

The constant rise in both production and demand for energy means that no region of the world can fully satisfy its energy needs. Developing novel techniques for energy storage is of greater interest to researchers than enhancing the effectiveness of existing traditional approaches. The purpose of this paper was to investigate several forms of conservative storage technologies and their substitute approaches for enhancing performance [8]. The following specifications should be met by an efficient energy storage technology.

- High energy storage density.
- Low energy dissipation to the surrounding.
- Storage media should be easily available at low cost.
- Capacity to store for long duration.

Developing an effective storage system using modern optimization tools is necessary to monitor and operate the performance with respect to the changes in demand, generation, and weather pattern. Although the development of regulatory frameworks and design standards is needed to produce best-performing storage systems. The best storage system must ensure that reliability, grid stability, and supply security at low cost are essential [28].

Machine Learning for Sustainable Energy Solutions, First Edition. Edited by Zafar Said and Prabhakar Sharma.

13.1.1 Conventional Energy Storage

The globe has been using the energy storage technologies displayed below for a considerable amount of time Figure 13.1

13.1.2 Mechanical Energy Storage

This type of energy storage allows for an ample amount of energy to be stored as potential and kinetic energy, which can then be used to produce power if the primary energy source is unavailable. Although there is a significant energy loss during charging and discharging, this form of storage is only practical in locations where other methods are impractical. Pumped hydro energy storage is one of the methods of mechanical energy storage, which is high in investment and energy losses. However, it is best at hydroelectric power plant locations during off-peak hours when electricity demand is low. The extra energy generated there is then used to pump water back upstream and to empty it again during peak hours, thereby storing energy in water as potential energy. When water flows from upstream to downstream, the water turbine produces electricity. When there is surplus energy, the same turbine can also be used as a pump to release water from downstream into upstream. Benefits of this technology include a large storage capacity and a lengthy operating life.

Another mechanical energy storage method is compressed air energy storage, which uses the excess energy generated during off-peak load in any power production method to produce compressed air, which is subsequently utilized to run a gas turbine during peak load to produce electricity. Since a compressor is used to save energy, this method's efficiency is not very noteworthy. Compressor uses the

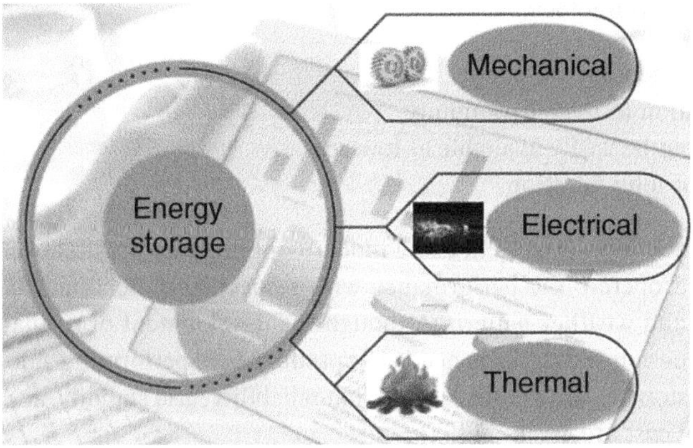

Figure 13.1 Energy storage.

greatest amount of excess energy for operation; there is not enough energy in storage to be used during peak hours. Because of this, using this technology to store energy is not practical [19].

The flywheel mechanical energy storage method is inefficient due to energy loss in the storing and releasing process because of the use of mechanical components like bearings and flywheel. This energy storage uses the motor shaft's rotation to store energy as kinetic energy for cycles of charging and discharging. Recent improvements in the design of magnetic bearings and composite materials for flywheel manufacture have increased energy efficiency and decreased energy loss. Due to its superior energy conversion efficiency, quick response time, and high instantaneous energy delivery, this technology is the best choice for urgent power needs [2, 30].

13.1.3 Electrical Energy Storage

This method uses the electrostatic or magnetic energy storage methods to store energy in the form of electrical current (without any conversion). The electrostatic storages using capacitors and supercapacitors for storing energy and superconducting magnets are used in magnetic energy storage [29]. The ability to store energy for longer periods of time has been aided by the development of capacitor materials. Supercapacitors are similar to batteries in that they store energy by means of conductor electrodes, an electrolyte, and a separator. This type of energy storage is not used very often because it is costly, has a shorter storage period, and has smaller storage capacity [2, 5]. Moreover, the high self-discharge rate of electrical energy storage limited its usage further because of low overall effectiveness [14, 31].

13.1.4 Electrochemical Energy Storage

This technique uses an electrochemical battery's mutual conversion process to store and release energy. This technique uses an electrochemical battery's mutual conversion process to store and release energy.

Lithium-ion battery: Industry utilizes these batteries mainly because of their worthy qualities and long service life. The high initial expenditure required for big installations is an obstacle to the commercial use of lithium-ion batteries [18].

Lead-acid battery: The high concentration of lead in these batteries poses a threat to the environment. The usage of these batteries in energy storage applications has the benefits of low cost, safety, and recycling choices.

Sodium–Sulfur battery: The choice of these batteries for the operation was largely influenced by their high energy density, no self-discharge, and extended operating life. Battery structure and selection of materials are challenges for the construction issues of batteries. The selection of materials by understanding their crystal structure has been an essential factor for storing energy in electrochemical processes [2, 32, 33]. Edwin et al. [33] highlighted that creating tailored material with respect to their operation requirements facilitates improvement in energy storage. Gharahcheshmeh et al. [34] highlight that the development of conjugated polymers by the methods of solution processing, electrochemical deposition, and layer-by-layer assembly exhibits high energy storage density through redox reactions for changing their bonds in delocalized electrons.

13.1.5 Thermal Energy Storage

This approach stores energy with the ability of material to store it for both short and extended periods by taking advantage of the latent heat, sensible heat, and thermochemical properties of materials [1, 3]. It is designed to store energy in both hot and cold forms depending on demand by utilizing materials that have the right energy density, heat capacity, and operating temperature[4]. Additional classifications for thermal energy storage are displayed below. Apart from energy storage, thermal energy storage systems are employed in green building technologies by storing latent heat in floors and ceilings. Further, the employment of TES helps provide sustainable solutions to balance energy demand and consumption [35]. Figure 13.2.

13.1.6 Characteristics of Thermal Energy Storage

The performance of thermal energy storage systems depends upon the following important factors: Working fluid, storage method, storage tanks, and charging/discharging time. Water serves as a versatile energy storage medium for heating, ventilation, and water heating applications, operating within a temperature range of 0–120 °C. The energy storage medium is also specified as heat carrier and HTF (heat transfer fluid), and the following factors are to be considered while selecting heat storage materials.

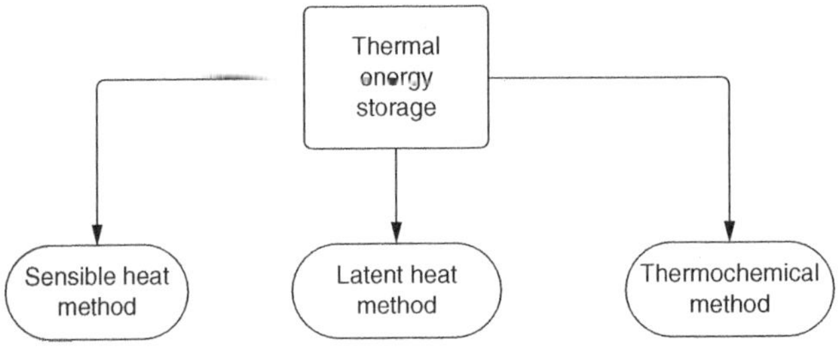

Figure 13.2 Classification of thermal energy storage.

- Low cost and easily available.
- Noncorrosive, noninflammable, and nontoxic.
- No energy loss during charging and discharging.
- Long service life.

In addition to the above specifications, the material for energy storage is chosen based on its high heat storage capacity, high energy density, and latent heat of fusion—the energy released from the material during a phase transition, such as melting or freezing. The substance with a high heat-storing capacity decreases the amount of space used to store sensible heat. Also, thermal conductivity and thermal diffusivity are the terms defining the energy charging and discharging time of material. The behavior of the charging and discharging cycle will be affected if undesired particles are present. Thermal energy storage can be classified as either indirect or direct depending on the method of storage. Since the HTF is maintained at a low temperature and partial pressure, the design and manufacturing of the vessel become less expensive when using the direct method. The high pressure of the HTF requires complex engineering and construction of pressure vessels, making the indirect approach less cost-effective than the direct method.

Based on the quantity of storage tanks, thermal energy storage systems are divided into two types: single-tank and two-tank designs. The temperature distribution of a material in a single storage tank is not uniform due to the energy loss between the different regions of the storage tank [47]. This issue can be solved by adopting two tank storage methods. The thermal energy storage system's performance is mostly dictated by its charge/discharge time, which is primarily based on its thermal power and thermal capacity.

$$\text{Charge/Discharge Time} = \text{Thermal Power in kW/Thermal Capacity kWh}$$

13.1.7 Sensible Heat Storage

In this method, heat energy is stored in the material without changing its phase and by differentiating the temperature. The difference in temperature directly relates to the amount of energy stored. Sensible heat storage materials are classified as sensible solid storage or sensible liquid storage depending on the condition of the energy storage medium. The quantity of sensible heat energy (Q) stored in the material is calculated by

$$Q = m \cdot C_p \cdot (T_2 - T_1)$$

where C_p is the specific heat capacity at constant pressure (J/kg K); T_2 is the temperature of the material after storing energy (K); T_1 is the temperature of the material before storing energy (K).

Phase transfer and chemical processes for the purpose of storing energy inside the material are typically not permitted in a sensible heat storage device. The following methods: aquifer storage, hot water storage, gravel water storage, cavern storage, and molten salt storage belong to sensible liquid heat storage. Borehole storage and packed bed storage belong to sensible solid storage. For the purpose of storing low-temperature solid thermal energy, both produced solids (ceramics) and natural materials (rocks and pebbles) are utilized [36]. Solids consisting of silica, alumina, magnesia, and iron oxide are employed at high temperatures. Usage of graphite is preferred in thermal energy storage due to its high thermal diffusivity, and metals are preferred for their high thermal conductivity. This method of energy storage has been a cost-effective method compared to other technologies [34]. Figure 13.3.

Sensible liquid storage is categorized as a low-temperature system (below 100 °C) or a high-temperature system (above 100 °C) based on the operating temperature. The high-temperature system causes the water to flash by reducing the pressure at the discharge while maintaining a constant temperature. This results in the production of steam at a temperature higher than 100 °C.

13.1.8 Aquifer Thermal Energy Storage

This technique is mostly applied to household heating and cooling systems in cold climates to provide comfortable living spaces. To build an aquifer storage, individual wells for hot and cold water storage and pumps are needed Figure 13.4.

Water from the cold well is utilized to keep the building colder during the summer season. As the water circulates throughout the building, heat is absorbed and then returned to the hot well to store it. The hot well is adequately insulated to prevent heat loss to the environment and to boost the system's long-term heat storage capacity. The building receives circulation of hot water from the well throughout the winter months to maintain a comfortable temperature for human habitation. When the water gets cooler, it is returned to the cold well for summer

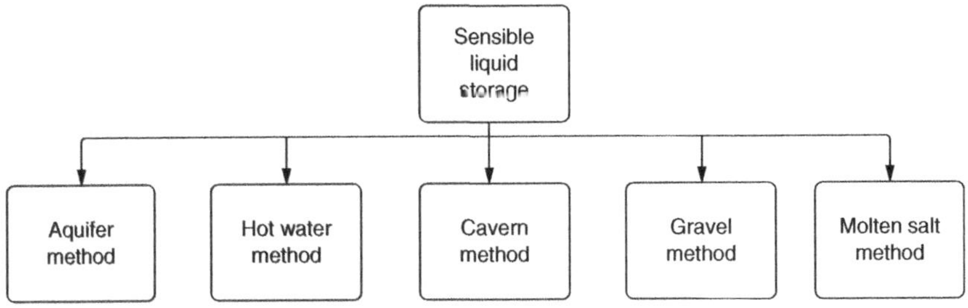

Figure 13.3 Sensible liquid storage.

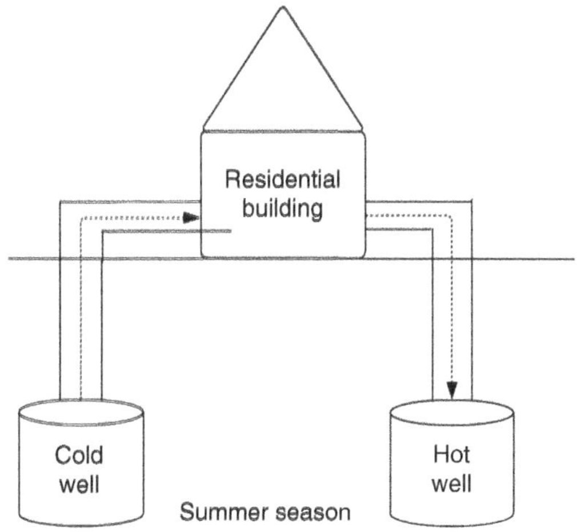

Figure 13.4 Aquifer storage (summer).

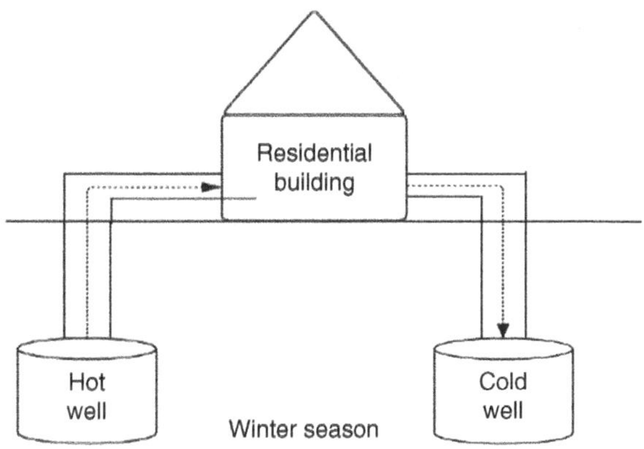

Figure 13.5 Aquifer storage (winter).

operations. This system's ability to function is mostly dependent on the site's geology. Tas et al. [37] conducted distance-based global sensitivity analysis (DGSA) to analyze energy storage parameters such as hydraulic conductivity and hydraulic gradient Figure 13.5.

13.1.9 Hot Water Energy Storage

During the summer, this method uses solar energy gathering to provide hot water. The produced hot water is kept in high-capacity tanks that are either buried entirely or partially in the ground, or they can be left standing on the ground.

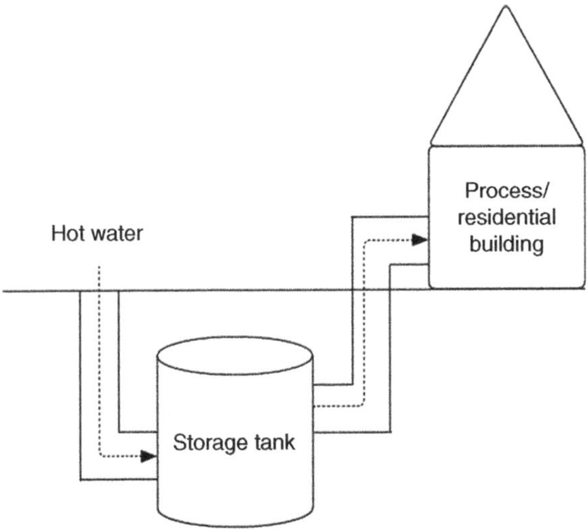

Figure 13.6 Hot water storage.

The tanks are constructed with adequate insulation to prevent heat loss and maintain thermal energy for the extended periods of time required to store temperatures between 80 and 90 °C. Figure 13.6.

13.1.10 Cavern Energy Storage

It is one of the technologies to store energy for integrating renewable energy by utilizing the naturally available geological caverns for storing energy, compressing air for future usage. If the caverns are manmade, then this method is unfeasible in terms of an economic point of view. Huang et al. [38] highlighted that the development of cavern storage helps mitigate the problems associated with a high amount of energy after removing the risk of leakage.

13.1.11 Gravel Energy Storage

It is also an underground energy storage by employing gravel water mixture to store the maximum energy with a temperature range of 80–90 °C by constructing a huge tank underground with proper insulation. The heat energy will be transferred to the gravel by either direct or indirect means. Using inexpensive materials for energy storage is the primary benefit of the gravel energy storage method, which has a lower specific heat capacity than other methods. The primary benefit of this approach is the use of inexpensive material for energy storage.

13.1.12 Molten Salt Energy Storage

This technique stores the thermal energy in a liquid that is created by the inorganic salt. Both direct and indirect methods are available for storing thermal energy. The same substance is utilized in the direct technique to both store and absorb heat energy from the source. The indirect approach uses molten salt to absorb heat energy from a source and uses a heat exchanger to transmit the energy to another medium of storage material. Due to the molten salt's high heat storage capacity, low viscosity, low vapor pressure, and high boiling temperature, this method works well at temperatures above 100 °C. The resistance element for choosing this method is the high freezing point of molten salts. Adding an extra heating system to the transportation line will help prevent freezing.

Wang et al. [39] conducted an analysis on molten thermal energy storage by nitrate molten salt and observed changes in their physical properties, thermal stability, energy storage density, and economics. Wang et al. concluded that the changes observed are less than 10% in storage media. Also, the authors suggested that the nitrate salt can be a suitable energy storage medium because of its melting temperature of 102.3 °C and decomposition temperature of 651.2 °C. Figure 13.7.

13.1.13 Borehole Energy Storage

In this underground energy storage system, thermal energy is stored and transferred to the storage material through a U-shaped tube composed of high-density polyethylene. The fluid used in borehole energy storage is a mixture of additives and antifreeze liquids, such as ethylene glycol and ethanol. The performance of this energy storage depends upon the material of the pipe, the number of pipes, and the conductivity of the storage material. The expense associated with building the system limits the use of this technology.

13.1.14 Packed-Bed Energy Storage

Using this method, thermal energy from solar radiation is captured by the air or water and transferred to a packed bed of sand or rock, which serves as a storage

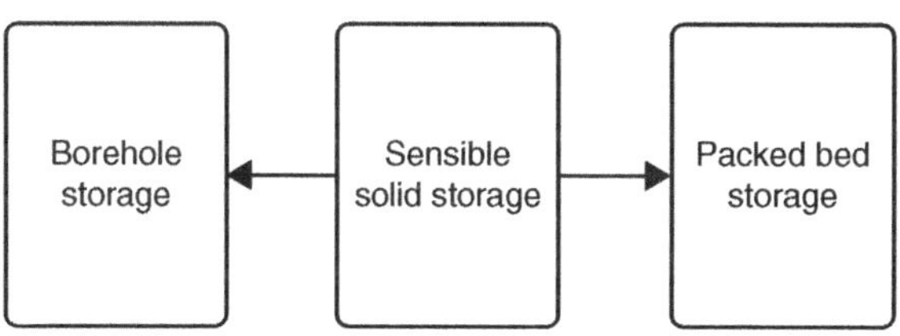

Figure 13.7 Sensible solid storage.

medium. During the need, the heat energy stored in these materials is recovered by circulating cold water or low-boiling liquids across the packed bed. The performance of this system is influenced by several factors, including the porosity, size, shape, and rate of heat transfer of the storage material. It is mostly used for space heating applications. Eddemani et al. [40] conducted a comparative study on packed-bed materials and concluded that quartzite has high potential for use in large-scale applications. Rabi et al. [41] investigated the impact of changing diameter, aspect ratio, and void fraction of packed-bed materials.

13.1.15 Latent Heat Storage

Latent heat storage method stores energy inside a material by changing its phase while retaining a constant temperature solid to liquid, liquid to solid, solid to gas, gas to solid, or gas to liquid. Latent heat is defined as the energy emitted or absorbed when a phase transition happens at a fixed temperature. It is possible to extract significant latent heat from solid–gas or liquid–gas conversion; however, the primary reason to avoid this phase transition is contamination. Solid–liquid phase change is typically chosen for heat storage because it offers moderate late heat generation and manageable volume changes [6]. Based on the phase change process and the type of storage material employed, latent heat storage systems are categorized. Latent heat storage systems are further categorized based on the phase transition process: (a) solid–solid, (b) solid–liquid, and (c) solid–gas.

13.1.15.1 Latent Heat Energy Storage by Phase Change Material

Due to the advantages of having a high energy storage density and capacity over other technologies, thermal energy storage systems using phase change materials (PCMs) are recommended for most applications [42]. Another reason to use this method instead of others is that the material releases energy at the same temperature. Generally, the energy storage material is categorized as organic and nonorganic. Despite having a high energy density, organic phase transition materials still release stored energy at low temperatures (below 120 °C); therefore, adequate insulation is necessary for long-term storage. Because of their longer carbon-hydrogen bonds, paraffin waxes are typical organic material types with high fusion temperatures and large energy storage capacities.

In operating temperature ranges of 120–1000 °C, nonorganic phase transition materials are appropriate for preventing heat loss during storage. These materials come in the form of metallic and salt hydrates, and they both have the crucial characteristics of having a high thermal conductivity and latent heat of fusion. The energy storage capacity of these materials is 5–14 times higher than that of sensible method storage materials. By including components like graphite, silicon carbide, and ceramics, PCMs can attain excellent thermal conductivity.

A good PCM must possess the following characteristics to be selected for energy storage: (a) No freezing property at operation temperature, (b) should have high phase change enthalpy, (c) low vapor pressure at the operation temperature, and (d) should have melting and solidification properties at defined temperatures. But paraffin has poor thermal conductivity, energy storage density, and potential fire hazard. Duquesne et al. [43] classified active, passive, and hybrid according to the applications.

13.1.15.2 Encapsulation of PCM

Encapsulation is used in PCMs to improve the transfer of heat, and this approach also eliminates the PCM leakage issue by encapsulating the PCM in thin-layered capsules within the application area. In general, polymer-based containers are employed in low-temperature applications, whereas metal layers tend to be used for concentrated solar power applications. Designing capsules for salt-based storage systems requires taking corrosive variables into consideration. The design of the capsule must take into account the excessive volume in order to accommodate the material's volume variations during the phase transition process. Liao et al. [44] state that encapsulating PCMs by NaCl—Al_2O_3 or Sic—Al_2O_3 provides additional benefits like an increase in thermal conductivity and prolonged thermal stability at high operating temperature ranges of 850 °C for more than 1000 hours of operation. Rahi et al. [45] highlight that the provision of encapsulation layer over PCMs reduces the chances of flammability and helps mitigate the issues of phase separation.

13.1.15.3 Latent Heat Energy Storage by Salt Hydrates

The cheap cost, less corrosive, and less poisonous properties of salt hydrates make them an excellent choice for latent heat storage as an energy storage material. When salt hydrates go through a solid–liquid phase transition, the energy-storing and -releasing processes are reversible. The solid salt hydrates release water of crystallization with a small volume change and a large energy release during the phase change process. As mentioned by Thomas Bauer et al., the following reactions indicate the process of energy storage.

$$X(Y) \cdot n\mathrm{H_2O} \rightarrow X(Y)i + n\mathrm{H_2O}$$
$$(X_9Y)i \cdot n\mathrm{H_2} \rightarrow X(Y)I \cdot 0\mathrm{H_2O} + (n-0)\mathrm{H_2O}$$

From the above equation, salt hydrates produce anhydrous salt and water during the melting process. Due to a lack of water, some solid particles are unable to come into contact with water particles during the solidification process. This can be solved by adding external agents to hold the solid particles together. Li et al. [46]

suggested that the method of adding potassium chloride and other thickening agents such as carboxylmethyl cellulose helps improve the properties of PCMs.

13.1.16 Thermochemical Energy Storage

It is an indirect method of energy storage based on reversible thermochemical reactions, which utilizes chemical bonds of materials to store the heat energy. The amount of energy stored by the thermochemical materials is based on their type, quantity, and enthalpy. It also relies on the mass and rate at which the storage material's chemical reactions occur. During an endothermic reaction, heat energy is stored inside the individual components of thermochemical materials. The stored energy is liberated during an exothermic reaction when individual components combine with each other. Thermochemical energy storage is the preferred approach for long-term energy storage because it has a higher energy storage density than other thermal energy storage methods.

Endothermic Reaction:

$$XY \rightarrow X + Y(\text{Energy Abosrption})$$

Exothermic Reaction:

$$X + Y \rightarrow XY\ (\text{Energy Liberation})$$

13.2 Artificial Intelligence (AI)

Artificial Intelligence (AI) is a contemporary tool that simulates human thought processes in machines so they may act and behave. Other benefits like problem-solving and decision-making will be helpful to take quick response with respect to the difficulty in operating conditions, and these features facilitate error minimization and human interventions [17]. Implementation of AI in the energy sector helps make a smarter energy management system to provide the following features: (a) forecasting of energy generation and production, (b) modeling of the energy system, and (c) improving efficiency by effective energy management [7, 9, 10]. Also, the use of AI ensures the safety of the energy storage system.

13.2.1 AI in Energy Sector

The application of AI tools in design, modeling, and analysis of energy storage systems helps overcome the issues in the transformation of energy sector from conventional storage systems. The modeling and design of thermal latent heat energy storage using PCM are complicated and involve complex mathematics [11]. These issues are solved by adopting AI tool, which will simplify the complexity of modeling. Omojola et al. [46] confirm that the implementation of AI helps monitor the battery performance, safety, and durability by the preset predictive maintenance [16].

13.2.1.1 Artificial Neural Network (ANN)

ANN is an efficient computing system using nodes (A group of interconnected neurons) working similar to biological neurons of humans to transmit signal from one node to another connected node. The frequency of the signal transmitted is adjusted based on the weight and biases of the node transmitting the signal to others [12]. ANN is mostly used to derive solutions for nonlinear engineering problems. Zhang et al. [48] conducted research on energy storage systems driven by data control model and confirmed that the system has exhibited accurate control with an enhanced efficiency of 31.3% and improved heat storage capacity by reducing operational costs to 40.6%. Sudhir et al. [49] experimented machine learning in thermal energy storage systems and found a prediction error in melting fraction time of 10%. Lack of standardization and insufficient data are limiting the application of AI in a wide range of thermal energy storage systems.

13.2.1.2 Fuzzy Logic (FL)

Fuzzy Logic is favorable to solve engineering problems, which have insufficient data and information. FL has sections called rule base, inference engine, and fuzzification. Rule base has expert conditions like IF-THEN. Based on the rule base conditions, inference engine monitors the FL and input rule. FL uses graphical information to display the output in the form of 0 and 1. Fuzzy logics are highly suitable for renewable energy integration because of their varying nature conditions [51, 52]. Ibrahim et al. [50] analyzed the uses of FL control in battery aging control and found a remarkable reduction in operating costs by managing sources effectively to enhance system's efficiency. Usanova et al. [20, 53] applied an FL control system for energy management and concluded that the system has shown enhancement in state of charge due to optimized cycle of operation.

13.2.1.3 Adaptive Neuro Fuzzy Inference System (ANFIS)

It is a combination technique of FL and ANN, which is widely used for solving nonlinear problems of engineering. ANFIS models the input—output data based on simulations by employing neural networks to study the system performance with respect to the operating conditions and analyzes using FL to manage the energy supply and demand. Peng et al. [21, 53] conducted research on AI-controlled energy storage systems using FL control and concluded that FL control stabilizes the system to provide a constant voltage supply [22].

13.2.1.4 Particle Swarm Optimization (PSO)

Particle swarm optimization based on the concept of bird flocking randomized approach. In this approach, the solution is called a particle (or) bird, which occupies a place in the problem domain [24, 25]. The optimization starts from assuming an approximate solution for the problem, and then iteration will happen for each particle to get the best solution [23, 26]. Generally, PSO uses weight-based hybrid algorithms in energy storage systems to provide cost-efficient and stabilized operation. Pan et al. [54] confirmed through the research that PSO enhances energy storage by frequency modulation through bridging the state of charge and power.

13.2.1.5 Support Vector Machine (SVM)

Support vector machine is used to solve the linear quadratic problem, which is an unusual method but helps minimize the error in finding solutions. SVM plays a critical role in forecasting and management applications of energy storage systems [27]. Feng et al. [55] conducted research on the optimization of energy storage system using SVM by controlling and optimizing nonlinear behavior of energy storage systems. Shrivastava et al. [56] confirmed that the application of SVM helps reduce operating costs by optimizing the charging and discharging cycle of energy storage systems.

13.2.1.6 Implementation of AI in Energy Storage

Generally, AI is used in energy storage for predicting, optimizing, and controlling operations. Energy storage system effectiveness, efficiency, and user-friendliness will all be improved with the integration of AI tools. ANN application is essential for determining the proportion of liquid and solid in PCM thermal energy storage. Also, finding the heating time and water temperature in the heating application of an energy storage system will also be useful in order to conserve money and energy.

FL is familiar for predicting the efficiency of the thermal collectors and temperatures of the heat transfer fluid and heat carrier to maximize the energy transfer for increased heat exchange effectiveness. ANFIS helps in the effective design of solar-powered energy storage using molten salt for the enhancement of maximum energy storage [13, 15]. The PSO technique is utilized for the estimation of solar collector efficiency for space heating applications.

13.3 Conclusion

Traditional storage needs to be updated in all areas pertaining to current technologies in order to expand storage capacities and achieve lower losses, lower transportation and energy storage costs, more cycles of charging and discharging, and environmental friendliness. The mechanical energy storage method compresses air storage and pumped hydro storage monopolize the maximum energy for the operation of pump and compressor, this energy contemplate only a small portion for future usage.

Systems for storing thermal energy in hot water, caverns, and gravel are suitable for home use and water heating, but they are not designed to store large amounts of energy for extended periods of time. Thermal energy storage devices have individual storage capacities that correspond to their specific design specifications, but they do not adapt to changes in usage or design.

The current global energy difficulties necessitate innovations in flexible storage, which is achieved by combining additives with storage material to alter the material's properties relevant to the operating requirements. This allows energy to be reserved for long-term use. In the near future, advances in energy storage will be made possible by the application of technologies like machine learning and AI to assess and track system progress. Also, the development of dual thermal energy storage media by combining different forms of sensible, latent heat together will possibly yield a good impact on the energy sector.

Implementation of AI, machine learning, and deep learning in the development of Smart energy storage systems will provide advantages of prediction, control, and modeling for efficient energy storage with flexibility in operation, cost saving, and control. Renewable energy sources are generally fluctuating in nature, and these difficulties are overcome by integrating AI tools in energy storage systems. In order to increase the reliability of energy storage, SEEs assist us in forecasting weather patterns, energy generation from sources, and future energy usage. Also, the quality of energy supplied to the grid from renewable energy sources is maintained by the smart energy storage systems.

The development of green energy needs the support of an energy management system, so the cost of implementing such energy into the commercial market requires a huge investment for the investors. As a result, the rapid development of energy storage systems as well as renewable energy generation is lagging. On the other hand, modern technology and tools provide benefits in addressing design complications, reducing manufacturing cost, and solving the problems in the control of energy storage systems. In addition to the long operation cycle, AI technologies help reduce the usage of fossil fuels, coal, petrol, and diesel. The reduction of fossil fuel usage reduces the impact of harmful pollutants on the environment, which will be helpful to save the earth and human life.

Integration of modern tools in energy storage systems offers wide opportunities for researchers to focus on and develop effective energy storage systems. Hybrid energy storage systems like PCM-based latent heat energy storage performance are optimized through SVM, FL, and ANFIS to extend energy storage capacity. AI techniques also provide benefits of reduction in space, decreasing cost, and minimizing error to increase efficiency.

References

1 Bauer, T., Wolf-Dieter, D.L., and Tamme, R. (2012). Thermal energy storage materials and systems. *Annual Review of Heat Transfer* 15: 131–177.

2 Kumar, A. and Shukla, S.K. (2015). A review on thermal energy storage unit for solar thermal power plant application" Science Direct. *Energy Procedia* 74: 462–469.

3 Chavan, S., Rudrapati, R., and Manickam, S. (2022). A comprehensive review on current advances of thermal energy storage and its applications. *Alexandria Engineering Journal* 61: 5455–5463.

4 Sadeghi, G. (2022). Energy storage on demand: thermal energy storage development, materials, design and integration challenges" Elsevier. *Energy Storage Materials* 46: 192–232.

5 Mitali, J., Dhinakaran, J., and Moahmad, A.A. (2022). Energy storage and saving" KeAi. *Energy Storage and Saving* 1: 166–216.

6 Agyenim, F., Hewit, N., Eames, P., and Smyth, M. (2010). A review of materials, heat transfer and phase change problem formulation for latent heat thermal energy storage systems (LHTESS). *Renewable and Sustainable Energy Reviews* 14 (2): 615–628.

7 Olabi, A.G., Abdelghafar, A.A., Maghrabie, H.M. et al. (2023). Application of artificial intelligence for prediction, optimization, and control of thermal energy storage systems. *Thermal Science and Engineering Progress* 39: 101730.

8 A.G. Olabi, C. Onumaegbu, T. Wilberforce, M. Ramadan, M.A. Abdelkareem, A. H. Al–Alami, "Critical review of energy storage systems", *Energy* 214 (2021), 118987 accessed date January 2024.

9 J. Frankenfield. (2021). Artificial intelligence. https://www.investopedia.com/terms/a/artificial-intelligence-ai.asp (January 16, 2024).

10 Nilsson, N.J. (2014). *Principles of Artificial Intelligence*. Morgan Kaufmann.

11 M. van Someren and I. d. I. e. I. Artificial. MLnet (1993). Workshop on learning and problem solving: artificial intelligence. Research Institute IIIA-CSIC, Blanes, Spain. (23 September 1993).

12 K.D. Foote. (2021). Artificial neural network: an overview. http://www.dataversity.net/artificial-neural-networks-overview/ (accessed January 16, 2024).

13 Milidonis, K. et al. (2021). Review of application of AI techniques to solar tower systems, solar energy. *Solar Energy* 224: 500–515.

14 Waag, W. et al. (2014). Critical review of the methods for monitoring of lithium-ion batteries in electric and hybrid vehicles. *Journal Power Sources* 258: 321–339.

15 Ahmed, N.A., Nazir, M.S., Tao, H. et al. (2021). Integration of energy storage system and renewable energy sources based on artificial intelligence: An overview. *Journal of Energy Storage* 40: 102811.

16 Xu, B. et al. (2015). Application of phase change materials for thermal energy storage in concentrated solar thermal power plants: a review to recent developments. *Applied Energy* 160: 268–307.

17 Youssef, A. et al. (2017). The role of artificial intelligence in photo-voltaic systems design and control: a review. *Renewable and Sustainable Energy Reviews* 78: 72–79.

18 Vasantharaj, S., Indragandhi, V., Subramaniyaswamy, V. et al. (2021). Efficient control of DC microgrid with hybrid PV—fuel cell and energy storage systems. *Energies* 14 (11): 3234.

19 Alirahmi, S.M., Mousavi, S.B., Razmi, A.R., and Ahmadi, P. (2021). A comprehensive technoeconomic analysis and multi-criteria optimization of a compressed air energy storage (CAES) hybridized with solar and desalination units. *Energy Conversion and Management* 236: 114053.

20 Zeinalnezhad, M., Chofreh, A.G., Goni, F.A., and Kleme˘s, J.J. (2020). Air pollution prediction using semi-experimental regression model and adaptive neuro-fuzzy inference system. *Journal of Cleaner Production* 261: 121218.

21 Sonmez, A.Y., Kale, S., Ozdemir, R.C., and Kadak, A.E. (2018). An adaptive neuro-fuzzy inference system (ANFIS) to predict of cadmium (Cd) concentrations in the Filyos River, Turkey. *Turkish Journal of Fisheries and Aquatic Sciences* 18 (12): 1333–1343.

22 Jang, J.-S. (1993). ANFIS: adaptive-network-based fuzzy inference system. *IEEE Transactions on Systems, Man, and Cybernetics* 23 (3): 665–685.

23 Kennedy, J., Eberhart, R.1995). Particle swarm optimization. *Proceedings of ICNN'95-international Conference (27 November–1 December 1995) on Neural Networks* IEEE.

24 Jaberipour, M., Khorram, E., and Karimi, B. (2011). Particle swarm algorithm for solving systems of nonlinear equations. *Computers and Mathematics with Applications* 62 (2): 566–576.

25 Dashtdar, M. et al. (2022). Improving the power quality of island microgrid with voltage and frequency control based on a hybrid genetic algorithm and PSO. *IEEE Access* 10: 105352–105365.

26 Kuo, B.-C., Ho, H.-H., Li, C.-H. et al. (2013). A kernel-based feature selection method for SVM with RBF kernel for hyperspectral image classification. *IEEE Journal of Selected Topics in Applied Earth Observations and Remote Sensing* 7 (1): 317–326.

27 Shao, M., Wang, X., Bu, Z. et al. (2020). Prediction of energy consumption in hotel buildings via support vector machines. *Sustainable Cities and Society* 57: 102128.

28 Ölmez, M.E., Ari, I., and Tuzkaya, G. (2024). A comprehensive review of the impacts of energy storage on power markets. *Journal of Energy Storage* 91: 111935. https://doi.org/10.1016/j.est.2024.111935.

29 Njema, G.G., Ouma, R.B.O., and Kibet, J.K. (2024). A review on the recent advances in battery development and energy storage technologies. *Journal of Renewable Energy* 2024: 1–35. https://doi.org/10.1155/2024/2329261.

30 Bustamante, J. S., Singh, A., Soto Crisanto, M. I. et al. (2023). Comparative analysis of mechanical energy storage systems. *Proceedings of the* (6–8 August 2023). Long Beach California, USA 10.1115/power2023-108530.

31 Andújar, J.M., Segura, F., and Rey Luengo, J. (2023). Electrical storage. In: *Energy Storage Systems: Fundamentals, Classification and a Technical Comparative*, 49–57. Springer Science+Business Media https://doi.org/10.1007/978-3-031-38420-2_4.

32 Chen, M. and Li, Y. (2024). Fundamentals of electrochemical energy storage technologies. In: *Towards Next Generation Energy Storage Technologies: From Fundamentals to Commercial Applications* (ed. M. Chen), 3–30. https://doi.org/10.1002/9783527845316.ch2.

33 Mombeshora, E.T., Muchuweni, E., Davies, M. et al. (2024). *Electrochemical Capacitors: Basic Concepts and Emerging Nanomaterials for Electrodes*, 83–118. Elsevier BV https://doi.org/10.1016/b978-0-323-96125-7.00030-7.

34 Gharahcheshmeh, M. and Chowdhury, K. (2024). Fabrication methods, pseudocapacitance characteristics, and integration of conjugated conducting polymers in electrochemical energy storage devices. *Energy Advances* 3: https://doi.org/10.1039/d4ya00504j.

35 Singh, A. K., & Lakeh, R. B. (2024). Thermal energy storage conceptual design using reclaimed minerals as heat storage material. *Proceeding of the* (15–17 July 2024). Anaheim, California, USA. 10.1115/es2024-131375.

36 Katekar, V.P., Rao, A.B., and Sardeshpande, V. (2023). An experimental investigation to optimise pebbles-based sensible heat storage system: an exploration to improve thermal efficiency of solar devices. *Journal of Energy Storage* 73: https://doi.org/10.1016/j.est.2023.108964.

37 Tas, L., Hartog, N., Bloemendal, M. et al. (2025). Efficiency and heat transport processes of low-temperature aquifer thermal energy storage systems: new insights from global sensitivity analyses. *Geothermal Energy* 13 (1): https://doi.org/10.1186/s40517-024-00326-1.

38 Huang, S., Li, Y., Shi, X. et al. (2024). The role of underground salt caverns in renewable energy peaking: a review. *Energies* 17 (23): 6005. https://doi.org/10.3390/en17236005.

39 Wang, Y., Lu, Y., Wang, Y. et al. (2024). Investigation on thermal performance of quinary nitrate/nitrite mixed molten salts with low melting point for thermal energy storage. *Solar Energy Materials & Solar Cells* 270: https://doi.org/10.1016/j.solmat.2024.112803.

40 Eddemani, A., El Baamrani, H., Aharoune, A. et al. (2024). Comparative study of the performance of packed beds using different types of thermal storage materials. *Solar Energy and Sustainable Development Journal* 14: 176–184. https://doi.org/10.51646/jsesd.v14isi_msms2e.395.

41 Rabi, A.M., Radulović, J., and Buick, J.M. (2024). Packed bed thermal energy storage system: parametric study. *Thermo* 4 (3): 295–314. https://doi.org/10.3390/thermo4030016.

42 Abdulsitar, A.S. and Abdullah, A.Y. (2024). Heat transfer improvement in latent heat thermal energy storage: a review paper. *International Journal of Science and Research Archive* 13 (1): 1820–1835. https://doi.org/10.30574/ijsra.2024.13.1.1857.

43 Duquesne, M. and Jomaa, W. (2024). Latent heat storage systems. In: *Heat and Cold Storage 1: Sensible and Latent Storage* (ed. P. Odru and E.P. del Barrio), 109–149. Wiley https://doi.org/10.1002/9781394312528.ch6.

44 Liao, S., Zhou, X., Chen, X. et al. (2024). Development of macro-encapsulated phase-change material using composite of NaCl-$Al2O_3$ with characteristics of self-standing. *Processes* 12 (6): 1123. https://doi.org/10.3390/pr12061123.

45 Rahi, M., Anand, P., Jagirdar, M.N.S. et al. (2024). *Development of Microencapsulated Phase Change Materials*, 41–67. Informa https://doi.org/10.1201/9781003331957-3.

46 Omojola, A.F., Ilabija, C.O., Onyeka, C.I. et al. (2024). Artificial intelligence-driven strategies for advancing lithium-ion battery performance and safety. *Int. J. Adv. Eng. Manag* 6 (10): 452–484. https://doi.org/10.35629/5252-0610452484.

47 Li, B., Ren, B., Yu, J., and Wang, Y. (2024). Preparation of a novel sodium acetate trihydrate-based composite phase change material and thermal performance of its integration in a coil-type thermal energy storage unit for heat pump system. *Journal of Energy Storage* 88: https://doi.org/10.1016/j.est.2024.111648.

48 Zhang, Y., He, Z., Guo, W., and Zhang, P. (2024). Data-driven optimization of packed bed thermal energy storage heating system with encapsulated phase change material. *Journal of Energy Storage* 79: https://doi.org/10.1016/j.est.2023.110017.

49 Sudhir, P. S., & Banerjee, D. 12th October to 2023 to 2nd November 2023). Exploring efficacy of machine learning (artificial neural networks) for enhancing reliability and resilience of thermal energy storage platforms utilizing phase change materials for sustainability and mitigating food-energy-water (FEW) nexus. *Proceeding of the* New Orleans, Louisiana, USA. 10.1115/imece2023-117109.

50 Ibrahim, O., Abdul Aziz, M.J., Ayop, R. et al. (2024). Fuzzy logic-based particle swarm optimization for integrated energy management system considering battery storage degradation. *Results in Engineering* 24: 102816. https://doi.org/10.1016/j.rineng.2024.102816.

51 Ali, H.S.F. and Ali, M.S.F. (2024). Fuzzy logic-based power management in grid-connected solar energy systems. *European Journal of Social Sciences Studies* 10 (4): https://doi.org/10.46827/ejsss.v10i4.1843.

52 Usanova, K.I., Rao, D., Pandey, S. et al. (2024). Fuzzy logic-based energy management in sustainable management for renewable integration. *E3S Web of Conferences* 537: 08003. https://doi.org/10.1051/e3sconf/202453708003.

53 Peng, T., Wan, R., Xu, Y., et al. (2023). An energy control strategy based on adaptive fuzzy logic for onboard hybrid energy storage system. (19 November 2023). Singapore IEEE 1–6 https://doi.org/10.1109/iecon51785.2023.10312025.

54 Pan, Y. and Y, Y. (2024). Optimizing adaptive particle swarm for combined fire and storage control FM energy reserve. *The Journal of Combinatorial Mathematics and Combinatorial Computing* 123 (1): 449–459. https://doi.org/10.61091/jcmcc123-32.

55 Feng, L., Li, S., Yang, J. et al. (2024). Research on energy storage configuration optimization of microgrid based on GCC-PHAT and ARIMA-SVM model. *2024 IEEE 6th International Conference on Power, Intelligent Computing and Systems (ICPICS)* (24 December 2024). Shenyang, China IEEE. 10.1109/icpics62053.2024.10796607.

56 Shrivastava, A., Raturi, A., Gupta, H.,et al. (2023). Battery energy management systems in active distribution network. *2023 3rd International Conference on Pervasive Computing and Social Networking (ICPCSN)* (04 October 2023). Salem, India IEEE. 10.1109/icpcsn58827.2023.00123.

14

Application of IoT and Machine Learning to Improve Biogas Production Through Anaerobic Digestion

Akshay Jain[1] *and Bhaskor Jyoti Bora*[2]

[1] *Department of Mechanical Engineering, National Institute of Technology, Hamirpur, India*
[2] *Rajiv Gandhi Institute of Petroleum Technology, Sivasagar Campus, India*

Abbreviations

AD	Anaerobic digestion
AI	Artificial intelligence
ANN	Artificial neural network
BG	Biogas
CCF	Centered composite design
COA	Cuckoo optimization algorithm
ELM	Extreme learning machine
ERWCA	Evaporation-rate water cycle algorithm
DET	Direct electron transfer
DNN	Deep neural network
GA	Genetic algorithm
HRT	Hydraulic retention time
IoT	Internet of Things
KNN	K-nearest neighbors
LCA	Leagues championship algorithm
ML	Machine learning
MECs	Microbial electrolysis cells
MVO	Multiverse optimization algorithm
MLP	Multilayer perceptron
NARX	Nonlinear autoregressive exogenous
RF	Random forest
RNN	Recurrent neural network
SMY	Specific methane yields

Machine Learning for Sustainable Energy Solutions, First Edition. Edited by Zafar Said and Prabhakar Sharma.

SFLA	Shuffled frog-leaping algorithm
SMA	Slime mold algorithm
SMOTE	Synthetic minority oversampling technique
SMOTER	SMOTE for regression
SVM	Support vector machines
TFT	Temporal fusion transformer
TLBO	Teaching learning-based optimization
VSNs	Variable selection networks

14.1 Introduction

In numerous global regions, bioenergy plays a vital part in nurturing local financial development, indicating climate change, and enhancing energy security [1]. Given the imperative to replace substantial quantities of fossil energy with renewable alternatives, bioenergy is poised to serve as a crucial supplement to other renewable sources, including hydropower, solar, and wind, in achieving this transition [2]. Anaerobic digestion (AD) technology is attracting significant interest due to its environmentally friendly nature and high efficiency [3]. It can transform organic wastes into biogas (BG), mainly consisting of carbon monoxide and hydrogen, as well as digestate, a byproduct generated by a variety of bacteria during the AD process [4].

14.2 Biogas Production

The process of BG generation, a well-established biofuel production method, presents a promising opportunity for the efficient utilization of diverse raw materials to address contemporary energy requirements. This method not only profits individual users but also pays for the inclusive welfare of the community while ensuring the conservation of natural resources. Essentially, BG serves as a sustainable alternative for managing the entire life cycle of organic waste, aligning with the principles of circular economy and biorefinery through open-cycle recycling, where waste materials are repurposed as valuable resources [5]. BG power generation, which utilizes organic waste from humans and animals, is an essential component of the circular economy and a source of renewable energy. Global energy requirement is anticipated to rise by two- to threefold in the current century, placing significant strain on resources. Fossil fuels, however, are capable of meeting only around 88% of this heightened demand for global energy, as per the

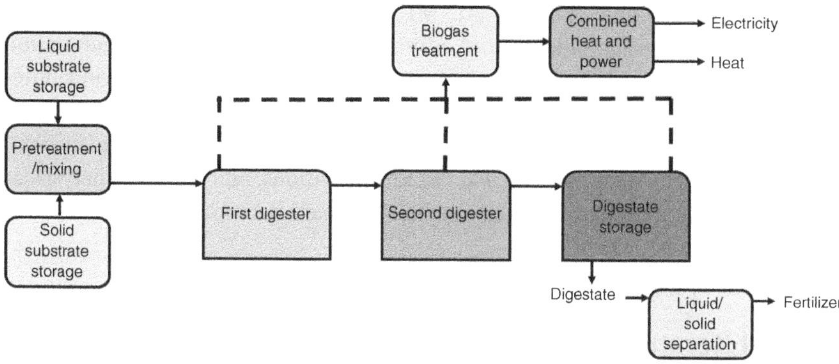

Figure 14.1 Processes followed in biogas plant. *Source:* Mathur [9]/with permission of Elsevier.

International Energy Agency [6]. The natural production of methane (CH_4) from the blend of BG is a key aspect of the chemical reaction occurring during AD. The hurdles in the successful establishment of BG facilities can be considered in the following tiers:

i) data, marketing, financial, societal, technical infrastructure, and monitoring challenges (tier one),
ii) in-depth understanding of the method (tier two), and
iii) various analytical tools for assessing the performance of the BG facility (tier three) [7].

Co-digestion, which involves simultaneous AD of numerous waste streams, is becoming increasingly popular due to its capacity to enhance BG and CH_4 production when compared with single-feedstock AD. Co-digestion serves as a technique for optimizing the carbon-to-nitrogen (C/N) ratio in the digestion process, leading to improved overall efficiency and performance [8]. Figure 14.1 portrays the pictorial representation of the process followed in the BG plant.

14.3 Techniques for Biogas Production Enhancement

In order to enhance AD performance and ensure real-time monitoring, process stability, optimization, and prediction of BG production processes are being facilitated through the use of machine learning (ML) methods [10]. The incorporation of cutting-edge technologies, such as artificial intelligence (AI), ML, and the Internet of Things (IoT), is essential for harnessing the power of IoT sensor data analysis. This integration facilitates instantaneous decision-making and process enhancement. By amalgamating AI, ML, and IoT strategies, we can greatly enhance bioprocess surveillance and administration, resulting in the generation of high-value chemicals and products from food waste. This, in turn, contributes to the enhancement of environmental sustainability and food security [11]. Researchers have explored the integration of Industry 4.0 techniques and circular economy concepts, including AI, IoT, cloud applications, and big data analysis, as a foundation for sustainable operations management decisions [12, 13].

14.4 Literature Review

Deep learning (DL) has enhanced the forecast accuracy of renewable energy. Although the application of DL in this field is in its early stages, there are numerous deep-learning models that have scope for research [14]. For instance, the analysts [15] used deep neural network (DNN) and support vector machine (SVM) algorithms for nonlinear time-series datasets to forecast wind force and direction for wind power generation. Researchers have also employed ML for predicting the direction of wind, providing valuable insights for the installation of wind turbines [16]. Research also endeavors to harness these techniques for sustainable BG energy production systems. The storage of data in the cloud using an IoT-based monitoring system can be analyzed using ML algorithms. However, there are certain challenges associated with automating the BG plant and implementing smart farming techniques [17–19]. The utilization of AI has been proposed as a more effective approach for optimizing and monitoring BG plant operations [20]. The field of ML has gained widespread popularity, particularly in circular economy literature, due to its predictive capabilities in the management, optimization, and retrieval of valued resources from organic waste. The application of ML is expected to significantly enhance the reuse, transparency, scalability, accuracy, and productivity of nutrient recovery, dissemination, and logistics [21]. ML can be used for responsibilities including association, estimation, clustering, and classification to analyze data, identify patterns, make predictions, or derive rules. The implementation of ML involves the following steps: preprocessing of data, development of the model, data collection and analysis, and implementation of the model. Through the stage of data collection, inputs and the expected outcomes are identified to form the basis for the ML process [22]. Challenges such as automated sorting systems, predictive maintenance, real-time monitoring, optimized collection routes, and public awareness and engagement can impede waste management efficiency [23]. Figure 14.2 provides a visual representation of the different categories of ML, and Figure 14.3 highlights the key algorithms specific to each group.

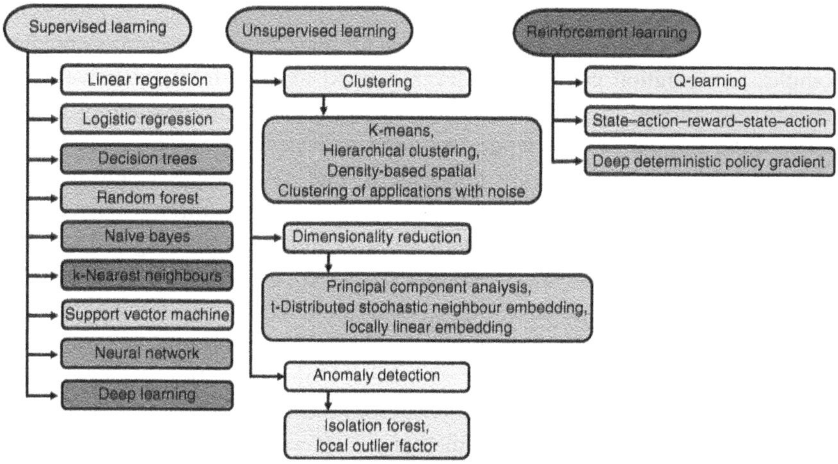

Figure 14.2 Elements of ML. *Source:* Munir [23]/with permission of Elsevier.

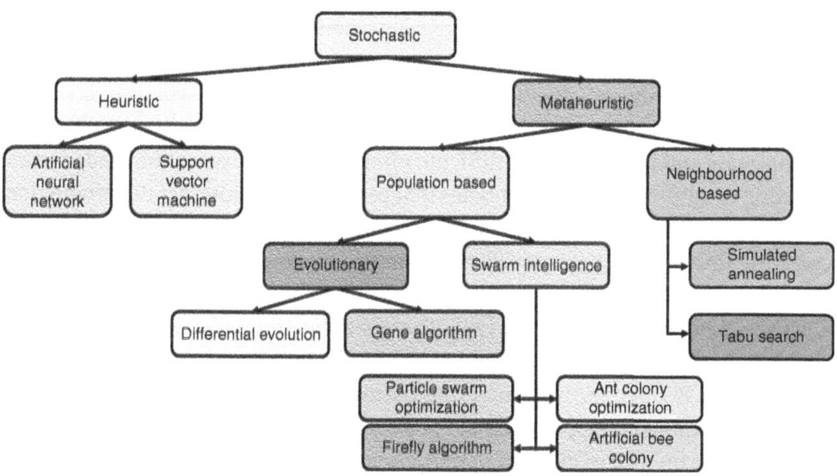

Figure 14.3 Classification of algorithms. *Source:* Mathur [9]/with permission of Elsevier.

Hybrid ML technique was utilized to enhance the estimation of BG production. This method combined long short-term memory (LSTM) and random forest (RF) analysis. The results revealed that the performance of this hybrid model surpassed its closest competitor, achieving a 20% increase in accuracy for predicting BG production [6]. The extreme learning machine (ELM) is an innovative algorithm designed for a feed-forward neural network with one hidden layer. ELM operates by randomly assigning weights to the input layer, which simplifies the process of optimization [24]. DNNs are highly esteemed for their ability to accurately predict complex fermentation processes by leveraging AI to learn from extensive datasets [25]. According to Yildirim and Ozkaya [26], maintaining stability in the AD presents a significant challenge due to the intricate nature of the system. Variation of feedstock, temperature fluctuations, and pH changes resulting from microbial processes can lead to instability, necessitating unremitting control and monitoring. By implementing increased unremitting monitoring and IoT utilization within the framework of Industry 4.0 in AD, stability of process control and prior involvement can be achieved. Integration of ML applications into AD facilities will enable real-time process control, thereby preventing low-efficiency BG production and maintaining process stability [26]. AD is a time-honored process for managing various organic waste sources while simultaneously producing renewable energy in the form of biomethane. However, AD is reliant on specific microbial communities and is thus susceptible to process disturbances that can impact BG production and methane yield. Consequently, there is a growing focus on process monitoring as a means of maintaining process stability and maximizing BG yield [27]. The conditions and intricacies of the AD process make it challenging for plant operators to obtain accurate measurements of many biochemical constituents within the digester tank. While parameters like pH and temperature are continuously monitored due to their ease of measurement, routine sampling inside the reactor is impractical. As a result, quantitatively assessing the digester's internal state and its efficiency is difficult, often relying on expert knowledge to set operating conditions. However, models that can effectively forecast BG production rates offer a solution to this challenge. Assuming that these models accurately represent the internal biochemistry of AD, it may help identify improved operational setpoints. Additionally, these models can provide high-quality multiday forecasts of expected BG production rates [28]. The quest for suitable predictive analytics models is of utmost importance in optimizing anaerobic fermentation processes in large-scale BG plants. Implementing cutting-edge ML tools can significantly improve operational efficiency and aid project operators in making informed decisions [29]. The predictive abilities of traditional ML algorithms are hindered when dealing with imbalanced data [30]. This poses a significant challenge when applying ML to forecast the performance of large-scale BG plants, where transmission failure, missed readings due to failure of sensor, and irregularities in operation are common issues [31]. To address this, the synthetic minority oversampling technique (SMOTE) may be employed to resolve the imbalanced data problem by creating additional data from minority groups [32]. Furthermore, Torgo et al. [33] have developed the SMOTE for regression (SMOTER), an algorithm based on the SMOTE approach, specifically tailored for regression tasks. SMOTER is a technique that facilitates the neural network's ability to comprehensively understand the potential correlations among various production parameters. Also, the genetic algorithm (GA) is a sophisticated optimization algorithm that leverages genetic traits to find optimal solutions [34]. Known for high scalability and faster optimization compared to traditional algorithms, researchers commonly integrate GA with ML models to enhance prediction accuracy [35]. GA-ANN may be utilized to simulate and optimize the AD process for aquatic weeds and potato waste [36]. The method accomplishes this by enhancing the distribution of imbalanced plant data and refining the neural network's ability to learn extreme parameters [24]. Further, Sappl et al. [28] have proposed the temporal fusion transformer (TFT), a new ML architecture designed for forecasting BG production rate. The TFT can capture fluctuations caused by various factors such as ambient temperature or public holidays, leading to a more comprehensive understanding. Moreover, the TFT incorporates an inherent attention mechanism that highlights the individual features crucial for forecasting. This is especially advantageous compared to black box models like ANNs, as it provides transparency into the decision-making process [28]. The BG production rate was evaluated using ANN that was developed with variable selection carried out using advanced optimization algorithms such as leagues championship algorithm (LCA), teaching learning based optimization (TLBO), multiverse optimization algorithm (MVO), cuckoo optimization algorithm (COA), evaporation rate water cycle algorithm (ERWCA), and stochastic fractal search (SFS), to synthesize the multilayer perceptron (MLP). The study found that certain hybrid techniques used in this process yielded better prediction outputs than traditional MLP rankings. Furthermore, after additional investigation, it was determined that the ERWCA outperformed the other algorithms [37].

14.5 Implementation of Mathematical Techniques for Biogas Production Enhancement

AD comprises the conversion of organic biomass into BG, consisting mainly of CH_4 and CO_2 [38, 39]. The production of methane is directly related to the degradation of organic matter, following specific chemical reactions represented by equations (14.1) and (14.2) [40–43]:

$$C_aH_bO_c + x\,H_2O \rightarrow y\,CH_4 + z\,CO_2 \tag{14.1}$$

$$CO_2 + 4\,H_2 \leftrightarrow CH_4 + 2\,H_2O \tag{14.2}$$

The ANN configuration developed by Avinash and Mishra [44] employs the feed-forward backpropagation algorithm for training multilayer neural networks using the trainlm algorithm. The network consists of five input factors in the input

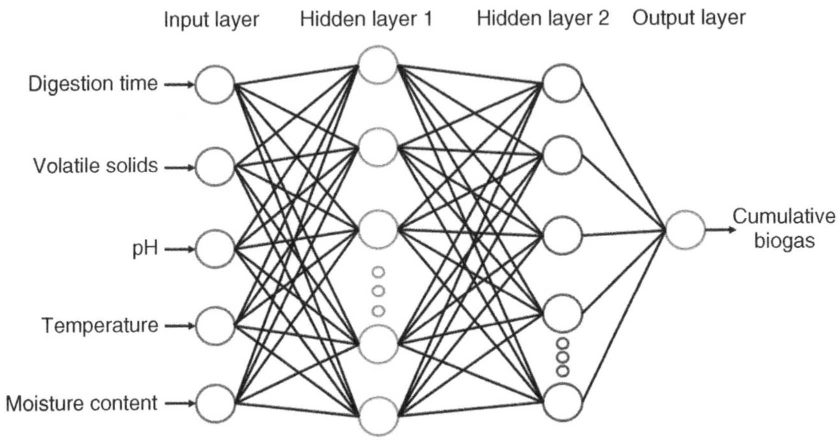

Figure 14.4 ANN-based architecture for BG prediction. *Source:* Avinash [44] / with permission of Elsevier.

layer (which includes moisture content, temperature, pH, volatile solids, and digestion time), two hidden layers, and one output layer for the output variable, as illustrated in Figure 14.4.

Savand-Roumi et al. [45] utilized a BG-specific e-nose to analyze BG samples collected from the headspace of a mesophilic BG digester. The study sought to classify the samples based on different hydraulic retention times (HRTs). The HRTs investigated included 30, 25, 20, and 15 days. The results indicated that HRTs of 30 days were identified as optimal for the process [45]. Sonwai et al. [46] employed ML techniques to forecast specific methane yields (SMY) using readings derived from the operational parameters of fully mixed reactors. The analysis revealed that the RF model outperformed other models in predicting SMY [46]. Zhang et al. [47] considered environmental factors as the input layer and microorganism levels as the second layer to build a third-layer H_2O AutoML framework for their work on a multilayer automated ML framework [47]. Hunter et al. [48] utilized a full-factorial, face-centered composite design (CCF) to conduct in silico training and testing of various model structures to ascertain the ideal artificial neural network (ANN) configuration. Leveraging a CCF design enables the elucidation of the interrelationships between hyperparameters [48]. Jeong et al. [49] introduced a novel hybrid deep learning architecture that combines a dual-stage attention-based long short-term memory network with variable selection networks. They conducted hyperparameter optimization to improve the model's predictability. The model's accuracy was authenticated with data from long-term anaerobic co-digestion monitoring [49]. Furthermore, the increasing amount of data has led to the use of data-driven ML models for modeling various processes and reactions. Offie et al. [50]

devised a comprehensive framework that encompasses several key stages, including data gathering and imputation, the implementation of a sensitivity analysis, the recurrent neural network (RNN)/nonlinear autoregressive exogenous (NARX) model, and the shuffled frog-leaping algorithm (SFLA) optimization model. The efficacy of this context was verified through its claim to a real micro AD plant. The NARX model was specifically tailored to predict BG yields based on historical feeding data, with its lag times adjusted using the SFLA. Furthermore, the SFLA was employed to identify the optimal daily feeding pattern for maximizing BG production. Results indicate that this framework can enhance BG productivity through optimal operational strategies. Sensitivity analysis revealed that the optimized BG production is predominantly impacted by the elements of catering waste and oats and optimal day for introducing feed when compared to other feed parameters [50]. Xiao et al. [51] developed several data-driven models for simulating microbial electrolysis cells (MECs) due to the ambiguity of traditional models in comprehending the mechanisms. They introduced a new general data-driven NARX-BP hybrid neural networks to foresee CH_4 generation in the biocathode MECs via direct electron transfer (DET). Compared to traditional one-stage models, this new model demonstrated excellent predictive capabilities for methane production via DET and revealed the mechanisms of BG upgrading. This systematic modeling approach holds the potential to enhance versatility and applicability by incorporating significant intermediate variables [51]. Alrowais et al. [52] compared the effects of thermophilic and mesophilic anaerobic co-digestion of wheat straw and waste activated sludge at various mixing ratios. They employed RNN with a slime mold algorithm (SMA) to simulate and optimize the process. This entailed determining the most effective structure of the RNN model to optimize the connection weights, number of feedback connections, and activation functions. The outcomes showed that co-digestion of straw and sludge resulted in enhancements to the C/N ratio and increased production of BG. Furthermore, the RNN model accurately predicted BG production, while the SMA efficiently identified the most suitable structure for the RNN model [52].

14.6 Conclusion

The traditional linear production process often leads to resource wastage. Utilizing renewable sources of energy reduces waste and protects the planet's resources. BG manufacturing encounters challenges such as high input costs and the lack of standardized manufacturing procedures. As a result, there is a shortage of predictive models using numerous algorithms to analyze critical input waste factors and forecast output in BG production. The lack of analysis of key waste factors can hinder production process optimization and stability, ultimately impacting the widespread adoption of BG production. This study explores various AI and ML techniques to redefine key factors and improve the estimation of BG production. By employing a hybrid ML approach, we can enhance parameter adjustment, identify crucial factors, and provide recommendations to users. These methods enable precise control over the final output and manufacturing process, leading to cost reductions and increased adoption of BG technology. This literature provides a comprehensive inspection of mathematical methodologies for modeling aerobic digestion and delves into diverse optimization approaches. It also investigates the application of ML and soft computing methods for controlling aerobic digestion facilities and presents suggestions for prospective research endeavors in this domain.

References

1 Tjutju, N.A.S., Ammenberg, J., and Lindfors, A. (2024). Biogas potential studies: a review of their scope, approach, and relevance. *Renewable and Sustainable Energy* 201: 114631. https://doi.org/10.1016/j.rser.2024.114631.

2 Gielen, D., Boshell, F., Saygin, D. et al. (2019). The role of renewable energy in the global energy transformation. *Energy Strategy Reviews* 24: 38–50. https://doi.org/10.1016/j.esr.2019.01.006.

3 Alharbi, R.M. (2024). Anaerobic co-digestion of cow manure and microalgae to increase biogas production: a sustainable bioenergy source. *Journal of King Saud University, Science* 36: 103380. https://doi.org/10.1016/j.jksus.2024.103380.

4 Li, B., Yun, S., Xing, T. et al. (2021). A strategy for understanding the enhanced anaerobic co-digestion via dual-heteroatom doped bio-based carbon and its functional groups. *Chemical Engineering Journal* 425: 130473. https://doi.org/10.1016/j.cej.2021.130473.

5 Maria, M.P., Torres, N.H., Nascimento, V.R.S. et al. (2023). Current advances in the brewery wastewater treatment from anaerobic digestion for biogas production: a systematic review. *Environmental Advances* 13: 100394. https://doi.org/10.1016/j.envadv.2023.100394.

6 Chiu, M.-C., Wen, C. Y., Hsu, H.-W., and Wang, W.-C. (2022). Key wastes selection and prediction improvement for biogas production through hybrid machine learning methods. *Sustainable Energy Technologies and Assessments* 52: 102223. https://doi.org/10.1016/j.seta.2022.102223.

7 Ilangovan, P., Sharmila Begum, M., and Srividhya, P.K. (2023). Development of online monitoring device and performance evaluation of biogas plants using enhanced methane prediction algorithm (EMPA). *Sustainable Energy Technologies and Assessments* 56: 103041. https://doi.org/10.1016/j.seta.2023.103041.

8 Ahmad, A., Yadav, A.K., Singh, A., and Singh, D.K. (2024). A comprehensive machine learning-coupled response surface methodology approach for predictive modeling and optimization of biogas potential in anaerobic co-digestion of organic waste. *Biomass and Bioenergy* 180: 106995. https://doi.org/10.1016/j.biombioe.2023.106995.

9 Mathur, P. and Singh, S. (2022). Analyze mathematical model for optimization of anaerobic digestion for treatment of waste water. *Materials Today Proceedings* 62: 5575–5582. https://doi.org/10.1016/j.matpr.2022.04.606.

10 Sharma, V., Sharma, D., Tsai, M.-L. et al. (2023). Insights into the recent advances of agro-industrial waste valorization for sustainable biogas production. *Bioresource Technology* 390: 129829. https://doi.org/10.1016/j.biortech.2023.129829.

11 Said, Z., Sharma, P., Thi Bich Nhuong, Q. et al. (2023). Intelligent approaches for sustainable management and valorisation of food waste. *Bioresource Technology* 377: 128952. https://doi.org/10.1016/j.biortech.2023.128952.

12 de Sousa, L., Jabbour, A.B., Jabbour, C.J.C. et al. (2018). Industry 4.0 and the circular economy: a proposed research agenda and original roadmap for sustainable operations. *Annals of Operations Research* 270: 273–286. https://doi.org/10.1007/s10479-018-2772-8.

13 Tseng, M.-L., Tan, R.R., Chiu, A.S.F. et al. (2018). Circular economy meets industry 4.0: can big data drive industrial symbiosis? *Resources, Conservation and Recycling* 131: 146–147. https://doi.org/10.1016/j.resconrec.2017.12.028.

14 Wang, H., Lei, Z., Zhang, X. et al. (2019). A review of deep learning for renewable energy forecasting. *Energy Conversion and Management* 198: 111799. https://doi.org/10.1016/j.enconman.2019.111799.

15 Chen, J., Zeng, G.-Q., Zhou, W. et al. (2018). Wind speed forecasting using nonlinear-learning ensemble of deep learning time series prediction and extremal optimization. *Energy Conversion and Management* 165: 681–695. https://doi.org/10.1016/j.enconman.2018.03.098.

16 Demolli, H., Dokuz, A.S., Ecemis, A., and Gokcek, M. (2019). Wind power forecasting based on daily wind speed data using machine learning algorithms. *Energy Conversion and Management* 198: 111823. https://doi.org/10.1016/j.enconman.2019.111823.

17 Zhang, R. and Ma, J. (2008). An improved SVM method P-SVM for classification of remotely sensed data. *International Journal of Remote Sensing* 29: 6029–6036. https://doi.org/10.1080/01431160802220151.

18 Lim, H.R., Khoo, K.S., Chia, W.Y. et al. (2022). Smart microalgae farming with internet-of-things for sustainable agriculture. *Biotechnology Advances* 57: 107931. https://doi.org/10.1016/j.biotechadv.2022.107931.

19 Tham, P.E., Ng, Y.J., Vadivelu, N. et al. (2022). Sustainable smart photobioreactor for continuous cultivation of microalgae embedded with Internet of Things. *Bioresource Technology* 346: 126558. https://doi.org/10.1016/j.biortech.2021.126558.

20 Cinar, S., Cinar, S.O., Wieczorek, N. et al. (2021). Integration of artificial intelligence into biogas plant operation. *Processes* 9: 85. https://doi.org/10.3390/pr9010085.

21 Soo, A., Wang, L., Wang, C., and Shon, H.K. (2023). Machine learning for nutrient recovery in the smart city circular economy—a review. *Process Safety and Environment Protection* 173: 529–557. https://doi.org/10.1016/j.psep.2023.02.065.

22 Kavitha, S., Ravi, Y.K., Kumar, G. et al. (2024). Microalgal biorefineries: advancement in machine learning tools for sustainable biofuel production and value-added products recovery. *Journal of Environmental Management* 353: 120135. https://doi.org/10.1016/j.jenvman.2024.120135.

23 Munir, M.T., Li, B., and Naqvi, M. (2023). Revolutionizing municipal solid waste management (MSWM) with machine learning as a clean resource: opportunities, challenges and solutions. *Fuel* 348: 128548. https://doi.org/10.1016/j.fuel.2023.128548.

24 Zhang, Y., Li, L., Ren, Z. et al. (2022). Plant-scale biogas production prediction based on multiple hybrid machine learning technique. *Bioresource Technology* 363: 127899. https://doi.org/10.1016/j.biortech.2022.127899.

25 Sadoune, H., Rihani, R., and Marra, F.S. (2023). DNN model development of biogas production from an anaerobic wastewater treatment plant using Bayesian hyperparameter optimization. *Chemical Engineering Journal* 471: 144671. https://doi.org/10.1016/j.cej.2023.144671.

26 Yildirim, O. and Ozkaya, B. (2023). Prediction of biogas production of industrial scale anaerobic digestion plant by machine learning algorithms. *Chemosphere* 335: 138976. https://doi.org/10.1016/j.chemosphere.2023.138976.

27 Cruz, I.A., Andrade, L.R.S., Bharagava, R.N. et al. (2021). An overview of process monitoring for anaerobic digestion. *Biosystems Engineering* 207: 106–119. https://doi.org/10.1016/j.biosystemseng.2021.04.008.

28 Sappl, J., Harders, M., and Rauch, W. (2023). Machine learning for quantile regression of biogas production rates in anaerobic digesters. *Science of the Total Environment* 872: 161923. https://doi.org/10.1016/j.scitotenv.2023.161923.

29 De Clercq, D., Jalota, D., Shang, R. et al. (2019). Machine learning powered software for accurate prediction of biogas production: a case study on industrial-scale Chinese production data. *Journal of Cleaner Production* 218: 390–399. https://doi.org/10.1016/j.jclepro.2019.01.031.

30 Tao, X., Zheng, Y., Chen, W. et al. (2022). SVDD-based weighted oversampling technique for imbalanced and overlapped dataset learning. *Information Sciences* 588: 13–51. https://doi.org/10.1016/j.ins.2021.12.066.

31 Bourel, M., Segura, A.M., Crisci, C. et al. (2021). Machine learning methods for imbalanced data set for prediction of faecal contamination in beach waters. *Water Research* 202: 117450. https://doi.org/10.1016/j.watres.2021.117450.

32 Elreedy, D. and Atiya, A.F. (2019). A comprehensive analysis of synthetic minority oversampling technique (SMOTE) for handling class imbalance. *Information Sciences* 505: 32–64. https://doi.org/10.1016/j.ins.2019.07.070.

33 Torgo, L., Branco, P., Ribeiro, R.P., and Pfahringer, B. (2015). Resampling strategies for regression. *Expert Systems* 32: 465–476. https://doi.org/10.1111/exsy.12081.

34 Guo, H., Wu, S., Tian, Y. et al. (2021). Application of machine learning methods for the prediction of organic solid waste treatment and recycling processes: a review. *Bioresource Technology* 319: 124114. https://doi.org/10.1016/j.biortech.2020.124114.

35 Beltramo, T. and Hitzmann, B. (2019). Evaluation of the linear and non-linear prediction models optimized with metaheuristics: application to anaerobic digestion processes. *Engineering in Agriculture, Environment and Food* 12: 397–403. https://doi.org/10.1016/j.eaef.2019.06.001.

36 Jacob, S. and Banerjee, R. (2016). Modeling and optimization of anaerobic codigestion of potato waste and aquatic weed by response surface methodology and artificial neural network coupled genetic algorithm. *Bioresource Technology* 214: 386–395. https://doi.org/10.1016/j.biortech.2016.04.068.

37 Sun, Y., Dai, H., Moayedi, H. et al. (2024). Predicting steady-state biogas production from waste using advanced machine learning-metaheuristic approaches. *Fuel* 355: 129493. https://doi.org/10.1016/j.fuel.2023.129493.

38 Andrade Cruz, I., Silva Nascimento, V.R., Alves Felisardo, R.J. et al. (2023). Evaluation of artificial neural network models for predictive monitoring of biogas production from cassava wastewater: a training algorithms approach. *Biomass and Bioenergy* 175: 106869. https://doi.org/10.1016/j.biombioe.2023.106869.

39 Kumar, V., Thakur, C., and Chaudhari, P.K. (2022). Anaerobic biological treatment of dye bearing water in anaerobic sequencing batch reactor: performance and kinetics studies. *Journal of the Indian Chemical Society* 99: 100673. https://doi.org/10.1016/j.jics.2022.100673.

40 Bryant, M.P. (1979). Microbial methane production—theoretical aspects. *Journal of Animal Science* 48: 193–201. https://doi.org/10.2527/jas1979.481193x.

41 Buswell, A.M. and Mueller, H.F. (1952). Mechanism of methane fermentation. *Industrial and Engineering Chemistry* 44: 550–552. https://doi.org/10.1021/ie50507a033.

42 Gutiérrez-Martín, F., Rodríguez-Antón, L.M., and Legrand, M. (2020). Renewable power-to-gas by direct catalytic methanation of biogas. *Renewable Energy* 162: 948–959. https://doi.org/10.1016/j.renene.2020.08.090.

43 Kamravamanesh, D., Rinta Kanto, J.M., Ali-Loytty, H. et al. (2023). Ex-situ biological hydrogen methanation in trickle bed reactors: integration into biogas production facilities. *Chemical Engineering Science* 269: 118498. https://doi.org/10.1016/j.ces.2023.118498.

44 Avinash, L.S. and Mishra, A. (2024). Comparative evaluation of artificial intelligence based models and kinetic studies in the prediction of biogas from anaerobic digestion of MSW. *Fuel* 367: 131545. https://doi.org/10.1016/j.fuel.2024.131545.

45 Savand-Roumi, E., Salehiyoun, A.R., and Mohtasebi, S.S. (2024). Use of a biogas-specific e-nose with machine learning to identify biogas pattern changes linked to hydraulic retention times in an anaerobic digester: a case study. *Fuel* 357: 130013. https://doi.org/10.1016/j.fuel.2023.130013.

46 Sonwai, A., Pholchan, P., and Tippayawong, N. (2023). Machine learning approach for determining and optimizing influential factors of biogas production from lignocellulosic biomass. *Bioresource Technology* 383: 129235. https://doi.org/10.1016/j.biortech.2023.129235.

47 Zhang, Y., Jing, Z., Feng, Y. et al. (2023). Using automated machine learning techniques to explore key factors in anaerobic digestion: at the environmental factor, microorganisms and system levels. *Chemical Engineering Journal* 475: 146069. https://doi.org/10.1016/j.cej.2023.146069.

48 Hunter, S.M., Blanco, E., and Borrion, A. (2024). Predicting total biogas potential of food waste using the initial output of biogas potential tests as input data to train an artificial neural network. *Bioresource Technology Reports* 26: 101845. https://doi.org/10.1016/j.biteb.2024.101845.

49 Jeong, K., Abbas, A., Shin, J. et al. (2021). Prediction of biogas production in anaerobic co-digestion of organic wastes using deep learning models. *Water Research* 205: 117697. https://doi.org/10.1016/j.watres.2021.117697.

50 Offie, I., Piadeh, F., Behzadian, K. et al. (2023). Development of an artificial intelligence-based framework for biogas generation from a micro anaerobic digestion plant. *Waste Management* 158: 66–75. https://doi.org/10.1016/j.wasman.2022.12.034.

51 Xiao, J., Liu, C., Ju, B. et al. (2021). Estimation of in-situ biogas upgrading in microbial electrolysis cells via direct electron transfer: two-stage machine learning modeling based on a NARX-BP hybrid neural network. *Bioresource Technology* 330: 124965. https://doi.org/10.1016/j.biortech.2021.124965.

52 Alrowais, R., Said, N., Al-Otaibi, A. et al. (2023). Comparing the effect of mesophilic and thermophilic anaerobic co-digestion for sustainable biogas production: an experimental and recurrent neural network model study. *Journal of Cleaner Production* 392: 136248. https://doi.org/10.1016/j.jclepro.2023.136248.

Index

Machine Learning for Sustainable Energy Solutions, First Edition. Edited by Zafar Said and Prabhakar Sharma.

Printed and bound by CPI Group (UK) Ltd, Croydon, CR0 4YY
17/09/2026
14964331-0004